Einstein's Physics

Albert Einstein at Barnes Foundation in Merion PA, c.1947.
Photography by Laura Delano Condax; gift to the author from Vanna Condax.

Einstein's Physics

Atoms, Quanta, and Relativity
Derived, Explained, and Appraised

TA-PEI CHENG

University of Missouri–St. Louis
Portland State University

OXFORD
UNIVERSITY PRESS

Great Clarendon Street, Oxford, OX2 6DP,
United Kingdom

Oxford University Press is a department of the University of Oxford.
It furthers the University's objective of excellence in research, scholarship,
and education by publishing worldwide. Oxford is a registered trade mark of
Oxford University Press in the UK and in certain other countries

Impression: 2

Reprinted (with corrections 2014)

British Library Cataloguing in Publication Data

Data available

ISBN 978–0–19–966991–2

Printed and bound by
Clays Ltd, St Ives plc

To

Leslie

Preface

Einstein explained in equations

Albert Einstein's achievement in physics is proverbial. Many regard him as the greatest physicist since Newton. What did he do in physics that's so important? While there have been many books about Einstein, most of these explain his achievement only in qualitative terms. This is rather unsatisfactory as the language of physics is mathematics. One needs to know the equations in order to understand Einstein's physics: the precise nature of his contribution, its context, and its influence. The most important scientific biography of Einstein has been the one by Abraham Pais: *Subtle is the Lord ... The Science and the Life of Albert Einstein*: The physics is discussed in depth; however, it is still a narrative account and the equations are not worked out in detail. Thus this biography assumes in effect a high level of physics background that is perhaps beyond what many readers, even working physicists, possess. Our purpose is to provide an introduction to Einstein's physics at a level accessible to an undergraduate physics student. All physics equations are worked out from the beginning. Although the book is written with primarily a physics readership in mind, enough pedagogical support material is provided that anyone with a solid background in an introductory physics course (say, an engineer) can, with some effort, understand a good part of this presentation.

In historical context This is a physics book with material presented in the historical context. Although it is not a scholarly history and there is hardly any original work in the Einstein biography, historical material from secondary sources is used to make the physics material more comprehensible and interesting. For example, a more careful discussion of the results obtained by Hendrik Lorentz will precede Einstein's special relativity. Planck's and Einstein's work on blackbody radiation are presented only after reviewing first the thermodynamics and scaling results of Wilhelm Wien. Our opinion is that the history conveyed through standard physics textbooks sometimes misses the proper context of the discovery. The original Einstein story is actually more interesting and illuminating.

Post-Einstein development Also, we do not stop at Einstein's discovery, but carry the discussion onto some of the advances in physics that had been made because of Einstein's contribution. We discuss gauge symmetry leading to the Standard Model of particle physics as a legacy of Einstein's invariance-principle approach. As an example of Einstein's unified field theory we present the Kaluza–Klein unification of electromagnetism and gravitation in a space with an extra dimension. Such knowledge is needed to fully appreciate the profound influence that Einstein's physics had on subsequent development.

Can you answer these "Einstein questions"?

Physics students have already learnt aspects of Einstein's physics—from segments in their course work or from popular accounts. Here is a list of 21 Einstein questions. Can you answer them?[1]

[1]Brief answers are given in Appendix C, where the reader can also find the chapter and section numbers where the discussion of, and answer to, such Einstein questions are carried out in the text.

1. Einstein's research played a significant part in people's acceptance of the reality of the molecular constituents of matter. In one year, 1905, he showed three separate ways to deduce Avogadro's number from macroscopic measurements. What were the three areas in physics where these deductions were made? Surprisingly, one of these was the blackbody radiation.

2. Einstein's celebrated Brownian motion paper did not have the words "Brownian motion" in its title. How come?

3. Einstein's 1905 photoelectric paper, where the idea of light quanta was first proposed, and which was the work cited when he was awarded the Nobel Prize, was concerned mostly with a statistical study of blackbody radiation. If the papers on quantum theory by Planck and by Einstein were both concerned with blackbody radiation, what was their key difference?

4. In the classical theory we have an "ultraviolet catastrophe" for the blackbody radiation. How does the postulate of energy quantization cure this problem?

5. Einstein's quantum theory of specific heat is historically important because it is the first instance when the quantum idea was shown to be relevant to physical systems well beyond the esoteric case of blackbody radiation. His theory also clarified the questions about matter's molecular composition. How is that so?

6. The statement of wave–particle duality was made first by Einstein in his 1909 study of fluctuations of radiation energy. Einstein and Bohr had influenced each other's work, especially with respect to the idea of quantum transitions (the quantum jumps). How did quantum mechanics and quantum field theory accommodate, in one elegant framework, simultaneously waves, particles, and quantum jumps? Famously, this is not the resolution that Einstein was able to accept.

7. Einstein never accepted the orthodox interpretation of quantum mechanics. Was he just too set in his ways to appreciate the new advances in physics? How had Einstein's criticism influenced subsequent investigation on the meaning of quantum mechanics?

8. By the time Einstein proposed his special theory of relativity, the Lorentz transformation had already been written down. Einstein was unaware of this latest development, as he was working (in the Swiss Patent Office) outside an academic environment. Einstein's derivation of this transformation rule differed fundamentally from the way it was gotten by Lorentz and others. How?

9. While the Michelson–Morley measurement did not play a direct role in Einstein's motivation for special relativity, there were other results

(stellar aberration, Fizeau's experiment, and Fresnel's formula) that Einstein had acknowledged as having had an influence. In what ways were they relevant to Einstein's motivation? How were they explained in the final relativity theory?

10. The key element of special relativity is the new conception of time. Just about all the counter-intuitive relativistic effects spring from the relativity of simultaneity. What about the well-known result of "length contraction"? Does it have this connection with time also? If so, how?

11. What is the difference between the special and general theories of relativity? Special relativity is applicable to electromagnetism, mechanics, thermodynamics, etc. but not to gravity (why not?); on the other hand, general relativity is the field theory of gravitation. Why then is special relativity special and general relativity general? Why does the principle of general relativity automatically bring gravity into the consideration?

12. Einstein originally dismissed Minkowski's geometrical formulation of his relativity theory as "superfluous learnedness". What caused Einstein to change this appraisal later on? With respect to the role of mathematics in the discovery of physics theory, how did Einstein's view evolve? Was Einstein a great mathematician as well as a great physicist? What difference would it make?

13. What was the idea that Einstein called "my happiest thought"? Which moment of elation was characterized by his biographer Pais as "by far the strongest emotional experience in Einstein's scientific life, perhaps, in all his life"?

14. Einstein's general relativity is said to be a geometric theory of gravity. What does one mean by a "geometry theory"? How did Einstein get the idea that "a gravitational field is simply spacetime with curvature"? To what physical realm exactly does Einstein's theory extend Newtonian gravity?

15. One way to state the equivalence of inertia and gravitation is to say that gravity can always be transformed away locally (by going to a reference frame in free fall). Thus the essence of gravity is represented by its differentials (tidal forces). How does this feature appear in the field equation of general relativity (the Einstein equation)?

16. According to Einstein's gravity theory, shaking a mass gives rise to gravitational waves. Because gravity is such a weak force, it is extremely difficult to detect the resultant waves. At this moment there is no confirmed evidence for their direct detection. Nevertheless, consequences of gravitational wave emission have been measured and found in agreement with Einstein's prediction. What are these indirect effects?

17. It is well known that black holes are regions where gravity is so strong even light cannot escape. Black holes demonstrate the full power and glory of general relativity also because, inside black holes, "the roles of space and time are interchanged"! What does this mean? How does this come about?

18. Planck's discovery of Planck's constant allowed him to construct, from h, c, and G_N, a natural unit system of mass-length-time. Through the essential contribution by Einstein, we now understand each of these

fundamental constants as the "conversion factor" that connects disparate realms of physics. Can you name these areas? What are Einstein's works that made these syntheses possible?

19. The modern study of cosmology started with Einstein's 1917 paper. The story was often told that Einstein regarded his introduction of the cosmological constant as "the biggest blunder of my life". What is the source of this piece of anecdotal history? What role does Einstein's cosmological constant play in our present understanding of the universe?

20. Special relativity, photons, and Bose–Einstein statistics are crucial ingredients of modern particle physics. On the other hand, Einstein did not work directly on any particle theory. Yet, one can still claim that the influence of his ideas had been of paramount importance in the successful creation of the Standard Model of particle physics. What is the basis of this claim?

21. In the later years of his life, Einstein devoted the major part of his physics effort in the search for a unified field theory. Was this just a misguided chasing of an impossible dream? Based on our current understanding, what was the legacy of this somewhat less appreciated part of his research?

Clearly the story is a fascinating one. But to understand it properly one needs to know the relevant physics, to know some of the technical details. This, an undergraduate physics student, with some help, should be able to do.

Atoms, quanta, and relativity—Our presentation

The material is logically divided into five parts: atoms, quanta, special/general relativity, and later developments. Each of the 17 chapters has a detailed summary, in the form of a bullet list, placed at the beginning of the chapter. The reader can use these lists to get an overview of the contents and decide which part the book he or she wants to study in detail. For example, a reader may well wish to postpone Chapter 1 for a later reading; it discusses Einstein's doctoral thesis and concerns the subject of classical hydrodynamics, which may not be all that familiar to a present-day student.

Physics focus   Although many of Einstein's papers are discussed in this book, his physics is not presented in the exact form as given in his papers. For example, the derivation of the Lorentz transformation is different from that given in Einstein's 1905 paper, even though the assumption and result are the same. In finding the general relativity field equation, Einstein's original steps are not followed because, after Einstein's discovery, it had been shown by others that the same conservation law condition could be obtained much more simply by using the Bianchi identities. In other words, the focus of this book is Einstein's physics, rather than the strict historical details of his physics. It is hoped that our presentation (without the obsolete notation of the original papers) is more accessible to a modern-day reader.

As a textbook?   Since Einstein's legacy has permeated so many areas in physics, a wide range of topics will be covered in our presentation. It is hoped that after studying these lessons, a student will not only have learnt some history of physics and a better appreciation of Einstein's achievement, but, perhaps more importantly, will have enhanced their understanding of some of the basic areas in their physics curriculum (and a glimpse of more advanced topics): thermodynamics, hydrodynamics, statistical mechanics, Maxwell's equations, special and general relativity, cosmology, quantum mechanics, quantum field theory, and particle physics. Although this book is written for a general-interest physics readership, it can be used as a textbook as well—for a "Special Topics" course, or an "Independent Reading" course. One possibility is to have the book function as the basis of a "senior year project". Working through the book may well be an enjoyable experience for both the student and the instructor.

Acknowledgement

My editor Sonke Adlung has given me much encouragement and useful advice. Jessica White helped me with the whole process of submitting this book manuscript to OUP. This book grew from the lecture notes of a summer course I taught at the Portland State University. I thank Erik Bodegom, John Freeouf, Drake Mitchell, and Kim Doty-Harris for their support. Sam Paul, my student at UMSL, read through the entire manuscript and made useful comments. I am grateful to Cindy Bertram for doing all the line-drawing figures in this book.

Book website   As always, I shall be glad to receive, at tpcheng@umsl.edu, readers' comments and notification of errors. An updated list of corrections will be displayed at <http://www.umsl.edu/~tpcheng/einstein.html>.

St. Louis MO and Portland OR   *T.P.C.*
July, 2012

Contents

ATOMIC NATURE
OF MATTER

Molecular size from classical fluids

1

- The existence of atoms and molecules as constituents of matter was still being debated at the beginning of the twentieth century. Einstein's doctoral thesis provided another important piece of evidence for their reality. He was able to derive two independent equations relating the molecular size P and Avogadro's number N_A to properties of a macroscopic fluid, such as viscosity and the diffusion coefficient. The system studied by Einstein was a fluid with small suspended particles.

- The first equation, $\eta^* = \eta\left(1 + \frac{5}{2}\varphi\right)$, relates the change of viscosity coefficient $(\eta \to \eta^*)$ by the solute particles to the fraction of volume φ occupied by these suspended particles, which can be expressed in terms of P and N_A. This viscosity relation then becomes the equation for the product $P^3 N_A$.

- We obtain this $\eta^*(\eta)$ relation from a solution of the Navier–Stokes equation, the equation of motion for a viscous fluid. This equation is related to that for an ideal fluid (the Euler equation), viewed as a momentum conservation equation. The viscosity is then brought in through the appropriate modification of the fluid's energy–momentum tensor. Solving the Navier–Stokes equation for the velocity and pressure fields altered by the presence of solute particles, we then calculate the change in viscosity through the change of heat loss from the fluid.

- As is the case throughout our presentation, we provide some of the background material and all the calculational steps. Elementary material and some details are relegated to the supplementary material (SuppMat) sections at the end of the chapter. In SuppMat Section 1.4 we discuss the Euler equation as the statement of momentum conservation. In SuppMat Section 1.5 the calculational details of the effective viscosity due to the presence of solute particles are displayed.

- The second equation, $D = \mu k_B T$, is obtained by a consideration of balancing the forces on the solute particles in the fluid. The fluctuation of molecular motion gives rise to a diffusion force (D being the diffusion coefficient). It is countered by the frictional force on a particle moving in a viscous fluid (dissipation) obeying Stokes' law, with the mobility μ related to the viscosity $\mu = 1/(6\pi\eta P)$. This second relation, also known as the Einstein–Smoluchowski relation, is an equation in which the combination of PN_A enters.

- Finally one can solve these two equations to obtain P and N_A separately. This result allowed people to deduce the values of these crucial parameters from simple table-top experimental measurements. The agreement of these numbers with those obtained from completely different gaseous systems lent convincing support to the idea that matter was composed of microscopic particles.
- As mentioned in the Preface, this first chapter on fluid mechanics is likely to be a particularly difficult one for a present-day student. It is placed at the beginning of the book for historical and logical reasons. Still, one can skip this chapter (even the whole of Part I) without impeding the understanding of later chapters, and return to this material when one is so inclined at a later time.

1.1   Two relations of molecular size and the Avogadro number

Even before his famous 1905 papers Albert Einstein (1879–1955) had already published three papers, "Einstein's statistical trilogy", in *Annalen der Physik* on the foundations of statistical mechanics: they form important parts of the bridge going from Boltzmann's work to the modern approach to statistical mechanics (Einstein 1902, 1903, and 1904). Einstein completed his thesis "On the determination of molecular dimensions" in April 1905 and submitted it to the University of Zurich on July. His Alma Mater (Swiss) Federal Polytechnic Institute ETH, also in Zurich, did not yet grant doctoral degrees. Shortly after the thesis was accepted, he submitted a slightly shortened version to *Annalen der Physik* in August. In his thesis work (Einstein 1905b), he provided an important piece of supporting evidence for the idea that matter is composed of material particles (atoms and molecules). He deduced the molecular size P and Avogadro number N_A (which in turn fixes the mass of the individual molecule through the known molar mass) from a study of the effects due to small suspended particles[1] on the viscosity and diffusion coefficient of a liquid (a sugar solution, for example). This is in contrast to previous work where such molecular dimensions were deduced from the kinetic theory of gases.

The key idea of Einstein's thesis work is that particles suspended in a fluid behave just like molecules in the solution (they differ only in their sizes). He assumed the validity of classical hydrodynamics in the calculation of the effects of solute molecules on the viscosity of a dilute solution. When treating the molecules as rigid spheres with radius P, Einstein succeeded in obtaining two independent relations from which he could solve for the two unknowns P and N_A in terms of macroscopic quantities such as viscosity and the diffusion coefficient:

$$N_A P^3 = \frac{2}{5}\left(\frac{\eta^*}{\eta} - 1\right)\frac{3M}{4\pi\rho} \tag{1.1}$$

$$N_A P = \frac{RT}{6\pi\eta}\frac{1}{D} \tag{1.2}$$

[1] In our discussion, we often refer to such particles as "solute particles".

where T is the absolute temperature, M is the molar mass (molecular weight), ρ the mass density, R is the gas constant, D the diffusion coefficient, η the viscosity coefficient of the liquid without the solute molecules (e.g. water), and the effective viscosity coefficient η^* of the liquid with solute molecules (e.g. dilute solutions with sugar molecules). The second relation was the first example of a fluctuation and dissipation relation as it shows that there is a connection between diffusion and viscosity.[2] The calculated values of P and N_A were in accord with those that had been obtained from other methods. This result is important for the consistency of the atomic theory of matter because it shows that molecular dimensions deduced from widely disparate environments are in agreement with each other.

[2]This fluctuation–dissipation relation will be discussed further in the next chapter.

1.2 The relation for the effective viscosity

In this section we shall see how Einstein derived the relation (1.1). Basically, he used the Navier–Stokes equation: it allowed him to calculate the dependence of heat dissipation on the viscosity of the solution—with and without solute particles (see Fig. 1.1). He deduced a simple relation between the effective viscosity η^* and the viscosity η of the ambient fluid without solute particles:

$$\eta^* = \eta \left(1 + \frac{5}{2}\varphi \right), \tag{1.3}$$

where φ is the fractional volume occupied by the solute particles in the fluid, and can easily be related to the molecular dimension to yield Eq. (1.1). The assumptions Einstein used in this calculation are listed here:

- One can ignore (as small) the inertia of translational and rotational motion of the solute molecules.
- The motion of any particle is not affected by the motion of any other particles.
- The motion of a particle is due only to the stress it feels at its surface.
- One can take the boundary condition that the velocity field vanishes at the particles' surfaces.

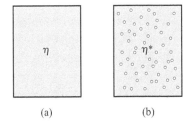

(a) (b)

Fig. 1.1 (a) A volume without solute particles having a viscosity η, vs. **(b)** a volume with many suspended particles having an effective viscosity η^*.

1.2.1 The equation of motion for a viscous fluid

Some of the basic elements of fluid mechanics are presented in SuppMat Section 1.4. We review how the conservation laws in a continuum system are written as the **equations of continuity**. In particular the equation of motion for an incompressible ideal fluid, the **Euler equation**,[3]

$$-\nabla p = \rho \frac{\partial \mathbf{v}}{\partial t} + (\mathbf{v} \cdot \nabla)\rho\mathbf{v} \tag{1.4}$$

can be viewed as a statement of fluid momentum conservation

$$\frac{\partial}{\partial t}(\rho v_i) + \frac{\partial}{\partial x_j}\tau_{ij} = 0 \tag{1.5}$$

[3]$\mathbf{v}(x,t)$ and $p(x,t)$ are, respectively, the velocity and pressure fields, and ρ is the mass density.

where we have followed the 'Einstein summation convention' of summing over repeated indices.[4] τ_{ij} is the energy–momentum tensor for an ideal fluid

$$\tau_{ij} = p\delta_{ij} + \rho v_i v_j. \tag{1.6}$$

With this view of the fluid equation of motion, we can extend it to the case of a viscous fluid in a straightforward manner.

Navier–Stokes equation for viscous fluids

In a viscous fluid, there is internal friction resulting in the irreversible transfer of momentum from points where the velocities are high to points where the velocity is low. The simplest way to implement this feature is to add another tensor, call it σ_{ij}, to the ideal fluid stress tensor τ_{ij} of (1.6). Since the viscosity stress tensor must vanish in the limit of constant velocity, we can assume it to be proportional to the first derivatives of the velocity $\partial v_i/\partial x_j$; and since the stress tensor should always be symmetric $\sigma_{ij} = \sigma_{ji}$, we arrive at the simple form:

$$\sigma_{ij} = \eta \left(\frac{\partial v_i}{\partial x_j} + \frac{\partial v_j}{\partial x_i} \right), \tag{1.7}$$

where η is the **viscosity coefficient**. Replacing τ_{ij} by $\tau_{ij} - \sigma_{ij}$ in the momentum continuity equation (1.5), one obtains, after some algebra, the **Navier–Stokes equation**. For an incompressible fluid [hence the condition $\nabla \cdot \mathbf{v} = 0$ as discussed in (1.36)] it reads as

$$-\nabla p + \eta \triangle \mathbf{v} = \rho \frac{\partial \mathbf{v}}{\partial t} + (\mathbf{v} \cdot \nabla) \rho \mathbf{v} \tag{1.8}$$

where $\triangle$ is the Laplacian operator $\triangle = \nabla \cdot \nabla$.

We will show how the (velocity and pressure field) solution of the Navier–Stokes equation allows us to calculate the energy dissipation in terms of the viscosity coefficients.

1.2.2 Viscosity and heat loss in a fluid

Our goal is to find the change of viscosity coefficient ($\eta \to \eta^*$) due to the presence of the solute particles in a fluid and relate this change to the fraction of volume φ occupied by these suspended particles. This viscosity relation then becomes the equation (1.1) for the product $P^3 N_A$.

We first calculate the change in viscosity through the change of heat loss from the fluid. The rate of energy transfer in an incompressible viscous fluid can be expressed as a volume integral

$$W = \frac{d}{dt} \int \left(\frac{1}{2} \rho v^2 \right) dV = \int \rho \mathbf{v} \cdot \left(\frac{\partial \mathbf{v}}{\partial t} \right) dV. \tag{1.9}$$

The integrand can then be rewritten by using the Navier–Stokes equation of (1.8)

$$\frac{\partial v_i}{\partial t} = -\mathbf{v} \cdot \nabla v_i - \frac{1}{\rho} \partial_i p + \frac{1}{\rho} \partial_k \sigma_{ki} \tag{1.10}$$

where σ_{ki} is the stress tensor of (1.7) and we have used the incompressible fluid condition $\partial_i v_i = 0$ of (1.36) to relate $\partial_k \sigma_{ki} = \eta \partial_k (\partial_k v_i + \partial_i v_k) = \eta \triangle v_i$. Substituting Eq. (1.10) into (1.9), we have the integrand

$$-\rho v_i \mathbf{v} \cdot \nabla v_i - \mathbf{v} \cdot \nabla p + v_i \partial_k \sigma_{ki} = -\rho \mathbf{v} \cdot \nabla \left(\frac{1}{2} v^2 + \frac{p}{\rho} \right) + \partial_k (v_i \sigma_{ki}) - \sigma_{ki} \partial_k v_i$$

where to reach the right-hand side (RHS) we have combined the first two terms of the left-hand side (LHS) into one and expanded the last one into two. The second term on the RHS is a divergence. We now show that the first term is also a divergence: $v_k \partial_k (\ldots) = \partial_k [v_k (\ldots)] - (\ldots) \partial_k v_k$ But the last term vanishes because of the incompressible fluid condition of (1.36). Thus the first two terms on the RHS are divergences and their volume integral turns into a surface integral, which vanishes at large distance because velocity vanishes at large distance. In this way one finds

$$W = - \int \sigma_{ki} \partial_k v_i dV. \tag{1.11}$$

We see that the energy loss is proportional to the viscosity stress tensor of (1.7).

Effect of solute molecules on heat loss

Einstein then considered the effect of solute molecules on the energy dissipation of the viscous fluid—starting with one solute particle in the form of a rigid sphere, then generalizing to a collection of such spheres. In this way he was able to derive a relation between the viscosity coefficients with and without solute particles as shown in Eq. (1.3). The steps are given below.

Dissipation in an ambient fluid We can view the velocity field of the fluid with one spherical particle at the center as composed of two components $\mathbf{v} = \mathbf{u} + \mathbf{v}'$, where $\mathbf{u}$ is the "background field", namely the velocity far from the particle where the induced velocity $\mathbf{v}'$ due to the presence of the solute particle should be unimportant (Fig. 1.2). For a region around the location of the solute particle so small that only the linear terms in a Taylor expansion of the velocity field $u_i(x)$ need to be considered

$$u_i = \alpha_{ij} x_j \tag{1.12}$$

where α_{ij} is a constant traceless symmetric matrix: it is traceless because we are dealing with an incompressible fluid $\partial_i u_i = \alpha_{ij} \delta_{ij} = \alpha_{ii} = 0$; it must be symmetric $\alpha_{ij} = \alpha_{ji}$ as it is directly related to the symmetric stress tensor

$$\sigma_{ij} = \eta (\partial_i u_j + \partial_j u_i) = 2\eta \alpha_{ij}. \tag{1.13}$$

From this we can also calculate the heat loss (per unit volume) in such a fluid (without the solute particle) by Eq. (1.11) to yield the rate of energy loss as

$$W_0 = -2\eta \alpha_{ij} \alpha_{ij} V. \tag{1.14}$$

Dissipation in a fluid with solute particles Einstein then considered the effect of solute molecules. After a somewhat lengthy calculation[5] he found that the above relation (1.14) is modified by an additional term

$$W = -2\eta \alpha_{ij} \alpha_{ij} V - 5\eta \alpha_{ij} \alpha_{ij} \varphi V \tag{1.15}$$

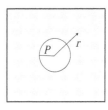

Fig. 1.2 A single sphere of radius P suspended in a fluid with an ambient field of $u_i = \alpha_{ij} x_j$ resulting in a net velocity field of $v_i = u_i + v_i'$. The origin of the coordinate system is taken to be the center of the submerged sphere.

[5]Details of this calculation are given in SuppMat Section 1.5.

where φ is the fraction of the volume occupied by the solute molecules. Given the assumptions listed at the top of this section, the fluid with solute particles still satisfies the Navier–Stokes equation and can be characterized by an effective viscosity coefficient η^*. The heat dissipation in a fluid with suspended particles (1.15), viewed this way, can also be written as

$$W = -2\eta^* \alpha_{ij}\alpha_{ij}V. \tag{1.16}$$

Equating this to the expression in (1.15), Einstein found a simple relation between η^* and η as shown in Eq. (1.3).

1.2.3 Volume fraction in terms of molecular dimensions

We now show that, after expressing the volume fraction φ in terms of molecular dimensions, Eq. (1.3) leads directly to the first relation claimed in (1.1). The volume fraction is the product of n, the number of solute particles per unit volume, and Φ, the volume occupied by each solute particle taken to be a rigid sphere of radius P,

$$\varphi = n\Phi \qquad \text{where} \qquad \Phi = \frac{4\pi}{3}P^3. \tag{1.17}$$

We then show that the number density can be expressed as

$$n = \frac{\rho}{M}N_A, \tag{1.18}$$

where ρ is the mass density of the solute molecules (mass per unit volume), N_A the Avogadro number (the number of particles in a mole volume), and M the molecular weight (the mass of one mole of molecules) of the solute particles; hence, $M = N_A m_1$, with m_1 being the mass of an individual molecule. To show the relation (1.18), let the total number of molecules in a volume V be $\mathcal{N}$ so that

$$n = \frac{\mathcal{N}}{V} = \frac{\mathcal{N}m_1}{Vm_1} = \frac{\text{total mass}}{Vm_1} = \frac{\rho}{m_1} = \frac{\rho}{M}N_A. \tag{1.19}$$

We now substitute this result into (1.17)

$$\varphi = n\Phi = \left(\frac{\rho}{M}N_A\right)\left(\frac{4\pi}{3}P^3\right) \tag{1.20}$$

so that Eq. (1.3) becomes

$$\frac{\eta^*}{\eta} - 1 = \frac{5}{2}\varphi = \frac{5}{2}\left(\frac{\rho}{M}N_A\right)\left(\frac{4\pi}{3}P^3\right). \tag{1.21}$$

This, we see, is the relation (1.1) claimed at the beginning of this chapter.

1.3 The relation for the diffusion coefficient

Here we shall work out the second relation of Einstein's thesis, Eq. (1.2), which also plays a key role in Einstein's Brownian motion paper to be discussed in Chapter 2. This relation between diffusion and viscosity was derived on the basis of thermal and dynamical equilibrium conditions, using a combined technique of fluid dynamics and the theory of diffusion. Einstein studied the viscous (frictional) force and the diffusion osmotic force acting on the solute molecules in a dilute solution. Most of us are familiar with the picture of an immersed sphere falling in a viscous fluid, reaching the terminal velocity when the viscous drag force is balanced by gravity. Here we have the frictional force balanced by, instead of gravity, the osmotic force.

We have the following situation: The uneven concentration, namely the presence of a density gradient, gives rise to a force, the osmotic force, which causes fluid particles to diffuse. These moving particles are quickly slowed down by the frictional force. This balance of osmotic and frictional forces allowed Einstein to derive the result

$$D = \mu k_B T \tag{1.22}$$

with T being the absolute temperature. Boltzmann's constant k_B is related to the gas constant and Avogadro's number as $k_B = R/N_A$, and the **mobility** μ is related to the molecular size and viscosity coefficient as $\mu = (6\pi P\eta)^{-1}$ through Stokes' law. Eq. (1.22) leads to the second relation (1.2) in which the combination $N_A P$ enters.

1.3.1   Osmotic force

The different concentration in the fluid gives rise to the osmotic force. One way to express this is through the phenomenological diffusion equation[6] (**Fick's first law**)

$$\rho v = -D \frac{\partial \rho}{\partial x} \tag{1.23}$$

which states that the diffusion flux (i.e. the diffusion current density) ρv must be proportional to the density gradient $\partial \rho/\partial x$, and the proportionality constant D is (defined to be) the **diffusion coefficient**.

To relate this to the osmotic force, we note that the pressure gradient has the physical significance of a "force density" [cf. Eq. (1.39) in SuppMat Section 1.4]—namely, when multiplied by the volume factor $-V\partial p/\partial x$ it is the total force.[7] We are going to compare this force to the Stokes formula [of (1.29) to be discussed below] for the frictional force on a single particle, so we need to divide this total force by the total number of particles $\mathcal{N}$. This yields the force per particle:

$$F_{os} = -\frac{V}{\mathcal{N}} \frac{\partial p}{\partial x} = -\frac{1}{n} \frac{\partial p}{\partial x} = -\frac{M}{\rho N_A} \frac{\partial p}{\partial x}, \tag{1.24}$$

where we have used the relation (1.18), with ρ being the mass density, M the molar mass (molecular weight), and N_A the Avogadro number.

To relate (1.24) to the diffusion equation (1.23), one needs to replace the pressure p by the density function ρ. For a dilute solution, Einstein assumed the validity of the so-called **van't Hoff analogy**—the behavior of solute molecules in a dilute solution is similar to those of an ideal gas. The ideal gas law then leads to

[6]Since we are interested in the force in one particular direction, we can simplify our writing to the one-dimensional (1D) case. Instead of the full gradient operator $\partial/\partial x_i$, we just need to work with $\partial/\partial x$.

[7]The total force is $\int(-\partial p/\partial x)\,dV$. Thus properly speaking what we should have is the average $-\overline{\partial p/\partial x}$. However, we are working in the framework of the phenomenological Fick's first law, and for consistency higher derivatives of density and pressure are assumed to be small. Thus we would have $\overline{\partial p/\partial x} = \partial p/\partial x$ anyway.

$$pV = \frac{\mathcal{N}}{N_A}RT \quad \text{or} \quad p = n\frac{RT}{N_A} = \frac{RT}{M}\rho. \tag{1.25}$$

The total number of molecules divided by the Avogadro number $\mathcal{N}/N_A$ is the "mole number", while the total number divided by the volume is the number density n. To reach the last expression we have used the relation (1.18). Thus, substituting this density–pressure relation into (1.24), the osmotic force on a solute molecule can now be written as

$$F_{os} = -\frac{RT}{\rho N_A}\frac{\partial \rho}{\partial x}. \tag{1.26}$$

1.3.2   Frictional drag force—the Stokes law

To study the drag force by a viscous fluid (of viscosity η) with a constant velocity field $\mathbf{u}$ acting on an immersed rigid sphere of radius P, we first need to find the induced velocity field. Namely, very much like the case when we calculated the effective viscosity in SuppMat Section 1.5, we have[8] $\mathbf{v} = \mathbf{u} + \mathbf{v}'$, where the induced velocity field vanishes when far from the immersed sphere. Just like the previous case we solve the Navier–Stokes equation for the induced velocity $\mathbf{v}'(r)$ and pressure $p'(r)$, with a similar boundary condition of $\mathbf{v} = 0$ at the surface of the immersed sphere $r = P$. These results can be translated into knowledge of the viscous stress tensor of Eq. (1.7):

$$\sigma'_{ij} = \eta\left(\frac{\partial v'_i}{\partial x_j} + \frac{\partial v'_j}{\partial x_i}\right). \tag{1.27}$$

Recall that when the inertia force can be ignored (the substantial derivative of the velocity field $Dv/Dt \simeq 0$), the Navier–Stokes equation reads (cf. Eq. 1.10),

$$\partial_i p = -\partial_k \sigma_{ki}.$$

Since the gradient of pressure is force per unit volume, the volume integral on the RHS also has the physical meaning of force. The volume integral can then be turned into a surface integral by the divergence theorem. We then have (as we expect the drag force to be in the direction of the velocity field $\mathbf{u}$, the magnitude is just the projection of the force in the $\mathbf{u}$ direction $F = \mathbf{F} \cdot \hat{\mathbf{u}}$) the drag force as

$$F_{dg} = -\oint \sigma'_{ij}\hat{u}_i n_j dS \tag{1.28}$$

where the surface integration is performed over the spherical surface of the immersed particle. Namely, we sum over all the force elements at every point of the surface, with $\mathbf{n}$ being the unit vector normal to the surface element, and $\hat{\mathbf{u}}$ is the unit vector in the direction of velocity $\hat{u}_i = u_i/u$.

To complete the calculation we naturally work with a spherical coordinate system with $\hat{\mathbf{u}}$ being the axis of the system. In the end we obtain[9] the drag force to be

$$F_{dg} = 6\pi \eta u P. \tag{1.29}$$

[8] Of course, instead of u_i = constant, we had in the previous case an ambient field $u_i = \alpha_{ij}x_j$.

[9] An outline of the calculation is provided in SuppMat Section 1.6.

which was first obtained by George Stokes (1819–1903). Clearly one would expect the frictional force to be proportional to the viscosity coefficient η and to fluid velocity u (or what amounts to the same thing, the velocity of the sphere in a stationary fluid). One can further deduce that, to obtain the correct dimension for the force, the product ηu has to be multiplied by a length factor.[10] Since the only length-scale available in this problem is the spherical radius P, this allows us to understand all the nontrivial factors in the Stokes formula. The remaining part is to calculate the coefficient, which turns out to be 6π.

We can also state the Stokes formula as a result for the mobility of the immersed particle. Under the action of some force F, a particle acquires some velocity u; the proportionality coefficient is defined to be the **mobility** of the particle

$$u = \mu F. \tag{1.30}$$

Thus in our case we can interpreted Stokes' formula (1.29) as yielding an expression for the mobility

$$\mu = (6\pi \eta P)^{-1}. \tag{1.31}$$

This viewpoint will be developed further when we discuss the fluctuation–dissipation relation in the next chapter.

When osmotic force is balanced by frictional force

As the frictional drag force (1.29) increases with velocity, until the osmotic force is balanced by the frictional force $F_{os} = F_{dg}$ and the velocity reaches its terminal value $u_{max} = \omega$, we have

$$6\pi \eta \omega P = -\frac{RT}{\rho N_A} \frac{\partial \rho}{\partial x} \quad \text{or} \quad \rho \omega = -\frac{1}{N_A P} \frac{RT}{6\pi \eta} \frac{\partial \rho}{\partial x}. \tag{1.32}$$

From the definition of the diffusion coefficient given in (1.23), we see that Eq. (1.32) gives, for our case:

$$D = \frac{1}{N_A P} \frac{RT}{6\pi \eta}, \tag{1.33}$$

which may be written in more compact form as (1.22) where we have used the expression (1.31) for the mobility μ and Boltzmann's constant $k_B = R/N_A$. This relation is just the "second relation" (1.2) that Einstein obtained[11] in his doctoral thesis. He then solved these two coupled equations (1.1) and (1.2) to obtain the parameters of molecular dimension, P and N_A.

[10] From (1.28) we have the dimension relation of [force] = $[\sigma_{ij}]$ [length]^2 while the dimension of the viscous stress tensor σ_{ij} in (1.27) has $[\eta]$ [time]$^{-1}$. Thus, we have the dimension relation of [force] = $[\eta]$ [velocity] [length].

[11] This relation was also obtained independently by Marian Smoluchowski; hence it is sometimes referred to as the Einstein–Smoluchowski relation.

1.4 SuppMat: Basics of fluid mechanics

In this section we provide a rapid introduction to some of the basic elements in fluid dynamics. We review the equation of motion for an ideal fluid, showing that it can also be interpreted as a statement of momentum conservation.

1.4.1   The equation of continuity

The conservation law of a continuum system (i.e. a field system) is written as the **equation of continuity**. For the simplest case when the conserved quantity is a scalar (e.g. energy, mass, or charge) the equation of continuity is

$$\frac{\partial \rho}{\partial t} + \nabla \cdot \rho \mathbf{v} = 0. \tag{1.34}$$

For definiteness, let us say we are discussing mass conservation. Performing a volume integration on both terms: the first term becomes the rate of change of mass in this volume $\int \rho dV$, the second one is, by the divergence theorem,[12] the mass flux $\rho \mathbf{v}$ passing through the surface surrounding this volume. Thus the equation is a conservation statement. We note that the flux density ρv_i in the ith direction[13] has the physical significance of

$$\rho v_i = \frac{\text{mass}}{\text{time} \times (\text{cross-sectional area} \perp \text{to } v_i)}. \tag{1.35}$$

[12]The divergence theorem (also known as Gauss' theorem) is presented in Appendix A. See Eq. (A.25).

[13]The index i ranges over $1, 2, 3$ for the x, y, z components.

Incompressible fluid

We shall assume that the fluid is incompressible; namely, the change of the mass density $d\rho$ is small when compared to the flux change of the velocity field, i.e. ρ can be treated as a constant. Equation (1.34) then turns this condition into the statement that the velocity field of an incompressible fluid has zero divergence:

$$\nabla \cdot \mathbf{v} = 0. \tag{1.36}$$

1.4.2   The Euler equation for an ideal fluid

The Euler equation as the equation of motion for an ideal fluid

The "$F = ma$" equation, the inertial force law, for an ideal fluid takes on the form

$$-\nabla p = \rho \frac{D\mathbf{v}}{Dt} \tag{1.37}$$

where we have on the RHS (mass density times acceleration) the **substantial derivative** of the velocity field[14] $\mathbf{v}(t, \mathbf{r})$:

$$\frac{D\mathbf{v}}{Dt} = \frac{\partial \mathbf{v}}{\partial t} + (\mathbf{v} \cdot \nabla) \mathbf{v}. \tag{1.38}$$

The gradient of the pressure field on the LHS can be viewed as the local force per unit volume. This interpretation is particularly clear, if one takes pressure as a vector field, i.e. the force vector perpendicular to the enclosure surface, so that ∇p can be replaced by the divergence $\nabla \cdot \mathbf{p}$. In this way, one sees that the volume integral of the divergence of the "pressure vector" turns into (via the divergence theorem) a surface integral

$$-\int \nabla \cdot \mathbf{p} dV = \oint p dS. \tag{1.39}$$

[14]While the first term on the RHS is the change of velocity during dt at a fixed point, the second term expresses the difference of velocities (at a given instance) at two points separated by the distance that the fluid element has moved during dt. Namely, these two terms together are the change of velocity of a fluid element as it moves about in space. The substantial derivative of the velocity thus expresses the true acceleration of the particle, hence is proportional to the applied force.

Since pressure is force per unit surface area, the surface integral on the RHS has the physical meaning of force. This then suggests that $-\nabla p$ should be viewed as force for unit volume, and the Euler equation (1.37) is the equation for the second law of motion for the density functions of an ideal fluid.

The Euler equation as the momentum conservation equation

We can look at the Euler equation (1.37) in another way. It is the equation of continuity expressing conservation of fluid momentum—just as Eq. (1.34) expresses mass conservation. For the momentum field, instead of a simple mass density ρ and a flux $\rho\mathbf{v}$, one needs to work with a momentum density $\rho\mathbf{v}$, which is a vector (a one-index tensor); and so a momentum flux, analogous to Eq. (1.35), must be a two-index tensor[15] (think of it as a 3×3 matrix)

$$T_{ij} = \frac{i\text{th component of momentum}}{\text{time} \times (\text{cross-sectional area} \perp \text{to } v_j)}. \tag{1.40}$$

We leave it as an exercise for the reader to show that it is a symmetric tensor $T_{ij} = T_{ji}$. The corresponding expression of momentum conservation, a generalization of (1.34), is a set of three continuity equations (one for each index value of i):

$$\frac{\partial}{\partial t}(\rho v_i) + \frac{\partial}{\partial x_j}T_{ij} = 0. \tag{1.41}$$

The momentum flux tensor T_{ij} for an ideal fluid is given by τ_{ij} defined by

$$\tau_{ij} = p\delta_{ij} + \rho v_i v_j. \tag{1.42}$$

This represents the completely reversible transfer of momentum, due simply to the mechanical transport of different fluid particles and the pressure forces acting on the fluid.[16]

Using this $T_{ij} = \tau_{ij}$ in Eq. (1.41),

$$v_i\frac{\partial\rho}{\partial t} + \rho\frac{\partial v_i}{\partial t} + \frac{\partial p}{\partial x_i} + \rho v_j\frac{\partial v_i}{\partial x_j} + v_i\frac{\partial\rho v_j}{\partial x_j} = 0,$$

namely,

$$-\frac{\partial p}{\partial x_i} = \rho\left(\frac{\partial v_i}{\partial t} + v_j\frac{\partial v_i}{\partial x_j}\right) + v_i\left(\frac{\partial\rho}{\partial t} + \frac{\partial\rho v_j}{\partial x_j}\right),$$

we see that this is just the Euler equation (1.37) because the first term on the RHS is the substantial derivative of velocity (1.38) and the last term vanishes because of mass conservation Eq. (1.34).

1.5 SuppMat: Calculating the effective viscosity

Here we provide some details of Einstein's calculation of heat loss in a viscous fluid with suspended particles, which are treated as rigid spheres with radius P. We will first calculate the effect of one solute particle; the total effect of all the solute particles is assumed to be a simple sum of individual molecules. A helpful textbook reference is Landau and Lifshitz (1959).

[15]This must be a quantity with two indices because, as seen in the following equation, it is the ratio of two quantities, each being a quantity with an independent index. This is in contrast to the flux density shown in (1.35) where the numerator is a scalar (without an index). We can also see from Eq. (1.5) that the two-index property is required as the second term must, like the first term, be a term with one index, after the divergence operation "takes away" one of the indices.

[16]The explicit form displayed in (1.6) can be easily understood from the definition of the stress tensor given in (1.40). For example, the result of $\tau_{11} = p + \rho v_x^2$ is consistent with the definition

$$T_{11} = \frac{x \text{ component of momentum}}{\text{time} \times (\text{area in } yz \text{ plane})}$$

which can be interpreted as "pressure" because

$$T_{11} = \frac{x \text{ component of force}}{(\text{area in } yz \text{ plane})}$$

as well as "mass density $\times v_x^2$" because

$$T_{11} = \frac{\text{mass} \times (x \text{ component of velocity})^2}{(\text{volume in } xyz \text{ space})}.$$

For further discussion see Section 11.5.

1.5.1   The induced velocity field v'

The Navier–Stokes equation (1.8) will be solved for the induced velocity v' due to the presence of a solute particle in the fluid. The LHS terms in the Navier–Stokes equation represent the inertia force, while the RHS represents pressure and viscous forces, respectively. As stated in the listed assumptions at the beginning of Section 1.2, we shall ignore the inertia force,[17] the terms on the LHS of Eq. (1.8). The equation we need to solve is now

$$\nabla p = \eta \triangle \mathbf{v}. \tag{1.43}$$

Taking the curl of this equation, the LHS being the curl of a gradient must vanish[18] so we have from the RHS

$$\triangle(\nabla \times \mathbf{v}) = 0. \tag{1.44}$$

Since all equations are linear, the induced velocity and pressure also satisfy these equations. In particular we also have the incompressible condition $\nabla \cdot \mathbf{v}' = 0$. This means that $\mathbf{v}'$ itself must be some curl $\mathbf{v}' = \nabla \times \mathbf{A}$, where $\mathbf{A}$ must be an axial vector related to the ambient velocity field u. It is not difficult to see that A_i must itself be a curl of the gradient, projected in the direction of the ambient velocity $u_i = \alpha_{ij}x_j$:

$$v'_i = \left[\nabla \times \nabla \times \boldsymbol{\alpha} \cdot \nabla f(r)\right]_i = \partial_i\left[\nabla \cdot (\boldsymbol{\alpha} \cdot \nabla f)\right] - \triangle(\boldsymbol{\alpha} \cdot \nabla f)_i \tag{1.45}$$

where the RHS is reached by using (A.12). The scalar function can be fixed by the Navier–Stokes equation (1.44) in the form of $f(r) = ar + b/r$, with (a, b) being constants.[19] Let us work out the terms on the RHS of (1.45) for such an $f(r)$:

$$\nabla \cdot (\boldsymbol{\alpha} \cdot \nabla f) = \partial_i \alpha_{ij} \partial_j f = \alpha_{ij} \partial_i \partial_j \left(ar + \frac{b}{r}\right)$$

$$= \alpha_{ij} \partial_i \left[x_j \left(\frac{a}{r} - \frac{b}{r^3}\right)\right] = \alpha_{ij} x_i x_j \left(-\frac{a}{r^3} + \frac{3b}{r^5}\right),$$

where we have used $\alpha_{ii} = 0$ to reach the last expression. Similarly,

$$\partial_i\left[\nabla \cdot (\boldsymbol{\alpha} \cdot \nabla f)\right] = \alpha_{ij} x_j \left(-\frac{2a}{r^3} + \frac{6b}{r^5}\right) + \alpha_{jk} x_i x_j x_k \left(\frac{3a}{r^5} - \frac{15b}{r^7}\right) \tag{1.46}$$

and

$$-\triangle(\boldsymbol{\alpha} \cdot \nabla f)_i = -(\boldsymbol{\alpha} \cdot \nabla)_i \triangle f = -(\boldsymbol{\alpha} \cdot \nabla)_i \frac{2a}{r} = \alpha_{ij} x_j \frac{2a}{r^3}, \tag{1.47}$$

which just cancels the first RHS term in (1.46). Substituting (1.46) and (1.47) and adding the ambient velocity field $u_i = \alpha_{ij}x_j$ into (1.45) we have for the full velocity field

$$v_i = u_i + v'_i = \alpha_{ij} x_j \left(1 + \frac{6b}{r^5}\right) + \alpha_{jk} x_i x_j x_k \left(\frac{3a}{r^5} - \frac{15b}{r^7}\right). \tag{1.48}$$

Imposing the boundary condition

The two constants (a, b) can then be fixed by the boundary condition that, at the surface of the solute particle $r = P$, the velocity vanishes $v_i = 0$. This can

[17]Technically speaking, we are in the regime of small **Reynolds number** (the pure number $\rho vP/\eta$)—a measure of nonlinearity. Otherwise turbulence would set in and pure streamline motion completely breaks down.

[18]Cf. Eq. (A.20). For a discussion of this and other subsequent vector calculus relations, see Appendix A, Section A.1.

[19]The constraint imposed by Eq. (1.44) turns into a differential equation on the scalar function of $\triangle^2 f = 0$. This leads, with the appropriate boundary condition, to $\triangle f = 2a/r$, which in turn has the solution $f = ar + b/r$.

happen only if both tensor coefficients vanish. We have two equations to solve for the two unknowns, leading to

$$a = -\frac{5}{6}P^3, \qquad b = -\frac{1}{6}P^5. \tag{1.49}$$

1.5.2 The induced pressure field p'

Knowing the induced velocity field v' as given in (1.48), we now solve for the induced pressure from (1.43), $\nabla p' = \eta \triangle v'$. Our strategy is to turn the RHS also into a gradient so we can identify p' directly. Recall that v' has the form as given in (1.45)

$$\triangle v' = \triangle \left[\nabla \times \nabla \times (\boldsymbol{\alpha} \cdot \nabla f) \right] = \nabla \times \nabla \times (\boldsymbol{\alpha} \cdot \nabla) \triangle f = \nabla \left[\nabla \cdot (\boldsymbol{\alpha} \cdot \nabla) \triangle f \right]$$

so that the RHS[20] is a pure gradient. The Navier–Stokes equation then gives us the result

$$p' = \eta \left[\nabla \cdot (\boldsymbol{\alpha} \cdot \nabla) \triangle f \right] = \eta \alpha_{ij} \partial_i \partial_j \frac{2a}{r} = \eta \alpha_{ij} x_i x_j \frac{6a}{r^5}. \tag{1.50}$$

[20] To reach the RHS we have used the cross product relation of Eq. (A.12) and dropped a term (proportional to $\triangle^2 f = 0$) from the result that follows from this identity.

1.5.3 Heat dissipation in a fluid with suspended particles

We are now ready to calculate the dissipation of energy in a viscous fluid with solute particles. Again we use the basic formula of Eq. (1.11).

$$W = -\int_V \sigma_{ki} \partial_k v_i dV = -\int_V \partial_k (\sigma_{ki} v_i) \, dV \tag{1.51}$$

where we have, consistent with our approximation of the Navier–Stokes equation of ignoring inertia force, dropped a term[21] involving $\partial_k \sigma_{ki}$. Up to this point we have concentrated on calculating the velocity and pressure field of a fluid with a single solute sphere, and hence the dissipation effect due to a single solute particle. We shall now attempt to broaden the result to include the case of a fluid with many solute spheres. Our volume integration will now cover a large volume V containing many solute particles. Using the divergence theorem Eq. (1.51) can be turned into a surface integral over the surface S enclosing V:

[21] The stress tensor σ_{ki} includes not only the viscosity term but also the pressure $p\delta_{ki}$. See the discussion following (1.53).

$$W = -\oint_S \sigma_{ki} v_i n_k dS = -2\eta \alpha_{ij} \alpha_{ij} V - \oint_S \sigma'_{ki} \alpha_{ij} x_j n_k dS \tag{1.52}$$

where $n_k = x_k/r$ is a unit vector normal to the surface element. On the RHS we have separated out the dissipation term due to a fluid without solute particles (with velocity field $\mathbf{u}$) as derived in (1.14); what remains is the energy transfer due to the presence of solute particles. Also on the RHS we have kept only the leading-order term (in η)[22] when we replaced the total velocity field v_i by the ambient field of $u_i = \alpha_{ij} x_j$ in the integrand. We recall that

[22] Since σ'_{ki} is $O(\eta)$, we only need to keep the velocity to $O(\eta^0)$ so that $v_i = u_i = \alpha_{ij} x_j$.

$$\sigma'_{ij} = \eta \left(\frac{\partial v'_i}{\partial x_j} + \frac{\partial v'_j}{\partial x_i} \right) - p'\delta_{ij}. \tag{1.53}$$

We will only keep the leading $O(r^{-2})$ term in v' from (1.48) because it will be evaluated over a surface of large distance. We have also added a pressure term[23] because the induced velocity also brings about extra pressure, as calculated above in Eq. (1.50):

$$v_i' = -\frac{5}{2}P^3\alpha_{lm}\frac{x_ix_lx_m}{r^5} \quad \text{and} \quad p' = -5P^3\eta\alpha_{ij}\frac{x_ix_j}{r^5}. \tag{1.54}$$

On the other hand, if the fluid with solute particles has an effective viscosity coefficient η^*, that leads us to identify the LHS of (1.52) with $W = -2\eta^*\alpha_{ij}\alpha_{ij}V$. We then have an expression for the extra dissipative effect due to the presence of solute particles:

$$2\left(\eta^* - \eta\right)\alpha_{ij}\alpha_{ij}V = \oint_S \sigma_{ki}'\alpha_{ij}x_jn_kdS. \tag{1.55}$$

Integrate over the surface area of all the immersed spheres

We will now take into account the effect of not only a single solute particle, but all the immersed spheres. Our assumption is that the solute particles are sufficiently far apart that they do not disturb each other's effect too much. Their total influence is a simple sum of the individual effects: We will first convert the surface integral in (1.55), which is over the large area S over the large volume V containing all the suspended spheres, into one over the surface area of all the immersed spheres—in such a way that one does not have to deal with the interior of these spheres, but only the fluid outside. We do this by subtracting, and then adding back, these individual area integrals, each over the spherical surface S_1 that encloses a spherical volume V_1:

$$\alpha_{ij}\oint_S \sigma_{ki}'x_jn_kdS = \alpha_{ij}\oint_{S-\Sigma S_1} \sigma_{ki}'x_jn_kdS + \alpha_{ij}\sum\oint_{S_1} \sigma_{ki}'x_jn_kdS. \tag{1.56}$$

Concentrating on the first term on the RHS, we apply the divergence theorem to get

$$\alpha_{ij}\oint_{S-\Sigma S_1} \sigma_{ki}'x_jn_kdS = \alpha_{ij}\int_{V-\Sigma V_1} \partial_k\left(\sigma_{ki}'x_j\right)dV = \alpha_{ij}\int_{V-\Sigma V_1} \sigma_{ij}'dV. \tag{1.57}$$

Again, consistent with our approximation as explained above, we have dropped a $\partial_k\sigma_{ki}'$ term. We can use (1.53) to convert this result back to a surface term[24]

$$\alpha_{ij}\int_{V-\Sigma V_1} \sigma_{ij}'dV = 2\eta\alpha_{ij}\int_{V-\Sigma V_1} \partial_iv_j'dV \tag{1.58}$$

$$= 2\eta\alpha_{ij}\int_S v_j'n_idS - 2\eta\alpha_{ij}\int_{\Sigma S_1} v_j'n_idS. \tag{1.59}$$

The first term on the RHS vanishes because $\mathbf{v}' = 0$ at the distant large surface S, and only small spherical surface S_1 terms (the second term) contribute. Combining with the second term of (1.56) we have from (1.55).

$$2\left(\eta^* - \eta\right)\alpha_{ij}\alpha_{ij} = \frac{\alpha_{ij}}{V}\sum\oint_{S_1}\left(\sigma'_{ki}x_jn_k - 2\eta v'_jn_i\right)dS \qquad (1.60)$$

$$= \frac{\alpha_{ij}}{V}\sum\oint_{S_1}\left[\eta\left(\partial_k v'_i + \partial_i v'_k\right)x_jn_k - p'x_jn_i - 2\eta v'_jn_i\right]dS.$$

To reach the final expression, we have used the expression for σ'_{ki} from (1.53).

Plug in the contribution due to each solute particle

We now substitute in (1.60) the solutions (1.54) of the Navier–Stokes equation (for a fluid with a single solute particle). It must be forewarned that the contributions from the two terms in $\left(\partial_k v'_i + \partial_i v'_k\right)$, although similar in appearance, are not equal and they must be calculated separately:

$$\partial_k v'_i = -\frac{5}{2}P^3\left(\alpha_{lm}\frac{\delta_{ik}x_lx_m}{r^5} + 2\alpha_{km}\frac{x_ix_m}{r^5} - 5\alpha_{lm}\frac{x_ix_kx_lx_m}{r^7}\right)$$

$$\partial_i v'_k = -\frac{5}{2}P^3\left(\alpha_{lm}\frac{\delta_{ik}x_lx_m}{r^5} + 2\alpha_{im}\frac{x_kx_m}{r^5} - 5\alpha_{lm}\frac{x_ix_kx_lx_m}{r^7}\right).$$

Thus

$$\left(\partial_k v'_i + \partial_i v'_k\right) = -5P^3\left(\alpha_{lm}\frac{\delta_{ik}x_lx_m}{r^5} - 5\alpha_{lm}\frac{x_ix_kx_lx_m}{r^7}\right)$$
$$- 5P^3\left(\alpha_{km}\frac{x_ix_m}{r^5} + \alpha_{im}\frac{x_kx_m}{r^5}\right)$$

and, with $n_k = x_k/r$,

$$\eta\left(\partial_k v'_i + \partial_i v'_k\right)x_jn_k = 15\eta P^3\alpha_{lm}\frac{x_ix_jx_lx_m}{r^6} - 5\eta P^3\alpha_{ik}\frac{x_kx_j}{r^4}, \qquad (1.61)$$

while the other two terms in (1.60) yield

$$-p'x_jn_i = 5\eta P^3\alpha_{lm}\frac{x_ix_jx_lx_m}{r^6} \qquad (1.62)$$

$$-2\eta v'_jn_i = 5\eta P^3\alpha_{lm}\frac{x_ix_jx_lx_m}{r^6}.$$

Altogether we have the result

$$\oint_{S_1}\left[\eta\left(\partial_k v'_i + \partial_i v'_k\right)x_jn_k - p'x_jn_i - 2\eta v'_jn_i\right]dS$$

$$= \oint_{S_1}\left[25\eta P^3\alpha_{lm}\frac{n_in_jn_ln_m}{r^2} - 5\eta P^3\alpha_{ik}\frac{n_kn_j}{r^2}\right]r^2d\Omega = 5\eta\frac{4\pi P^3}{3}\alpha_{ij} \qquad (1.63)$$

where to reach the last expression, we have used the well-known identities of integrating over the full solid angle[25]

$$\oint n_kn_jd\Omega = \frac{4\pi}{3}\delta_{kj} \qquad (1.64)$$

$$\oint n_in_jn_ln_md\Omega = \frac{4\pi}{15}\left(\delta_{ij}\delta_{lm} + \delta_{il}\delta_{jm} + \delta_{im}\delta_{jl}\right).$$

[25] One first 'guesses' the tensor structure, knowing that it must be constructed from the only invariant tensor, the Kronecker delta, with matching symmetry properties. The proportionality constant can then be fixed by selecting a particular index combination.

Summing over the contribution by all immersed spheres

We then obtain the final result by plugging (1.63) into (1.60):

$$2 \left(\eta^* - \eta \right) \alpha_{ij} \alpha_{ij} = 5 \eta \alpha_{ij} \alpha_{ij} \sum \frac{4\pi P^3 / 3}{V},$$

where the summation is over the contribution from all solute particles. We have φ as the fraction of the volume occupied by the immersed spheres

$$\varphi = \sum \frac{4\pi P^3 / 3}{V} \tag{1.65}$$

and the resulting expression is simply the relation quoted above in Eq. (1.3),

$$\eta^* = \eta \left(1 + \frac{5}{2} \varphi \right). \tag{1.66}$$

1.6   SuppMat: The Stokes formula for the viscous force

Here we outline the calculation of the frictional force on an immersed sphere in a viscous fluid. Since the calculational stages are similar to those used in SuppMat Section 1.5, we shall often just point out the corresponding steps. In both calculations we have a velocity composed of the ambient field and an induced field $\mathbf{v} = \mathbf{u} + \mathbf{v}'$. In the effective viscosity calculation we have $u_i = \alpha_{ij} x_j$ while in the Stokes case a constant u_i. To fix the induced velocity, both cases involve the Navier–Stokes equation in the form $\triangle (\nabla \times v') = 0$. In the previous case this leads to $v_i' = \left[\nabla \times \nabla \times \alpha \cdot \nabla f(r) \right]_i$, while the present case gives $v_i' = \left[\nabla \times \nabla \times \mathbf{u} f(r) \right]_i$. Both scalar functions have the same form $f(r) = ar + b/r$. Compared to the velocity and pressure fields of (1.48) and (1.50), here we have

$$v_i' = -u_i \left(\frac{a}{r} + \frac{b}{r^3} \right) + \mathbf{n} \cdot \mathbf{u} n_i \left(-\frac{a}{r} + \frac{3b}{r^3} \right) \tag{1.67}$$

and

$$p' = -\eta \mathbf{n} \cdot \mathbf{u} \frac{2a}{r^2}, \tag{1.68}$$

with the unit vector $n_i = x_i / r$. Because of the difference in ambient fields, the boundary conditions lead to different values for the constants: here we have

$$a = \frac{3}{4} P \qquad b = \frac{1}{4} P^3. \tag{1.69}$$

Setting the axis of the spherical coordinate system to be along the velocity $\mathbf{u}$, so that we have $\mathbf{n} \cdot \mathbf{u} = u_r = u \cos \theta$ and $u_\theta = -u \sin \theta$, leads to the full velocity components

$$v_r = u \cos \theta \left(1 - \frac{3P}{2r} + \frac{P^3}{2r^3} \right) \qquad \text{and} \qquad v_\theta = -u \sin \theta \left(1 - \frac{3P}{4r} - \frac{P^3}{4r^3} \right)$$

and the viscosity stress tensor elements

$$\sigma'_{rr} = 2\eta \frac{\partial v_r}{\partial r} = 0 \tag{1.70}$$

$$\sigma'_{r\theta} = \eta \left(\frac{\partial v_r}{r\partial \theta} + \frac{\partial v_\theta}{\partial r} - \frac{v_\theta}{r} \right) = \frac{3\eta}{2P} u \sin \theta.$$

Including the induced pressure (1.68) as well in Eq. (1.28), we have

$$F_{dg} = -\oint \left[\sigma'_{ij} + p' \delta_{ij} \right] \hat{u}_i n_j dS = -\oint \left[\sigma'_{ir} \hat{u}_i + p' \hat{u}_r \right] dS$$

$$= \oint \left[-\sigma'_{rr} \cos \theta + \sigma'_{r\theta} \sin \theta - p' \cos \theta \right]_{r=P} dS$$

$$= \oint \left(0 + \frac{3\eta u}{2P} \sin^2 \theta + \frac{3\eta u}{2P} \cos^2 \theta \right) dS$$

$$= \frac{3\eta u}{2P} 4\pi P^2 = 6\pi \eta u P,$$

which is Stokes formula, Eq. (1.29).

The Brownian motion

- Einstein advanced the first satisfactory theory of Brownian motion— the jiggling motion of suspended particles in a liquid as seen under a microscope. It provides us with direct visual evidence for the existence of the point-like structure of matter. The theory suggests that we can see with our own eyes the molecular thermal motion.

- The Brownian motion paper may be viewed as part of Einstein's doctoral dissertation work on the atomic structure of matter; it continues his pursuit of the idea that particles suspended in a fluid behave like molecules in solution. The motion of a Brownian particle is governed by the diffusion equation.

- Einstein was the first to provide a statistical derivation of the diffusion equation. From its solution one can calculate its variance, showing diffusion as fluctuations of a discrete system, like the prototype case of random walks. The mean-square displacement of a Brownian particle is related to the diffusion coefficient as $\langle x^2 \rangle = 2Dt$.

- The Einstein–Smoluchowski relation (already discussed in the previous chapter) between diffusion and viscosity, $D = k_B T/(6\pi \eta P)$ with η being the viscosity coefficient, is the first fluctuation–dissipation relation ever noted. This theory not only illuminates diffusion but also explains friction by showing that they both spring from the same underlying thermal process.

- Verification of Einstein's theory came about through the painstaking experimental work of Jean Perrin. This work provided another means to measure the molecular size P and Avogadro's number N_A. It finally convinced everyone, even the skeptics, of the reality of molecules.

Eleven days after Einstein completed his thesis on April 20, 1905, he submitted this "Brownian motion paper" to *Annalen der Physik* (Einstein 1905c). This paper can be regarded as part of Einstein's dissertation research and it represents the culmination of his study of atomic structure of matter (extending back at least to 1901) by explaining the Brownian motion. To many of us, before the advent of (field-ion) "atomic" microscopes in the 1960s, the most direct visual evidence for atoms' existence was viewing the jiggling motion of suspended particles (e.g. pollen) in a liquid, as seen under a microscope. This

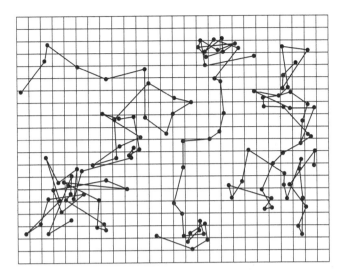

Fig. 2.1 The zigzag motion of Brownian particles as sketched by Perrin, at 30-second intervals. The grid size is 3.2μ and the radius of the particle is 0.53μ. Reproduction of Fig. 6 in Perrin (1909).

"Brownian motion" was originally discussed in 1827 by botanist Robert Brown (1773–1858). Although it was suggested soon afterwards that such Brownian motion is an outward manifestation of the molecular motion postulated by the kinetic theory of matter, it was not until 1905 that Einstein was finally able to advance a satisfactory theory.

Einstein pioneered several research directions in his Brownian motion paper. In particular he argued that, while thermal forces change the direction and magnitude of the velocity of a suspended particle on such a small time-scale that it cannot be measured, the mean-square displacement (the overall drift) of such a particle is an observable quantity, and can be calculated in terms of molecular dimensions. One cannot but be amazed by the fact that Einstein found a physics result so that a careful measurement of this zigzag motion (see Fig. 2.1) through a simple microscope would allow us to deduce Avogadro's number!

It is also interesting to note that the words "Brownian motion" did not appear in the title of Einstein's paper (Einstein 1905c), even though he conjectured that the motion he predicted was the same as Brownian motion. He was prevented from being more definitive because he had no access then to any literature on Brownian motion. One should remember that in 1905 Albert Einstein was a patent office clerk in Bern and did not have ready access to an academic library and other research tools typically associated with a university.

2.1 Diffusion and Brownian motion

Einstein argued that the suspended particles in a liquid, differing in their statistical and thermal behavior from molecules only in their sizes, should obey the same diffusion equation that describes the chaotic thermal motion of the liquid's constituent molecules. Here we follow Einstein in his derivation of the Brownian motion equation and show that it is just the diffusion equation.

2.1.1   Einstein's statistical derivation of the diffusion equation

Einstein assumed that each individual particle executes a motion that is independent of the motions of all the other particles; the motions of the same particle in different time intervals are also mutually independent processes, so long as these time intervals are chosen not to be too small. For simplicity of presentation, we will work in a one-dimensional (1D) model. One is interested in the displacement $x(t)$ after the particle makes a large number (N) of discrete displacement steps of size Δ in a time interval τ. The probability density $f_\tau(\Delta)$ is introduced as

$$f_\tau(\Delta)d\Delta = \frac{dN}{N}, \quad \text{or} \quad dN = Nf_\tau(\Delta)d\Delta. \tag{2.1}$$

The probability should clearly be the same whether the step is taken in the forward or backward direction: $f_\tau(\Delta) = f_\tau(-\Delta)$.

Let $\rho(x, t)$ be the number of particles per unit volume. One can then calculate the distribution at time $t + \tau$ from the distribution at time t. The change of particle density at the spatial interval $(x, x + dx)$ is due to particles flowing in from both directions (hence the $\pm\infty$ limits),

$$\rho(x, t + \tau) = \int_{-\infty}^{+\infty} \rho(x + \Delta, t)f_\tau(\Delta)d\Delta. \tag{2.2}$$

One then makes Taylor series expansions on both sides of this equation, the LHS being

$$\rho(x, t + \tau) = \rho(x, t) + \tau\frac{\partial\rho(x, t)}{\partial t} + \cdots \tag{2.3}$$

and the RHS being

$$\int_{-\infty}^{+\infty} \left[\rho(x, t) + \Delta\frac{\partial\rho(x, t)}{\partial x} + \frac{\Delta^2}{2!}\frac{\partial^2\rho(x, t)}{\partial x^2} + \cdots\right]f_\tau(\Delta)d\Delta$$
$$= \rho(x, t) + \left[\int_{-\infty}^{+\infty} \frac{\Delta^2}{2}f_\tau(\Delta)d\Delta\right]\frac{\partial^2\rho(x, t)}{\partial x^2} + \cdots \tag{2.4}$$

where we have used the conditions that the probability must add to unity and $f_\tau(\Delta)$ is an even function of Δ:

$$\int_{-\infty}^{+\infty} f_\tau(\Delta)d\Delta = 1, \quad \text{and} \quad \int_{-\infty}^{+\infty} \Delta f_\tau(\Delta)d\Delta = 0. \tag{2.5}$$

Equating the leading terms of both the LHS and RHS, we obtain the diffusion equation

$$\frac{\partial\rho}{\partial t} = D\frac{\partial^2\rho}{\partial x^2}, \tag{2.6}$$

with the diffusion coefficient being related to the probability density as

$$D = \frac{1}{2\tau}\int_{-\infty}^{+\infty} \Delta^2 f_\tau(\Delta)d\Delta. \tag{2.7}$$

In practice D can be obtained from experiment. We should remark that the fundamental assumption in this derivation is that $f_\tau(\Delta)$ depends only on Δ, not on previous history. Such a process we now call "Markovian" in the study of random processes. Einstein's Brownian investigation is one of the pioneering papers laying the foundation for a formal theory of stochastic processes.

In Chapter 1 we defined the diffusion coefficient D through (1.23) as opposed to its introduction in Eq. (2.6). Their equivalence can be demonstrated by taking the gradient of both terms in Fick's first law (1.23) and turning the equation into (2.6), sometimes called Fick's second law, by using the equation of continuity (written in 1D form again for simplicity[1])

$$\frac{\partial\rho}{\partial t} + \frac{\partial\rho v}{\partial x} = 0. \tag{2.8}$$

[1] Just as the continuity equation in 3D is written as $\partial_t\rho + \nabla \cdot \rho v = 0$, Fick's second law has the 3D form of $\partial_t\rho = D\nabla^2\rho$.

2.1.2   The solution of the diffusion equation and the mean-square displacement

The solution to the diffusion equation (2.6) is a Gaussian distribution (*Exercise*: check that this is the case)

$$\rho(x,t) = \frac{1}{\sqrt{4\pi Dt}}e^{-x^2/4Dt} \tag{2.9}$$

which is a bell-shaped curve, peaked at $x = 0$. Initially ($t = 0$) the density function is a Dirac delta function $\rho(x) = \delta(x = 0)$; as t increases, the height of this peak, still centered around $x = 0$, shrinks but the area under the curve remains unchanged. In other words, the probability of finding the particle away from the origin (as given by the density ρ) increases with time. There is, on the average, a drift motion away from the origin (cf. Fig. 2.2). One can easily check that it is properly normalized using the familiar result of Gaussian integrals (cf. Appendix A.2).

Clearly the curve in Fig. 2.2 is symmetric with respect to $\pm x$. This implies a vanishing average displacement $\langle x \rangle = 0$. But we have a nonzero mean-square displacement (the variance) that monotonically increases with time [cf. Eq. (A.44)]:

$$\langle x^2 \rangle = \int_{-\infty}^{\infty} x^2 \rho dx = 2Dt. \tag{2.10}$$

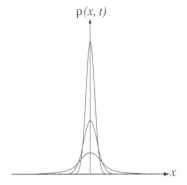

$\rho(x, t)$

Fig. 2.2 Density distribution plotted for various values of t. At $t = 0$ it is a delta function at $x = 0$; as t increases, the distribution becomes broader and spreads out.

We note that $\langle x^2 \rangle$ is just the width of the Gaussian distribution. Thus the broadening of the bell-shaped curve with time (as shown in Fig. 2.2) just reflects the increase of the mean-square displacement.

The basic point is that while the fluid density represents the overall coarse (i.e. averaged) description of the underlying molecular motion, still, once the density function is known, the fluid fluctuation of such motion, like the root-mean-square (i.e. the variance) displacement, can then be calculated. Such a fluctuation property shows up as an observable drift motion of the immersed Brownian particles.

2.2 Fluctuations of a particle system

To substantiate the claim that Brownian motion is evidence of the point-like structure of matter, we now directly connect this Gaussian distribution (2.9) and the mean-square displacement (2.10) to the fluctuations of a particle system.[2] Random walks being the prototype of a discrete system, we first discuss the fluctuation phenomenon associated with this system.

[2]This is to be contrasted with the fluctuation of a system of waves to be discussed in Section 6.5.

2.2.1 Random walk

Consider a particle moving in one dimension at regular time intervals in steps of equal size. At each instance it moves forward and backward at random. Let a walk of N steps end at k steps from the initial point. F is the number of forward steps and B the backward steps. Thus $F + B = N$ and $F - B = k$. After a large number of such walks (each ends at a different position), one is interested in the distribution of the end-point positions. Namely, we seek the probability $p(k, N)$ of finding the particle at k steps from the origin after making N steps. Since at each step one makes a two-valued choice, there are a total 2^N possible outcomes; the probability is evidently

$$p(k, N) = \frac{N!/F!B!}{2^N} = \frac{2^{-N}N!}{\left(\frac{N+k}{2}\right)!\left(\frac{N-k}{2}\right)!}. \tag{2.11}$$

Since we expect the result to have the form of an exponential, we proceed by first taking the logarithm of this expression,

$$\ln p = -N \ln 2 + \ln N! - \ln \left(\frac{N+k}{2}\right)! - \ln \left(\frac{N-k}{2}\right)!.$$

[3]See Section A.3 for a proof of Stirling's formula.

Using Stirling's formula[3] of $\ln X! \simeq X \ln X - X$ for large X, we have

$$\ln p = -N \ln 2 + N \ln N - N - \left(\frac{N+k}{2}\right) \ln \left(\frac{N+k}{2}\right) + \left(\frac{N+k}{2}\right)$$
$$- \left(\frac{N-k}{2}\right) \ln \left(\frac{N-k}{2}\right) + \left(\frac{N-k}{2}\right)$$
$$= -\frac{N}{2} \left[\left(1 + \frac{k}{N}\right) \ln \left(1 + \frac{k}{N}\right) + \left(1 - \frac{k}{N}\right) \ln \left(1 - \frac{k}{N}\right)\right].$$

In the limit of small k/N, the logarithm can be approximated by $\ln(1 + \epsilon) \simeq \epsilon - \epsilon^2/2$. The probability logarithm becomes $\ln p \simeq -k^2/2N$, which can be inverted and normalized (by a standard Gaussian integration) to yield

$$p(k, N) = \frac{1}{\sqrt{2\pi N}} e^{-k^2/2N}. \tag{2.12}$$

Similarly using the Gaussian integral we also have

$$\langle k^2 \rangle = \int k^2 p(k, N) dk = N, \tag{2.13}$$

which is the variance $\langle \Delta k^2 \rangle = \langle k^2 \rangle - \langle k \rangle^2$ because for our case $\langle k \rangle = 0$. In particular one often uses the fractional variance to characterize the fluctuation from the mean, leading to the well-known result

$$\frac{\sqrt{\langle \Delta k^2 \rangle}}{N} = \frac{1}{\sqrt{N}}. \tag{2.14}$$

This is the characteristic of fluctuations in a discrete system.

2.2.2 Brownian motion as a random walk

So far in this calculation no scales have been introduced. For the problem of calculating the displacement of a random walker, we will denote a displacement step by λ and the time interval by τ so that $x = k\lambda$ and $t = N\tau$. This allows us to translate the above probability density into the number density $p(k, N) \rightarrow \rho(x, t)$ with

$$\rho(x, t) = \frac{1}{\sqrt{4\pi Dt}} e^{-x^2/4Dt} \quad \text{with } D = \frac{\lambda^2}{2\tau}. \tag{2.15}$$

Thus we can interpret the solution (2.9) of the diffusion equation as representing, at some small scale, a Brownian particle executing a random walk. This exercise in random walks emphasizes the discrete nature of the molecular process underlying the diffusion phenomenon. We shall have occasion (in Section 6.1) to use this connection in the discussion of Einstein's proposal for a discrete basis of radiation—the quanta of light.

2.3 The Einstein–Smoluchowski relation

In the previous chapter on Einstein's doctoral thesis (Einstein 1905b) we derived this relation between the diffusion coefficient D and the viscosity η:

$$D = \frac{k_B T}{6\pi \eta P} \tag{2.16}$$

with P being the radius of the suspended particle, T the absolute temperature, and k_B Boltzmann's constant, related to the gas constant and Avogadro's number as $k_B = R/N_A$. This relation was also obtained by the Polish physicist Marian Smoluchowski (1872–1917) in his independent work on Brownian motion (Smoluchowski 1906). Hence, it is often referred to as the Einstein–Smoluchowski relation. Einstein in his Brownian motion paper (Einstein 1905c) rederived it and improved its theoretical reasoning in two aspects:

1. In his thesis paper, for the osmotic pressure $F_{os} = -n^{-1}\partial p/\partial x$ as shown in Eq. (1.24), Einstein assumed the validity of the **van't Hoff analogy**—the behavior of solute molecules in a dilute solution are similar to those of an ideal gas, and used the ideal gas law (1.25) to relate the pressure to mass density, obtaining the result:

$$F_{os} = -\frac{RT}{\rho N_A} \frac{\partial \rho}{\partial x}. \tag{2.17}$$

In the Brownian motion paper, he justified the applicability of this result on more general thermodynamical grounds. Consider a cylindrical volume with unit cross-sectional area and length $x = l$. Under an arbitrary virtual displacement δx, the change of internal energy is given by

$$\delta U = -\int_0^l F_{\text{os}}\rho\delta x dx \qquad (2.18)$$

and the change of entropy by[4]

$$\delta S = \int_0^l \frac{R}{N_A}\rho\frac{\partial\delta x}{\partial x}dx = -\frac{R}{N_A}\int_0^l \frac{\partial\rho}{\partial x}\delta x dx. \qquad (2.19)$$

[4]One may be more familiar with the entropy change under an isothermal expansion when written $\delta S = Nk_B\delta V/V = \rho k_B\delta V$. In the case here we have $\delta V = \delta x$ because of unit cross-sectional area.

We have performed an integration by parts in reaching the last expression. The relation (2.17) then follows from the observation that the free energy of a system of suspended particles vanishes for such a displacement, $\delta F = \delta U - T\delta S = 0$.

2. In his thesis paper, Einstein arrived at the result (2.16) by the balance of the osmotic force and the frictional drag force (described by Stokes' law) $F_{\text{os}} = F_{\text{dg}}$ on a single molecule. In the Brownian motion paper this would be obtained by a more general thermodynamical argument. Consider the flow of particles that encounters the viscous drag force F_{dg} reaching the terminal velocity ω. The mobility parameter μ is defined by $\omega = \mu F_{\text{dg}} = -\mu\partial U/\partial x$. The drift current density $j = \rho\omega$ produces a density gradient which in turn produces a counteracting diffusion current. This current is related to the diffusion coefficient by the diffusion equation in the form of Fick's first law (1.23)

$$D\frac{\partial\rho}{\partial x} = \rho\omega = -\rho\mu\frac{\partial U}{\partial x}. \qquad (2.20)$$

On the other hand, at equilibrium we must have the Boltzmann distribution

$$\rho(x) = \rho(0)e^{-U(x)/k_B T} \qquad (2.21)$$

so that

$$\frac{\partial\rho}{\partial x} = -\frac{\rho}{k_B T}\frac{\partial U}{\partial x}. \qquad (2.22)$$

Substituting this into (2.20),

$$D\frac{\rho}{k_B T}\frac{\partial U}{\partial x} = \rho\mu\frac{\partial U}{\partial x}, \qquad (2.23)$$

we have

$$D = \mu k_B T, \qquad (2.24)$$

which is the Einstein–Smoluchowski relation (2.16), upon using the Stokes' law of $\mu = (6\pi\eta P)^{-1}$ as derived in (1.29).

2.3.1 Fluctuation and dissipation

The Einstein–Smoluchowski relation (2.16) is historically the first example of a fluctuation–dissipation theorem, which would turn into a powerful tool in statistical physics for predicting the behavior of nonequilibrium thermodynamical systems. These systems involve the irreversible dissipation of energy into heat from their reversible thermal fluctuations at thermodynamic equilibrium.

As illustrated in our discussion of the Einstein–Smoluchowski relation the fluctuation–dissipation theorem relies on the assumption that the response of a system in thermodynamic equilibrium to a small applied force is the same as its response to a spontaneous fluctuation. Thus Browning motion theory not only illuminates diffusion but also explains friction by showing that they spring from the same underlying thermal process.

2.3.2 Mean-square displacement and molecular dimensions

Having demonstrated that the observable root-mean-square displacement of the Brownian particle was related to the diffusion coefficient D, as shown in (2.10), which in turn can be expressed in terms of molecular dimensions (molecular size P and Avogadro number N_A) through the Einstein–Smoluchowski relation (2.16), Einstein derived the final result of

$$x_{rms} = \sqrt{\langle x^2 \rangle} = \sqrt{2Dt} = \sqrt{\frac{2RT}{N_A} \frac{t}{6\pi \eta P}}. \tag{2.25}$$

This is what we meant earlier when we said "a careful measurement of this zigzag motion through a simple microscope would allow us to deduce the Avogadro number!"

2.4 Perrin's experimental verification

Precise observations of Brownian motion were difficult at that time. The results obtained during the first few years after 1905 were inconclusive. Einstein was skeptical about the possibility of obtaining sufficiently accurate data for such a comparison with theory.

But in 1908 Jean Perrin (1870–1942) entered the field and came up with an ingenious combination of techniques for preparing emulsions with precisely controllable particle sizes,[5] and for measuring particle numbers and displacements. For this series of meticulously carried out brilliant experiments (and other related work) Perrin received the Nobel Prize in physics in 1926. The Brownian motion work was summed up masterfully in his 1909 paper (Perrin 1909) from which we extracted Figs. 2.1 and 2.3.

In particular we have Fig. 2.3 in which Perrin translated 365 projected Brownian paths to a common origin. The end-position of each path is then projected onto a common plane, call it the x-y plane. The radial distance on this plane (labelled by σ) is marked by a series of rings with various σ values. The 3D version of the solution (2.9) of the diffusion equation reads as

[5] Recall that Einstein's calculation assumed equal size P for all suspended particles.

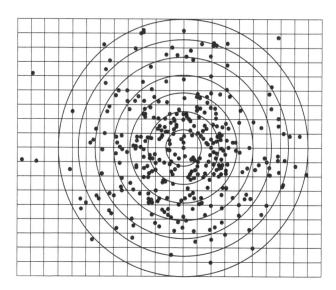

Fig. 2.3 In order to check the diffusion law, Jean Perrin parallel-transported 365 Brownian paths to a common origin. The end-position of each path is then projected onto a common plane. Reproduction of Fig. 7 in Perrin (1909).

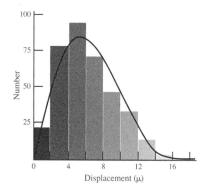

Fig. 2.4 Verification of the diffusion law in Brownian motion. The solid line is the theoretical curve. Each bar in the histogram represents the total number of paths with end-points at given plane-radial distance (σ) from the origin according to the data as shown in Fig. 2.3.

$$\rho(x,t) = \frac{1}{(4\pi Dt)^{3/2}} e^{-r^2/4Dt} \tag{2.26}$$

where r is the 3D radial distance from the origin $r^2 = \sigma^2 + z^2$. As an infinitesimal ring-shaped volume is $2\pi\sigma\,d\sigma\,dz$, one can calculate the number of particles within each of the rings in Fig. 2.3 by a simple integration over the vertical distance

$$\Delta N = \frac{2\pi\sigma\,\Delta\sigma}{(4\pi Dt)^{3/2}} \int_{-\infty}^{+\infty} e^{-(\sigma^2+z^2)/4Dt} dz. \tag{2.27}$$

A simple Gaussian integration yields

$$\Delta N = \frac{\sigma\,\Delta\sigma}{2Dt} e^{-\sigma^2/4Dt}, \tag{2.28}$$

which is the theoretical curve plotted in Fig. 2.4.

Thus experiments were able to confirm in detail the theoretical predictions by Einstein and Smoluchowski. This work finally convinced everyone, even the skeptics, of the reality of molecules.

QUANTUM THEORY

Part II

Blackbody radiation: From Kirchhoff to Planck

- We present in this chapter some historical background for Einstein's revolutionary proposal that under some circumstance radiation could be viewed as a collection of particles. This discovery was carried out in the context of blackbody radiation. We shall cover the era that started with Kirchhoff, showing (based on the second law of thermodynamics) that cavity radiation was independent of the shape and material of the cavity. Thus, the total energy density $u(T)$ as well as the radiation density per unit frequency $\rho(\nu, T)$ [related through $u(T) = \int \rho(\nu, T)\, d\nu$] must be universal functions.

- What are these universal functions of temperature T and frequency ν? Again electromagnetic and thermodynamical reasoning yielded the Stefan–Boltzmann law: $u(T) \propto T^4$, then Wien's displacement law: $\rho(\nu, T) = \nu^3 f(\nu/T)$; if one had the spectrum distribution at one temperature, one knew the distribution at any other temperature. The next step was to find the single-variable function $f(\nu/T)$.

- Wien's distribution $\rho(\nu, T) = \alpha\nu^3 \exp(-\beta\nu/T)$ gave a good fit to the then available data, until the new spectra in the infrared region became available in 1900. Planck postulated his distribution $\rho(\nu, T) = \alpha\nu^3/[\exp(\beta\nu/T) - 1]$ and found that it could fit both high- and low-frequency measurements.

- Given the phenomenological success, Planck immediately set out to find the physical basis of this distribution. Planck considered a cavity composed of a set of oscillators. He first related the oscillator energy to the radiation density $U(\nu, T) = \left(c^3/8\pi\nu^2\right)\rho(\nu, T)$, then found the entropy function S (corresponding to the Planck distribution) by integrating over the entropy–energy relation $dS = (dU)/T$. The next step in his program was to make a statistical analysis of this entropy so as to reveal the physics behind the distribution.

- Using Boltzmann's principle, $S = k_B \ln W$, with W being the microstate complexion, Planck did find such an account of the microstates. But, in order for it to match the entropy result, he had to postulate that the energy of an oscillator (or the energy transmitted by the oscillators) must be quantized—energy could only be transmitted in discrete packets $\epsilon = h\nu$. It is important to observe that, to Planck, the energy transmitted in the interaction between oscillators and radiation might be quantized; he did not propose or believe that the energy of radiation itself was intrinsically quantized.

- This thermal and statistical analysis then relates Planck's constant h and Boltzmann's constant k_B to the two constants (α, β) of Planck's distribution. In this way not only did Planck obtain h but he also obtained the first accurate determination of k_B.

- At various points of this chapter, we discuss electromagnetic radiation as a collection of oscillators. The ratio of the oscillator energy and frequency is an adiabatic invariant. From this we also provide a thermodynamical derivation of radiation pressure, being a third of the radiation energy density. These background materials are needed in various discussions of electromagnetic radiation.

[1] One can find an authoritative account of the quantum postulate resulting from the study of blackbody radiation in Kuhn (1978).

This is the first of the six chapters in Part II devoted to Einstein's contribution to the quantum theory. By and large, this involves the application of statistical ideas. We start with Einstein's 1905 "photoelectric paper" (Einstein 1905a) in which he first put forward the proposal of "light quanta" (later identified with point-like particles, photons). The Nobel Prize given to him in 1921 was based on this paper. As we shall see, this research was in the area of blackbody[1] radiation. This chapter gives the highlights of the relevant background, starting with the work by Gustav Kirchhoff (1824–87), Josef Stefan (1835–93), Ludwig Boltzmann (1844–1906), and especially the displacement law by Wilhelm Wien (1864–1928). Finally, all this led to the epoch-making research by Max Planck (1857–1947), who first arrived at the idea of the quantum of energy. Beyond the photoelectric paper, we shall also study Einstein's 1907 work on the quantum theory of specific heat, his first discovery of wave–particle duality in 1909, his 1916 papers on the absorption and emission of radiation, and his 1924 work on Bose–Einstein statistics and condensation. Finally we will also discuss the Einstein–Bohr debate, explaining why Einstein felt such aversion to the orthodox interpretation of quantum mechanics advocated by the Copenhagen School led by Niels Bohr (1885–1962).

The famous 1905 paper on light quanta (to be discussed in detail in Chapter 4) had the title *On a heuristic point of view concerning the production and transformation of light*. That in 1905 Einstein regarded the quantum idea as being "heuristic"—as a provisional method to learn and to study—is of course to be expected. But it is interesting to note that, all the way till the end of his life in 1955, he still viewed quantum theory as a provisional, that is, an incomplete theory.

3.1 Radiation as a collection of oscillators

In this Part II on quantum theory we shall at several instances make use of the understanding that radiation can be thought of as a collection of simple harmonic oscillators. This connection is valid both at the classical as well at the quantum levels. That the oscillator energy is proportional to the oscillator frequency under an adiabatic change (Section 3.5.1) and that radiation pressure is one-third of the radiation energy density (Section 3.5.2) are the classical radiation properties we shall use in deriving some of the basic properties of thermal radiation: the Stefan–Boltzmann relation and the Wien displacement law. This perspective of radiation as a collection of quantum oscillators also provides us with a simple understanding of some of the basic features of quantum field theory, as we shall discuss in Section 6.4.

We recall that the essential feature of oscillator dynamics is that the energy required to produce a displacement from the equilibrium position is proportional to the square of the displacement itself. Namely, the potential energy of an oscillator is $U = \frac{1}{2}kx^2$. This is also true for the electromagnetic field. The energy in any mode of oscillation is proportional to the square of the field strength (a 'displacement' from the normal state of field-free space). This is the physics of the statement that radiation is a collection of oscillators. Next we demonstrate this connection mathematically.

3.1.1 Fourier components of radiation obey harmonic oscillator equations

Maxwell's equations lead to the electromagnetic wave equation. For definiteness we shall write the wave equation[2] in free space in terms of the vector potential $\mathbf{A}$ (with the magnetic field given by $\mathbf{B} = \nabla \times \mathbf{A}$, and the electric field by $E = -\nabla\Phi + \frac{1}{c}\frac{\partial \mathbf{A}}{\partial t}$):

$$\frac{1}{c^2}\frac{\partial^2 \mathbf{A}}{\partial t^2} - \nabla^2 \mathbf{A} = 0. \tag{3.1}$$

We now make a Fourier expansion[3] of $\mathbf{A}(\mathbf{r}, t)$:

$$\mathbf{A}(\mathbf{r}, t) = \frac{1}{\sqrt{V}} \sum_{\mathbf{k},\alpha} \left(\hat{\epsilon}_{\mathbf{k}}^{(\alpha)} A_\alpha(\mathbf{k}, t) e^{i\mathbf{k}\cdot\mathbf{r}} + \hat{\epsilon}_{\mathbf{k}}^{(\alpha)*} A_\alpha^*(\mathbf{k}, t) e^{-i\mathbf{k}\cdot\mathbf{r}} \right) \tag{3.2}$$

where $\hat{\epsilon}_{\mathbf{k}}^{(\alpha)}$ is the polarization (unit) vector, α being the polarization index[4] $\alpha = 1, 2$; the volume factor V is required for normalization. $A_\alpha(\mathbf{k}, t)$ are the Fourier components for the wavevector $\mathbf{k}$ and polarization α. When we plug this decomposition (3.2) into the wave equation (3.1), the spatial derivatives have the effect of (the wavenumber k being the magnitude $|\mathbf{k}|$)

$$\nabla^2 \mathbf{A}(\mathbf{r}, t) \implies -k^2 A_\alpha(\mathbf{k}, t) = -\frac{\omega^2}{c^2} A_\alpha(\mathbf{k}, t).$$

Thus the wave equation (3.1) turns into

$$\frac{1}{c^2} \sum_{\mathbf{k},\alpha} \left(\frac{\partial^2}{\partial t^2} + \omega^2 \right) A_\alpha(\mathbf{k}, t) = 0. \tag{3.3}$$

Since Fourier components are independent, the above expression shows that each of them must obey the simple harmonic oscillator equation[5] (having angular frequency ω) with solution $A_\alpha(\mathbf{k}, t) = A_{\alpha,\mathbf{k}}e^{-i\omega t}$. Namely, $\mathbf{A}(\mathbf{r}, t)$ is a superposition of plane waves, $\exp[i(\mathbf{k}\cdot\mathbf{r} - \omega t)]$ with $\omega = ck$. Thus, already at

[2] See the further discussion in Section 9.2.1. For a compact derivation of the wave equation in the 4D spacetime formalism, see Section 16.4.1. To derive (3.1) we have taken the "Lorentz gauge" $\partial_\mu A^\mu = \frac{1}{c}\frac{\partial \Phi}{\partial t} + \nabla \cdot \mathbf{A} = 0$. In fact we can further pick the "radiation gauge": $\Phi = 0$ and $\nabla \cdot \mathbf{A} = 0$. Similarly we could have written the wave equation in terms of the fields $(\mathbf{E}, \mathbf{B})$. For our discussion of a field as a collection of oscillators, whether we work with $(\mathbf{E}, \mathbf{B})$ or $\mathbf{A}$ is quite immaterial.

[3] Recall that the solution to the wave equation can be any waveform $\mathbf{A}(\mathbf{r}, t)$ as long as the dependence on $(\mathbf{r}, t)$ is through the combination of $(\mathbf{k}\cdot\mathbf{r} \pm \omega t)$ with angular frequency $\omega = ck$, the light speed c times the wavenumber k. Fourier decomposition is just a convenient way to display the solution as a sum of sine and cosine functions.

[4] The radiation gauge condition $\nabla \cdot \mathbf{A} = 0$ requires $\hat{\epsilon}_{\mathbf{k}}^{(\alpha)} \cdot \mathbf{k} = 0$, which corresponds to the fact that there are only two independent polarizations.

[5] A familiar example is a particle moving under the restoring force $f = -\kappa x$ where κ is the spring constant. This leads to the equation of motion $m\ddot{x} = -\kappa x$ or $\left(\partial_t^2 + \omega^2 \right) x = 0$ (with $\omega^2 = \kappa/m$) just as in Eq. (3.3).

the classical level, we can say that electromagnetic waves are a collection of harmonic oscillators each characterized by $(\mathbf{k}, \alpha)$. In later chapters we shall quote this result in the simplified form of ignoring the polarization vector by saying that the field is a plane wave with amplitude A and phase factor $\phi = (\mathbf{k} \cdot \mathbf{r} - \omega t)$:

$$A(\mathbf{r}, t) = A e^{i\phi}. \tag{3.4}$$

3.2 Thermodynamics of blackbody radiation

The quantum idea originated from the study of blackbody radiation. When heated, a body (e.g. a tungsten filament) emits electromagnetic radiation at a rate that varies with the temperature. When in thermal equilibrium with its surroundings, a body emits and absorbs at the same rate. A convenient way to study such equilibrium radiation is to investigate the radiation inside a cavity heated to a given temperature. One observes the radiation through a small hole in such an enclosure. This situation, in which the radiation once absorbed into the cavity through the hole has little chance of being reflected back out again, is similar to that of a perfect absorber (a blackbody). Such cavity radiation has come to be called **blackbody radiation**.

3.2.1 Radiation energy density is a universal function

Let us denote the cavity radiation energy density by ρ namely it is the energy per unit volume, per unit radiation frequency ν. A priori we would expect it be a function of temperature T, and the frequency of radiation ν, as well as the cavity shape and material. However, in 1860 Kirchhoff used the second law of thermodynamics to show that such radiation was independent of the shape and material of the cavity, and so $\rho = \rho(T, \nu)$ only (Kirchhoff 1860). This simplification greatly stimulated people's interest in the problem of blackbody radiation.

Most of us know of Kirchhoff's law of electric circuits. Actually Kirchhoff was also one of the founding fathers of spectroscopy (three laws are named after him in that area as well). This prompted physicist/historian Abraham Pais (1918–2000) to remark that, if we regard Planck, Einstein, and Bohr as the fathers of quantum theory, then Kirchhoff can be said to be the grandfather. While Planck and Einstein pioneered the idea of quanta from the study of blackbody radiation, it was Kirchhoff's work on spectroscopy (passing through Balmer and Rydberg) that allowed Bohr to come up with his atomic model that was decisive in launching the quantum theory in 1913.

Here we outline Kirchhoff's proof that blackbody radiation energy densities are universal functions. Consider the cavities in Fig 3.1: one is large while the other is small, one is in the shape of a sphere while the other is in the shape of a cube, one is metallic while the other is ceramic—in other words, the two cavities differ in volume, shape, and material composition. Let these two cavities, call them A and B, be connected through a hole. When they reach thermal equilibrium, we must have $T_A = T_B$. In such a situation the energy densities

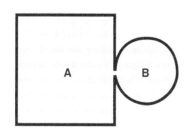

Fig. 3.1 Two cavities A and B (having different shapes and made of different material) are connected through a hole and reach thermal equilibrium with equal temperature $T_A = T_B$.

$u = \int \rho d\nu$ (i.e. the total density from summing over all frequencies) must be equal, $u_A = u_B$, because otherwise their radiation fluxes would be different, leading to $T_A \neq T_B$, which is forbidden by the second law of thermodynamics. In fact, a moment of reflection also makes it clear that the equality of $du_A(T) = du_B(T)$, being $\rho_A(\nu, T)d\nu = \rho_B(\nu, T)d\nu$, implies the equality of the density functions $\rho_A(\nu, T) = \rho_B(\nu, T)$ as well.

3.2.2 The Stefan–Boltzmann law

From Kirchhoff we learnt that energy densities u and ρ are universal functions $u(T)$ and $\rho(T, \nu)$. What are these universal functions? The first problem was solved in the form of the **Stefan–Boltzmann law**:

$$u(T) = a_{SB}T^4, \tag{3.5}$$

where a_{SB} is a constant, the Stefan–Boltzmann constant.[6] It was first obtained from experimental observation by Josef Stefan, and was then proven on theoretical grounds by his student Ludwig Boltzmann (Stefan 1879; Boltzmann 1884).

Consider a volume V of radiation at temperature T; the entropy S is related to the heat function Q by $dS = dQ/T$. When combined with the first law, $dQ = dU + pdV$, this leads to

$$dS = \frac{dU}{T} + p\frac{dV}{T}. \tag{3.6}$$

Since radiation energy U is the product of energy density (a function of T) and volume, $U = u(T)V$, we have $dU = udV + V(du/dT)dT$, and since radiation pressure is related[7] to energy density by $p = u/3$, Eq. (3.6) becomes

$$dS = \frac{u}{T}dV + \frac{V}{T}\frac{du}{dT}dT + \frac{u}{3T}dV = \frac{4u}{3T}dV + \frac{V}{T}\frac{du}{dT}dT. \tag{3.7}$$

The entropy S being a function of V and T,

$$dS = \left(\frac{\partial S}{\partial V}\right)_T dV + \left(\frac{\partial S}{\partial T}\right)_V dT, \tag{3.8}$$

a comparison of this to the expression for dS in Eq. (3.7) leads to

$$\left(\frac{\partial S}{\partial V}\right)_T = \frac{4u}{3T}, \tag{3.9}$$

$$\left(\frac{\partial S}{\partial T}\right)_V = \frac{V}{T}\frac{du}{dT}. \tag{3.10}$$

The second derivative $\left(\partial^2 S/\partial V \partial T\right)$ can be obtained by differentiating: i.e. taking $\partial/\partial T$ of (3.9) or $\partial/\partial V$ of (3.10). Equating these two expressions, we have

$$\frac{4}{3T}\frac{du}{dT} - \frac{4}{3}\frac{u}{T^2} = \frac{1}{T}\frac{du}{dT}$$

[6] The Stefan–Boltzmann law can eventually be derived from Planck's distribution with the Stefan–Boltzmann constant a_{SB} expressed in terms of Boltzmann's constant k_B and Planck's constant h.

[7] See SuppMat Section 3.5.2 for a proof.

or

$$\frac{du}{dT} = 4\frac{u}{T} \quad or \quad \frac{du}{u} = 4\frac{dT}{T}. \tag{3.11}$$

Integrating both sides, we obtain the result $\ln u = \ln T^4 +$ constant, or

$$\ln\left(uT^{-4}\right) = \text{constant}, \tag{3.12}$$

[8]It is also useful to get a qualitative understanding of this T^4 result from the picture of blackbody radiation as a gas of photons. At temperature T, most of the photons have energy $\simeq kT$. Namely, most of the photons have momentum $p' \simeq kT/c$. In the photon momentum space these photons are contained in a spherical volume (with radius $p' \propto T^3$. In this way we see that the total photon energy in the phase space is a quantity proportional to T^4 times the (position space) volume, or the energy per unit volume $\propto T^4$.

which is the Stefan–Boltzmann law[8] (3.5).

3.2.3   Wien's displacement law

All the attention was now on the density distribution $\rho(T, \nu)$. The first breakthrough was made in 1893 by Wilhelm Wien in the form of Wien's displacement law (Wien 1893). It was again based on electromagnetism and thermodynamics; but the reasoning relied on dimensional analysis as well. This displacement law Eq. (3.21) allowed one to reduce the function of two variables $\rho(T, \nu)$ to a function of one variable $f(\nu/T)$, thus greatly simplifying the task of eventually discovering the correct form of the radiation density.

Derivation of Wien's displacement law

As we have already shown, even at the classical Maxwell theory level, radiation (i.e. electromagnetic waves) can be regarded as a collection of simple harmonic oscillators. For a given temperature the density ρ is defined as

$$du = \rho(\nu, T)d\nu = \frac{N(\nu)U(\nu, T)}{V}d\nu \tag{3.13}$$

where N is the number of oscillators having frequency ν, U is the energy of a radiation oscillator, and V is the volume (with no loss of generality, we take the volume to be a cube with side-length L). Now consider how various terms scale under an adiabatic change of volume $V \longrightarrow V'$, thus $L \longrightarrow L'$.

1. We first consider the scaling property of the density ρ with respect to the radiation frequency. Since the total number of oscillators does not change, we have:

$$N(\nu)d\nu = N'\left(\nu'\right)d\nu'. \tag{3.14}$$

Because frequency is the inverse of wavelength which scales with length $\nu \sim L^{-1}$, from (3.14) we must have $N \sim L$,

$$\frac{N}{N'} = \frac{L}{L'}. \tag{3.15}$$

Similarly, because radiation energy (whether classical or quantum) scales with frequency,[9] we also have

[9]That the ratio of radiation energy to frequency is an adiabatic invariant is shown in SuppMat Section 3.5.1.

$$\frac{U}{U'} = \frac{L'}{L}. \tag{3.16}$$

Thus the combination NU is an adiabatic invariant and we can then conclude that the density ρ scales as the cubic power of the frequency,

$$\frac{\rho}{\rho'} = \frac{UN/V}{U'N'/V'} = \left(\frac{L'}{L}\right)^3 = \left(\frac{\nu}{\nu'}\right)^3. \tag{3.17}$$

2. We now consider the temperature dependence of $\rho(\nu, T)$. The energy density differential du should scale in the same way as u, which by the Stefan–Boltzmann law (3.5) has a fourth-power temperature-dependence:

$$\frac{du}{du'} = \left(\frac{T}{T'}\right)^4. \tag{3.18}$$

On the other hand, from $u = \int \rho d\nu$ and (3.17) we have

$$\frac{du}{du'} = \frac{\rho d\nu}{\rho' d\nu'} = \left(\frac{\nu}{\nu'}\right)^3 \left(\frac{\nu}{\nu'}\right) = \left(\frac{\nu}{\nu'}\right)^4. \tag{3.19}$$

Comparing Eqs. (3.18) and (3.19), we see that the ratio ν/T is invariant under this adiabatic change:

$$\frac{\nu}{T} = \frac{\nu'}{T'}. \tag{3.20}$$

Finally, in order for the density function $\rho(\nu, T)$ to satisfy both (3.17) and (3.20), we must have **Wien's displacement law**

$$\rho(\nu, T) = \nu^3 f\left(\frac{\nu}{T}\right) \tag{3.21}$$

where $f(x)$ is some one-variable function. As we can see that by using (3.20),

$$\frac{\rho(\nu, T)}{\rho(\nu', T')} = \frac{\nu^3 f(\nu/T)}{\nu'^3 f(\nu'/T')} = \frac{\nu^3}{\nu'^3},$$

we have a result consistent with (3.17). This "displacement law" allows one to deduce the radiation distribution at all other temperatures once one knows the distribution at one temperature.

For a consistency check, we should confirm that the Stefan–Boltzmann law (3.5) can be recovered from the displacement result (3.21):

$$u(t) = \int \rho(\nu, T) d\nu = \int \nu^3 f\left(\frac{\nu}{T}\right) d\nu = a_{SB} T^4 \tag{3.22}$$

with the **Stefan–Boltzmann constant** given by the integral

$$a_{SB} = \int x^3 f(x) \, dx. \tag{3.23}$$

The peak of the radiation distribution scales linearly as temperature

Now for each value of temperature, $\rho(\nu, T)$ can be plotted as a curve varying with frequency ν. It is observed (see Fig. 3.2) that each curve peaks at some

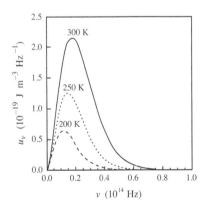

Fig. 3.2 Blackbody radiation spectrum distribution, peaked at $\nu_{max} \sim T$. The three density curves for temperatures of 200 K, 250 K, and 300 K have maxima around the frequencies 1.2, 1.5, and 1.8×10^{14} Hz, respectively.

frequency ν_{max}. Wien's displacement law (3.21), as we shall immediately show, has the corollary of

$$\frac{\nu_{max}}{T} = \text{constant}. \tag{3.24}$$

The constant here simply means some definite number. Thus, when the temperature rises, the position of the maximum of the distribution is displaced towards higher frequencies in proportion to T. The proof of (3.24) goes as follows. The first derivative of ρ vanishes at $\nu = \nu_{max}$; from Wien's displacement law (3.21), we have

$$3\nu_{max}^2 f\left(\frac{\nu_{max}}{T}\right) + \frac{\nu_{max}^3}{T} f'\left(\frac{\nu_{max}}{T}\right) = 0$$

with $f'(x) = df/dx$. This is a differential equation $xdf/dx + 3f = 0$ of one variable ($x = \nu_{max}/T$), having some solution as displayed in (3.24).

The measured Wien's displacement constant of (3.24) is commonly expressed in terms of the wavelength at the peak (namely the dominant component of the radiation)

$$\lambda_{max} T = 0.2897 \text{ cm} \cdot \text{K} \simeq 3 \times 10^{-3} \text{ m} \cdot \text{K}.$$

- The 3 K cosmic background radiation peaks in the microwave range $\lambda_{max} \simeq 10^{-3}$ m, hence is often referred to as the cosmic microwave radiation.
- The temperature on the surface of the sun is about $T_\odot \simeq 6 \times 10^3$ K; thus it peaks around $\lambda_{max} \simeq 0.5 \times 10^{-6}$ m $= 500$ nanometer. Our eyes have evolved to be sensitive to this range of the electromagnetic spectrum—the visible range.
- The basic feature that underlies the greenhouse effect is the following: Visible light at a few hundred nanometers easily passes through the glass of a greenhouse. It is absorbed and re-emitted at room temperature around 300 K, corresponding to a wavelength of $\lambda_{max} \simeq 10^{-5}$ m, or a thousand nanometers or so. This is in the infrared range and cannot easily get through the greenhouse glass windows.

3.2.4   Planck's distribution proposed

Since 1895 Max Planck had moved the focus of his research from thermodynamical investigation to the problem of resonant scattering of an electromagnetic wave by an oscillator and furthermore to blackbody radiation. He had written a whole series of papers on these subjects. The ultimate goal of his program was an explanation of irreversibility for conservative systems and, as a by-product, an understanding of the blackbody radiation spectrum distribution.

Following his derivation of his displacement law (3.21), Wien proposed in 1896 a specific distribution that fitted the then known experimental data rather well:

$$\rho(\nu, T) = \alpha \nu^3 \exp\left(-\beta \frac{\nu}{T}\right), \tag{3.25}$$

where α and β are two constants.[10] However, by 1900 H. Rubens (1865–1922) and F. Kurlbaum (1857–1927) were able to extend the experimental observation down to the infrared regime, finding that Wien's distribution no longer gave a good match to the newly acquired data.[11] As soon as Planck was made aware of this new development, this discrepancy, he proposed his distribution:

$$\rho(\nu, T) = \frac{\alpha \nu^3}{\exp(\beta \nu / T) - 1}. \tag{3.26}$$

Although some of his previous investigations had suggested this as the next simplest possibility (after Wien's distribution) for radiation entropy increase, what Planck did was basically guesswork—only with the constraint that it reduced to Wien's distribution for large ν/T. Remarkably, this new distribution (3.26), the Planck distribution, fitted all the known blackbody radiation data perfectly. Planck first presented this new radiation law in a meeting of the German Physical Society on October 19, 1900 (Planck 1900a).

3.3   Planck's investigation of cavity oscillator entropy

After this phenomenological triumph, Planck immediately set out to discover the physical significance of this success. For this purpose he needed to study the cavity-wall molecules absorbing and emitting radiation in achieving thermal equilibrium. Recalling Kirchhoff's theorem stating that the cavity radiation energy density is independent of cavity size, shape, and material composition, Planck considered the case of a cavity wall being a collection of oscillators, which absorb and emit electromagnetic radiation. This simple model of the cavity molecules allowed him to study the blackbody radiation as resulting from the oscillator/radiation interaction.

3.3.1   Relating the oscillator energy to the radiation density

We first present a result obtained by Planck (1900b) relating the radiation energy density $\rho(\nu, T)$ in an enclosure to the time-averaged energy $U(\nu, T)$ of a cavity oscillator[12] with the same frequency ν bathed in the radiation field:

$$U(\nu, T) = \frac{c^3}{8\pi \nu^2} \rho(\nu, T). \tag{3.27}$$

This is sometimes call Planck's law. It was obtained in 1899 by Planck through an electromagnetic calculation in the context of a specific model of resonant scattering of an oscillator. Remarkably the result, connecting the radiation density ρ to the average energy of the cavity oscillator, is independent of any properties (its mass, charge, etc.) of the oscillator. Plugging the distribution of (3.26) into (3.27), we have

$$U = \frac{c^3}{8\pi} \frac{\alpha \nu}{\exp(\beta \nu / T) - 1}. \tag{3.28}$$

[10] As we shall see, the constants α and β are to be related to Boltzmann's constant k_B and Planck's constant h.

[11] Rubens and Kurlbaum (1900) in agreement with earlier data by Lummer and Pringsheim (1900).

[12] As we shall explain in Section 4.1.3, U should be correctly interpreted as the average energy of a radiation oscillator, rather than a cavity-wall oscillator as first done by Planck. We shall not discuss Planck's derivation and refer the interested reader to Pais (1982, p. 369) and Longair (2003, p. 305–314) on this subject.

This is Planck's distribution expressed in terms of the energy (per unit frequency) of an oscillator state.

3.3.2 The mean entropy of an oscillator

In order to understand the physical significance of (3.28), Planck used a thermal and statistical approach. This was in contrast to other workers in the field who had concentrated on the temperature dependence of the spectral energy. Based on his previous thermodynamical work with Wien's distribution, Planck found an expression for the entropy of a cavity oscillator that is equivalent to Planck's distribution given above. Keep in mind that entropy and energy are additive functions of state. We can work with a single oscillator as well as an ensemble of them, and the obtained properties can be easily translated back and forth between them.

Let us first simplify the notation by replacing the two constants α and β in (3.28) by their linear combinations, denoted by κ and ϵ:

$$\frac{c^3}{8\pi}\alpha\nu = \epsilon \quad \text{and} \quad \beta\nu = \frac{\epsilon}{\kappa} \tag{3.29}$$

so that the Planck distribution takes on the form

$$U = \frac{\epsilon}{\exp(\epsilon/\kappa T) - 1}. \tag{3.30}$$

This relation can be inverted as an expression for the temperature:

$$\exp\left(\frac{\epsilon}{\kappa T}\right) = \left(1 + \frac{\epsilon}{U}\right), \quad \text{or} \quad \frac{1}{T} = \frac{\kappa}{\epsilon}\ln\left(1 + \frac{\epsilon}{U}\right). \tag{3.31}$$

One wishes now to find the equivalent result expressed in terms of the entropy using the relation between entropy, energy, and temperature, $dS = (dU)/T$ when the volume is fixed. In this way we obtain the differential of the entropy

$$dS = \frac{1}{T}dU = \frac{\kappa}{\epsilon}\ln\left(1 + \frac{\epsilon}{U}\right)dU = \frac{\kappa}{\epsilon}[\ln(U + \epsilon)dU - \ln U dU].$$

Up to an inconsequential constant, the entropy can be deduced by a straightforward integration $\int \ln x \, dx = x \ln x - x$:

$$S = \frac{\kappa}{\epsilon}[(U + \epsilon)\ln(U + \epsilon) - U \ln U],$$

or

$$S = \kappa\left[\left(1 + \frac{U}{\epsilon}\right)\ln\left(1 + \frac{U}{\epsilon}\right) - \frac{U}{\epsilon}\ln\frac{U}{\epsilon}\right]. \tag{3.32}$$

To recapitulate, this result for entropy is equivalent to the Planck spectral law. One can easily check this by a simple differentiation $dS/dU = 1/T$ to show that the relations of (3.31) and (3.30) are recovered. The constants (κ, ϵ) are directly related to (α, β); in particular as in (3.29):

$$\kappa = \frac{c^3}{8\pi}\frac{\alpha}{\beta}, \tag{3.33}$$

and its product with the constant β is another constant,

$$h = \beta\kappa = \frac{c^3}{8\pi}\alpha \quad \text{so that} \quad \epsilon = h\nu. \tag{3.34}$$

The following discussion will reveal the physical significance of the parameters ϵ, h, and κ.

3.4 Planck's statistical analysis leading to energy quantization

3.4.1 Calculating the complexion of Planck's distribution

Recall that the entropy constructed in the above discussion is that for a single oscillator. Planck now proceeded to a thermodynamical and statistical study by considering a system of N independent copies of such an oscillator. Since U of (3.30) and the entropy S of (3.32) are average values, the total energy U_N and entropy S_N of the whole system can be obtained by a simple multiplication:

$$U_N = NU \quad \text{and} \quad S_N = NS. \tag{3.35}$$

Having this thermodynamic, i.e. macroscopic, description of the system in terms of U_N and S_N, we proceed to decipher the physical meaning of such a depiction.

Boltzmann's principle

One of Boltzmann's great achievements was his showing that the second law of thermodynamics was amenable to precise mathematical treatment—the macroscopic notion of entropy S_N could be related to a counting of the corresponding microscopic states, the "complexions" W_N:

$$S_N = k_B \ln W_N \quad \text{or} \quad S = \frac{1}{N}k_B \ln W_N \tag{3.36}$$

where k_B is Boltzmann's constant.[13]

Planck now asked: what would be the microscopic statistical description that corresponded to the entropy just obtained for the radiation distribution? The hope was that an expression for the statistical weight W_N that matched the entropy of (3.32) would provide him with a deeper understanding of the physical meaning of Planck's distribution. Planck did find such an account of the microstates (Planck 1900b, 1901). But, in order for it to match the entropy result, he had to undertake, what he later described (Planck 1931) as, a "desperate act", to postulate that the energy of an oscillator (or the energy transmitted by the oscillators) was quantized—energy could only be transmitted in discrete packets. As it turns out, the packet of energy is just the quantity ϵ that first appeared in Eq. (3.32). If we denote the total number of energy packets (energy quanta) in this system of N oscillators by P, then

$$U_N = NU = P\epsilon. \tag{3.37}$$

[13] An interesting bit of history: Boltzmann actually never wrote down this famous equation $S = k_B \ln W$, although such a relation was implied in his study of ideal gases. It was first done by Planck, and it was Einstein who first called it, in 1905, the "Boltzmann principle". Also, the constant k_B was first introduced by Planck, and not by Boltzmann.

[14] This is the description given by Planck in the first edition of his book *The Theory of Heat Radiation*.

Planck computed the complexion as the number of ways these P packets of energies $(\epsilon's)$ can be distributed among N oscillators. In Planck's calculation,[14] "no regard is given to which energy packets, but only how many energy packets are assigned to a given oscillator". One can picture this as computing the different ways of distributing P identical balls into N boxes. It is just the number of permutations of P balls ($\bullet$) and $N - 1$ partition walls ($\shortmid$):

$\bullet\bullet\bullet\shortmid\bullet\bullet\shortmid\bullet\bullet\bullet\bullet\bullet\bullet\bullet\shortmid\bullet\ldots$, which leads to the expression of the probability

$$W_N = \frac{(P + N - 1)!}{P\,!(N - 1)!} \simeq \frac{(P + N)!}{P\,!N!}. \tag{3.38}$$

This gives rise to an entropy according to (3.36):

$$S = \frac{1}{N}\,k_B \ln W_N = \frac{1}{N}k_B\left\{\ln[(P + N)!] - \ln(P\,!) - \ln(N\,!)\right\}.$$

Applying Stirling's formula for large X,

$$\ln(X!) = X \ln X - X, \tag{3.39}$$

we have

$$S = \frac{1}{N}k_B\left\{(P + N)\ln(P + N) - (P + N) - P \ln P + P - N \ln N + N\right\}$$

$$= k_B\left\{\left(\frac{P}{N} + 1\right)\left[\ln\left(\frac{P}{N} + 1\right) + \ln N\right] - \frac{P}{N}\left[\ln\frac{P}{N} + \ln N\right] - \ln N\right\}.$$

All terms proportional to $\ln N$ cancel; the expression for the entropy is then simplified to

$$S = k_B\left[\left(1 + \frac{P}{N}\right)\ln\left(1 + \frac{P}{N}\right) - \frac{P}{N}\ln\frac{P}{N}\right], \tag{3.40}$$

which reproduces Planck's entropy of (3.32) when we identify $\kappa = k_B$, and the number of quanta per oscillator by the relation (3.37): $P/N = U/\epsilon$. This was how Planck derived his distribution law. In this way we see that the parameters we encountered in this discussion have the physical interpretation of

$$\epsilon = \text{energy quantum}$$
$$\kappa = \text{Boltzmann's constant } k_B$$
$$h = \text{Planck's constant.}$$

Energy quantization and Planck's distribution

[15] We shall often use the common notation of $\hbar \equiv h/2\pi$. In this way the quantum of energy becomes $\epsilon = \hbar\omega$, with ω being the angular frequency $\omega = 2\pi\nu$.

With the identification of $\beta\kappa$ as Planck's constant h, Eq. (3.29) can be seen as the famous Planck relation for energy quantization[15]

$$\epsilon = h\nu. \tag{3.41}$$

Planck's distribution (3.30) can now be written as

$$U = \frac{h\nu}{\exp(h\nu/k_B T) - 1} \tag{3.42}$$

and the radiation density function (3.26) in the more familiar form[16]

$$\rho = \frac{8\pi h}{c^3} \frac{\nu^3}{\exp(h\nu/k_B T) - 1}.$$ (3.43)

Planck's formula can fit all the blackbody radiation data perfectly. One of the most impressive encounters between theory and experimental observation was the discovery of the cosmic microwave background radiation (CMBR), which is the relic radiation from the era of thermal equilibrium between radiation and matter soon after the universe's big bang beginning. The CMBR was observed to follow perfectly the Planck distribution with a temperature of $T = 2.725 \pm 0.002$ K, to the extent that, when the observational data were plotted on paper, the error bars were less than one-hundredth of the width of the distribution curve itself (Mather *et al.* 1994), see Fig. 3.3.

Planck's analysis was at variance with Boltzmann's statistical method

It should be pointed out that Planck's statistical derivation of the complexion W_N differs from what Boltzmann would have done.[17] First of all, Planck did not suggest population from which a probability could be defined. His identification of the statistical weight (3.38) was stated without first giving an independent definition of this quantity. Namely, he simply defined his probability as the expression (3.38). The proper Boltzmann statistical analysis will be reviewed in Section 7.2.1 when we introduce the Bose–Einstein statistics. The ultimate justification of Planck's derivation would not come about until the advent of modern quantum mechanics, with its notion of identical particles and Bose–Einstein quantum statistics. We shall return to this question "why did Planck find the right result" in Section 7.6.

[16]One should thus keep in mind that the often quoted expression (3.43) of Planck's distribution is composed of two parts: the average oscillator energy as given in (3.42) and the radiation density distribution of (3.26) through their relation as given by (3.27).

[17]Another problematic aspect of Planck's derivation concerns its logical consistency. Energy quantization was the conclusion he drew from his distribution. Its derivation relied on the energy density relation (3.27), which was derived using classical radiation theory with the implicit assumption of continuous energy variation.

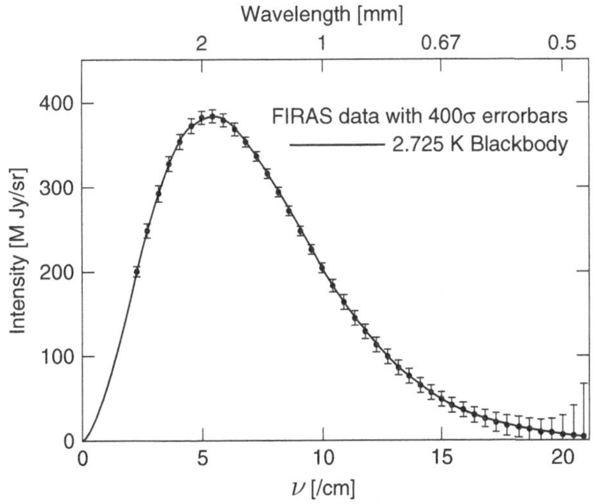

Fig. 3.3 Measurement by the FIRAS instrument on the COBE satellite showing that the cosmic background radiation spectrum fitted the theoretical prediction of blackbody radiation perfectly. The horizontal axis is the wavenumber or 1/[wavelength in cm]. The vertical axis is the power per unit area per unit frequency per unit solid angle in megajanskies per steradian. 1 jansky is 10^{-26} watts per square meter per hertz. In order to see the error bars they have been multiplied by 400; but the data points are consistent with the radiation from a blackbody with $T_0 = 2.725$ K. Data from Fixsen *et al.* (1996).

We note that Planck was not the first one to invoke a discrete energy unit. Boltzmann, for example, had resorted to such a calculational device in his discussion of an ideal gas, but the energy element ϵ was always set to zero at the end of the calculation to recover the continuous energy distribution. Planck, however, found that his entropy formula could fit the experimental data only if the energy packet was nonvanishing and proportional to the radiation frequency ν as given in (3.41).

3.4.2   Planck's constant and Boltzmann's constant

Although at first Planck was reluctant to assign physical significance to his energy quantum, he immediately attached great importance to the new constant h. Namely, while he was not ready to suggest that a quantum revolution was in the making, he was convinced the quantum h signified a new important development in physics. He noted that the radiation spectral distribution led to an accurate determination of Boltzmann's constant k_B.

Boltzmann's constant and Avogadro's number

By fitting the experimental data to his distribution, Planck was able to find the numerical values of two constants: Planck's constant and Boltzmann's constant

$$(\alpha, \beta) \xrightarrow[\text{Eqs. (3.33)+(3.34)}]{} (h, k_B).$$

In fact Planck's determinations of k_B and Avogadro's number N_A (through the relation $k_B N_A = R$, the gas constant) were by far the most precise values compared to those obtained by the kinetic theory of gases throughout the nineteenth century. Because of this success, Planck was convinced that his quantization relation was of fundamental significance, even though he thought Eq. (3.41) was only a formal relation and saw no need for a real quantum theory of radiation and matter of the kind that Einstein was to propose. In fact, Planck did not believe the photon idea for at least 10 years after its initial introduction in 1905.

Planck's constant and an absolute system of units

Planck's constant h has the unit of **action**, i.e. (distance $\times$ momentum), or (energy $\times$ time). Thus, h is often referred to as the **quantum of action**. The product hc having the dimension of (energy $\times$ length) is a convenient factor to use in converting different units.[18]

[18]As an example, we can use this conversion factor to arrive at Planck's mass: Gravitation potential energy being $G_N \times (\text{mass})^2/r$, the product of (energy $\times$ distance) over Newton's constant has the dimension of mass squared; we see that Planck's mass can be defined as the square root of hc/G_N.

Planck noted in particular that this new constant h, when combined with Newton's constant G_N and the velocity of light c, led in a natural way to defining the units of mass, length, and time, independent of any anthropomorphic origin. We have the **Planck system of units**

$$\text{Planck mass} \quad m_P = \sqrt{hc/G_N}$$
$$\text{Planck length} \quad l_P = \sqrt{hG_N/c^3}$$
$$\text{Planck time} \quad t_P = \sqrt{hG_N/c^5}.$$

Incorporating Boltzmann's constant k_B, one can also define the natural unit for temperature, Planck's temperature $T_P = k_B^{-1} \left(hc^5/G_N \right)^{1/2}$. This system of units, truly independent of any body or substance, or culture, deserves the name 'natural units'. As physics has developed (to be explained in the subsequent chapters of this book), we now understand that these three fundamental constants, h, c, and G_N, are also the basic conversion factors that allow us to connect apparently distinct types of physical phenomena: wave and particle, space and time, energy and geometry. Albert Einstein made essential contributions to such unification results through his work on quantum theory and special and general relativity.[19]

3.4.3 Planck's energy quantization proposal—a summary

Let us reiterate this discussion of Planck's discovery of energy quantization. Based on Wien's theoretical work and on the then newly obtained blackbody radiation data, Planck made a conjecture of the density distribution (3.26) that turned out to be a good fit to all the known experimental data. He then proceeded to seek out the physical significance of this distribution. Through a thermal and statistical investigation of the cavity oscillators interacting with the radiation, he was compelled to make the radical proposal that energy was quantized.

As we have noted, Planck's statistical steps were problematic. His argument leading up to (3.38) was not justified in the then known theoretical framework of Boltzmann's statistical mechanics. Its ultimate rationalization would not arrive until the advent of modern quantum statistics for identical photons (Bose–Einstein statistics). Furthermore, Planck's derivation of the relation (3.27) was based on a specific model in the framework of classical electromagnetism and was not logically consistent with the energy quantization result he eventually obtained. Rayleigh took the first step towards a correct approach by counting radiation wave states (as discussed in Section 4.1.3). It was not until 1923 that Bose derived (3.27) by counting photon states (as discussed in Section 7.2.2).

What Planck envisioned was not so much the energy quantization of any (cavity wall) material oscillator but rather the total energy of a large number of oscillators as a sum of energy quanta of $\epsilon = h\nu$. This counting device ϵ could not be set to zero if one was to have the correct spectrum distribution. In fact, for many years Planck himself did not believe the reality of energy quantization, and said nothing about the radiation emitted by these material oscillators. He firmly believed that the emitted radiation was the electromagnetic waves described by Maxwell's equations. Even after Einstein's put forward the idea of light quanta (Einstein 1905a), Planck resisted it for a long time after its proposal.

3.5 SuppMat: Radiation oscillator energy and frequency

In Section 3.1 we presented a calculation showing that even the classical radiation field can be regarded as a collection of simple harmonic oscillators. Here we first show that each oscillator has energy proportional to its frequency,

[19] See the index entry under 'fundamental constants' and further discussions in Sections 6.1, 11.4, and 14.4. Our present-day interpretation is that these Planck units are the natural physical scales for quantum gravity as the most basic physics theory. We note the large size of the Planck energy $m_Pc^2 = 1.22 \times 10^{19}$ GeV and, correspondingly, the small sizes of the Planck length and time, all reflect the weakness of the gravitational force. For further discussion see Chapters 15 and 17.

or more accurately we show that the ratio of the oscillator energy and the oscillator frequency is constant under an adiabatic change. From such considerations one can also show that radiation pressure is one-third of the radiation energy density.

3.5.1   The ratio of the oscillator energy and frequency is an adiabatic invariant

The electromagnetic field is a set of simple harmonic oscillators with some frequency $\omega = 2\pi\nu$. We now argue that under an adiabatic deformation—infinitesimal change of boundary conditions (e.g. slow change of volume of the container of radiation) the ratio of energy to frequency of each of these oscillator modes remains unchanged; it is an adiabatic invariant. This is an example of the general result for any oscillatory system. A rigorous proof would involve some rather advanced topics in classical physics, such as the Hamilton–Jacobi equation and action angle variables. Here we shall illustrate the result with a concrete example of an adiabatic variation of a simple pendulum[20] so that the deformation and the resultant change can be easily understood.

[20]Our presentation follows that given in Section 1.5 in Tomonaga (1962).

The simple pendulum

The pendulum has a mass m at the end of a weightless string of length l making an angle θ with the vertical (Fig. 3.4). Pivoting at the other end of the string, it has a moment of inertia $I = ml^2$. The oscillation is driven by gravity, with a torque $\tau = lm\mathsf{g}\sin\theta$ where g is the acceleration of gravitation. The equation of motion $I\ddot{\theta} = \tau$ can be written out for small angles

$$l^2\ddot{\theta} = -\mathsf{g}l\sin\theta \simeq -\mathsf{g}l\theta. \tag{3.44}$$

We have a minus sign because the torque restores the swing back towards the $\theta = 0$ direction. This is a simple harmonic oscillator equation $\ddot{\theta} = -\omega^2\theta$ with angular frequency

$$\omega = \sqrt{\frac{\mathsf{g}}{l}}. \tag{3.45}$$

It has solution $\theta = \Theta\sin(\omega t + \phi)$. The system has kinetic energy

$$K = \frac{1}{2}I\dot{\theta}^2 = \frac{1}{2}ml^2\omega^2\Theta^2\cos^2(\omega t + \phi), \tag{3.46}$$

and the (gravitational) potential energy $P = mgh$, where the height h is measured against the lowest point of the pendulum swing (see Fig. 3.4) $h = l - l\cos\theta \simeq \frac{1}{2}l\theta^2$ so that

$$P = \frac{1}{2}m\mathsf{g}l\theta^2 = \frac{1}{2}m\mathsf{g}l\Theta^2\sin^2(\omega t + \phi), \tag{3.47}$$

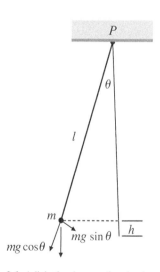

Fig. 3.4 Adiabatic change of a simple pendulum. As the string is slowly shortened (say, by a pulley at P), the oscillation energy and oscillation frequency will increase, but their ratio stays invariant.

leading to a constant total energy

$$U = K + P = \frac{1}{2}mgl\Theta^2. \tag{3.48}$$

Adiabatic change of oscillatory energy and frequency

We now consider an adiabatic variation of this simple pendulum. The pendulum is in a state of proper oscillation; the slow deformation of the dynamics system will be in the form of changing the string length gradually. For definiteness, consider the case of the string length being shortened $\delta l < 0$ by exerting a certain tension T on the string. We do it so slowly that the system remains in proper oscillation, but the frequency ω and oscillation energy U will change. We shall show that the rate of frequency change and the rate of oscillatory energy change are equal so that their ratio remains a constant under this adiabatic variation.

The frequency variation can be found by a straightforward differentiation of (3.45), yielding the result

$$\frac{\delta\omega}{\omega} = -\frac{\delta l}{2l}. \tag{3.49}$$

Namely, a shortening of the string length ($\delta l < 0$) results in higher frequency.

On the other hand, the variation of oscillatory energy cannot be obtained by a similar differentiation of (3.48) because we must find a way to separate out the variation of gravitational potential that's not oscillating. We do this by computing the work done by the applied tension, averaged over many oscillation cycles,

$$\delta W = -\delta l \langle T \rangle. \tag{3.50}$$

The tension is to counter the gravitational pull (along the string direction) and the centrifugal force $ma = m(l\dot\theta)^2 l^{-1}$:

$$T = mg\cos\theta + ml\dot\theta^2 \simeq mg - \frac{1}{2}mg\vartheta^2 + ml\dot\theta^2. \tag{3.51}$$

With its time dependence displayed,

$$T = mg - mg\Theta^2 \left(\frac{1}{2}\sin^2(\omega t + \phi) - \cos^2(\omega t + \phi) \right) = T_0 + T_{\mathrm{osc}},$$

where we have separated out the oscillatory part from the nonoscillatory force T_0 needed to pull up the weight. The shortening of the string takes place over many oscillations; the average values of $\langle \sin^2(\omega t + \phi) \rangle = \langle \cos^2(\omega t + \phi) \rangle = 1/2$ lead to $\langle T_{\mathrm{osc}} \rangle = \frac{1}{4}mg\Theta^2$. We are interested in the variation of oscillatory energy $\delta U = -\delta l \langle T_{\mathrm{osc}} \rangle = -\frac{1}{4}mg\Theta^2\delta l = -U\delta l/(2l)$, thus the variation rate of

$$\frac{\delta U}{U} = -\frac{\delta l}{2l}, \tag{3.52}$$

which is the same as the frequency variation (3.49):

$$\frac{\delta U}{U} = \frac{\delta\omega}{\omega}. \tag{3.53}$$

Performing the integration on both sides of the equation, we find $\ln(U/\omega)$ is a constant, or equivalently, U/ω is an adiabatic invariant.

Result for one-dimensional oscillators in general

We conclude this subsection by consideration of a one-dimensional oscillator having mass m and spring constant k so that the angular frequency is $2\pi \nu = \sqrt{k/m}$. The oscillator energy U in terms of some generalized position q and momentum p is

$$U = \frac{p^2}{2m} + \frac{1}{2}kq^2, \tag{3.54}$$

or, dividing both sides by U,

$$1 = \frac{p^2}{2Um} + \frac{q^2}{2U/k}. \tag{3.55}$$

We recognize this as the equation of an ellipse in 2D momentum–position phase space, with an area of

$$J = \pi\sqrt{2Um}\sqrt{2U/k} = 2\pi U\sqrt{\frac{m}{k}} = \frac{U}{\nu}. \tag{3.56}$$

Since the momentum–position phase space volume[21] is adiabatic invariant, so must be the ratio of oscillator energy to oscillator frequency.

[21] In more advanced language, the action angle variable $J = \oint p\,dq$ is an adiabatic invariant.

3.5.2   The thermodynamic derivation of the relation between radiation pressure and energy density

The relation between radiation energy density u and the average pressure $\bar{p}$ as exerted by radiation on the wall of its enclosure is

$$\bar{p} = \frac{1}{3}u. \tag{3.57}$$

As we shall see, the factor of 3 follows from the fact that our physical space has three dimensions. Another key input is that the radiation oscillator energy scales like frequency under an adiabatic change as shown in the previous section.

We start from the thermal relation (first law of thermodynamics) $dU_s = -p_s dV$ the system in a state s with pressure p_s, energy U_s, and volume V. Thus the mean pressure is

$$\bar{p} = \sum_s \bar{n}_s p_s = \sum_s \bar{n}_s \left(-\frac{\partial U_s}{\partial V}\right), \tag{3.58}$$

where $\bar{n}_s$ is the average number of modes (you can think of them as Fourier modes) in the s state. We shall also assume, without loss of generality, that the volume is a cube with side-length L; namely, $V = L^3$. The radiation energy of each oscillator mode changes in the same way, under an adiabatic change, as the inverse wavelength of radiation. The wavelength scales with the length dimension of the system $L = V^{1/3}$. We then have $U_s = C_s V^{-1/3}$, with C_s being some proportionality constant,

$$\frac{\partial U_s}{\partial V} = -\frac{1}{3} C_s V^{-4/3} = -\frac{1}{3} \frac{U_s}{V}, \tag{3.59}$$

which, after substituting into Eq. (3.58), leads directly to the final result:

$$\bar{p} = \sum_s \bar{n}_s \left(\frac{1}{3} \frac{U_s}{V} \right) = \frac{1}{3V} \sum_s \bar{n}_s U_s = \frac{1}{3V} U = \frac{1}{3} u. \tag{3.60}$$

We note that the key input in this derivation is the 3D nature of our physical space as well as the fact that the radiation energy scales inversely with respect to length change $U \sim L^{-1}$. This result of $\bar{p} = u/3$ will be derived again in Section 7.5, from the picture of radiation as a gas of photons in a manner entirely similar to the familiar derivation of the pressure/energy relation in the kinetic theory of gases.

4

Einstein's proposal of light quanta

- Einstein's 1905 "photoelectric paper", where he first proposed the idea of quanta of light, was concerned mainly with blackbody radiation.

- The radiation density distribution $\rho(\nu, T)$, according to Planck, was directly related to the average energy of cavity oscillators $\rho = 8\pi\nu^2 c^{-3}U$. Einstein applied the equipartition theorem of statistical mechanics to this average energy U and was the first one to correctly derive the Rayleigh–Jeans distribution law $\rho(\nu, T) = 8\pi\nu^2 c^{-3}k_B T$, which is the low-frequency limit of Planck's distribution. We explain why the sequence of events leading to the establishment of the Rayleigh–Jeans law could be characterized as "Planck's fortunate failure".

- Blackbody radiation is the third area in which Einstein derived Avogadro's number from macroscopic measurements.

- While Planck obtained the relation $\rho = 8\pi\nu^2 c^{-3}U$ through an electromagnetic model of material oscillators, Rayleigh derived it as an expression for the radiation oscillator energy U, by a counting of wave states in an enclosure.

- Planck's distribution gave an excellent account of the experimental data in their entire frequency range. After showing that its low-frequency limit had a firm theoretical foundation in classical physics, Einstein then concentrated on the high-frequency part, the Wien limit law $\rho = \alpha\nu^3 \exp(-\beta\nu/T)$, as representing the new physics.

- Like Planck, Einstein tried to find its physical significance through thermodynamic and statistical analysis by first translating this distribution into its corresponding entropy S. Einstein used an elegant method to find S by first relating the energy density per unit frequency $\rho(\nu, T)$ directly to $\varphi(\nu, T)$, the entropy density per unit frequency.

- Unlike Planck, Einstein did not try to interpret the meaning of the statistical weight $W = \exp(S/k_B)$ by a direct counting of the possible microstates. Rather, he investigated W's volume-dependence, showing that it had an analogous expression to that for the Joule expansion of an ideal gas. This allowed him to suggest that in the high-frequency Wien limit the blackbody radiation could be interpreted as a gas of quanta—each quantum of radiation (at frequency ν) having an energy of $h\nu$. Thus Einstein set himself against the then entrenched opinion of the physics community that electromagnetic radiation was composed of waves.

- At the end of his 1905 paper, he proposed to test this quantum of radiation idea by the properties of the photoelectric effect, the phenomenon first discovered by Hertz in 1887, which Lenard found in 1902 to have properties that were difficult to explain based on the wave theory of light. It took more than a decade of experimentation to confirm these predicted properties based on the quantum picture of light. In 1921 the Nobel Foundation awarded the physics prize to Einstein, citing his contribution to our understanding of the photoelectric effect.

As discussed in the previous chapter, stimulated by new observational data, Planck made essentially a stab-in-the-dark extension of Wien's distribution for the blackbody radiation—that turned out to be a good fit to all the experimental measurements. His attempt to find a deeper meaning of this result through a thermodynamical and statistical study ended up compelling him to make the energy quantization proposal. While Planck was reluctant to regard this as an actual description of reality, Einstein was the first one to take the quantum idea seriously, and to apply it to the physical system of electromagnetic radiation (Einstein 1905a), and to the theory of specific heat in the following year (Einstein 1907a). His quantum of light idea led him to explain various features of the photoelectric effect (which were first found experimentally by Philip Lenard (1862–1947) to have surprising properties), and to make predictions that were confirmed by detailed measurement by Millikan and others. In fact Einstein himself regarded his March 1905 photoelectric paper as being more revolutionary than his work on relativity (completed some three months later).

In the introduction section of his 1905 paper, Einstein wrote of his general motivation for this investigation. He was dissatisfied by the dichotomy of our description of matter as composed of particles, while radiation was waves. His seeking of a more symmetric world picture led him to the idea that under certain circumstances radiation could also be regarded as a gas of light quanta. It is interesting to note that Louis de Broglie (1892–1987) was inspired by these words to propose in 1923 that one could associate a wave to matter particles as well. This de Broglie wave idea received immediate and enthusiastic support from Einstein. Erwin Schrödinger (1887–1961) in turn found the reference to de Broglie's proposal in one of Einstein's papers, and this led to his famous equation for the matter wave.

4.1 The equipartition theorem and the Rayleigh–Jeans law

Recall that one of Planck's important contributions to blackbody radiation was his (electromagnetic) derivation of the relation[1]

$$\rho = \frac{8\pi \nu^2}{c^3} U, \tag{4.1}$$

[1] A proper interpretation of U as the average energy of a radiation oscillator was first proposed by Lord Rayleigh. The derivation of Eq. (4.1) by a counting of electromagnetic wave states will be given in Section 4.1.3. Its derivation by S. Bose (1924) by a counting of photon states will be discussed in Section 7.2.2.

where ν is the radiation frequency, c the speed of light, ρ the radiation density, and U is the average energy of a cavity-wall oscillator. Einstein pointed out that classical physics could make a definite statement about this average energy (Einstein 1905a).

4.1.1 Einstein's derivation of the Rayleigh–Jeans law

Einstein proceeded in his 1905 paper to apply the equipartition theorem (EPT) of classical statistical mechanics[2] to charged cavity oscillators. Corresponding to each degree of freedom, a physical system has a thermal energy of $\frac{1}{2}k_B T$. Since a 1D oscillator has two degrees of freedom (DOF)—kinetic and potential energy—each oscillator should have a thermal energy of

$$U = k_B T, \tag{4.2}$$

which can then be translated, through Eq. (4.1), into a result for the radiation density

$$\rho(\nu, T) = \frac{8\pi \nu^2}{c^3} k_B T. \tag{4.3}$$

This spectrum distribution, which came to be known as the **Rayleigh–Jeans law**, could only fit the low-frequency parts of the observational data, but not the high-frequency parts. In fact it implied an infinite energy density i.e. suffering from what Paul Ehrenfest (1880–1933) called the 'ultraviolet catastrophe':

$$u = \int_0^\infty \rho d\nu = \infty. \tag{4.4}$$

While the experimental validity was limited, the theoretical foundation of the Rayleigh–Jeans law was solid in the context of the known physics at the beginning of the twentieth century.

Supporting Planck's determination of k_B

The second section of Einstein's 1905 photoelectric paper had the title of "On Planck's determination of the elementary quanta". By "elementary quanta" Einstein meant "fundamental atomic constants". Recall (see Section 3.6.2) that an important part of Planck's result was his ability to deduce Boltzmann's constant k_B by fitting the blackbody radiation data to his distribution. But the theoretical foundation of Planck's distribution was uncertain. Here Einstein noted that the physical basis for the Rayleigh–Jeans law was firm and it could fit the low-frequency data well enough to extract k_B as well. Thus Planck's determination of Boltzmann's constant was on more reliable ground than Planck's deduction from his own distribution. Since k_B was the ratio of the gas constant R divided by Avogadro's number N_A, Einstein obtained $N_A = R/k_B = 6.02 \times 10^{23}$. We note that this was the third theoretical proposal made by Einstein in 1905 of finding Avogadro's number: his doctoral thesis involving the study of viscosity of a solution with suspended particles, his Browning motion paper, and now by way of blackbody radiation.

[2] See SuppMat Section 4.4 for some introductory comments on the equipartition theorem.

4.1.2 The history of the Rayleigh–Jeans law and "Planck's fortunate failure"

- June 1900, Rayleigh—Lord Rayleigh (John William Strutt, 1842–1919) wrote a paper (Rayleigh 1900) applying the equipartition theorem to radiation oscillators directly[3] and he obtained the result of $\rho = C_1 \nu^2 T$. The proportionality constant C_1 was not calculated. Although this distribution was consistent with Wien's displacement law,[4] it failed to account for all the observational data. He suggested that this should be a "limit law": valid only in the limit of low radiation frequency and high temperature (i.e. for small values of ν/T). To avoid the ultraviolet catastrophe (for large ν/T) he introduced an ad hoc exponential damping factor $\rho = C_1 \nu^2 T \exp(-C_2 \nu/T)$.
- October–December 1900, Planck—The Planck spectrum distribution was discovered; energy quantization proposed two months later. (Planck 1900a, b, 1901).
- March 1905, Einstein—What we now call the Rayleigh–Jeans law was derived correctly by Einstein. He noted the solid theoretical foundation of (4.3) and its consequential disastrous divergent radiation density.
- May 1905, Rayleigh again—Rayleigh returned to his $\nu^2 T$ result with a derivation of the proportional constant C_1. But he made a mistake, missing a factor of 8.
- June 1905, Jeans—Adding it as a postscript to a previously completed paper, James Jeans (1872–1946) corrected Rayleigh's error. There's an interesting bit of history: Jeans never accepted the idea that (4.3) was only a limit law. He explained away the incompatibility with experimental results by insisting that the observed radiation was somehow out of thermal equilibrium.
- Einstein's biographer Abraham Pais suggested that Eq. (4.3) should really be called the **Rayleigh-Einstein-Jeans law** (Pais 1982, p. 403).

The "popular history" of blackbody radiation as sometimes presented in textbooks may have left the impression that the Rayleigh–Jeans law preceded Planck's work. Somehow the failure of this solid piece of classical physics (EPT leading to the ultraviolet catastrophe) had forced Plank to make his revolutionary energy quanta proposal. Actually by temperament Planck was a very conservative man; the last thing he would want was to be any sort of revolutionary. The interesting point is that even though Rayleigh's 1900 paper did predate Planck's by six months, somehow Planck failed to take note of this work. It had been speculated that had he understood the significance of the Rayleigh–Jeans law (i.e. its solid classical physics foundation[5]), Planck would not have proceeded in proposing his distribution law. This sequence of events could be characterized as "Planck's fortunate failure".

The above-mentioned popular history of the quantum revolution would have been a better approximation of the actual history if Planck's name had been replaced by Einstein's. In fact it was Einstein who first understood the significance of the Rayleigh–Jeans law's inability to describe physical observation. He concluded that the failure of this piece of solid classical physics in its

[3] This is in contrast to Planck's material cavity oscillators. More discussion on this issue will be presented in the next subsection.

[4] It is in the form of $\rho = \nu^3 f(\nu/T)$ with $f(\nu/T) = C_1(\nu/T)^{-1}$.

[5] This should be qualified with the understanding that in those years the general validity of the equipartition theorem was in doubt. Before Einstein's 1906 work, as we shall discuss in Chapter 5, there was a great deal of confusion as to EPT's success in the area of specific heat.

confrontation with experiment meant that one had to alter the classical physics in some fundamental way. As we shall see, Einstein's method of arriving at $\epsilon = h\nu$ was quite independent of Planck's approach. (More on this point will be discussed in Section 5.1.)

4.1.3   An excursion to Rayleigh's calculation of the density of wave states

While this chapter mainly concerns Einstein's 1905 paper on blackbody radiation, we digress in this subsection to discuss the derivation of (4.1) as first given by Lord Rayleigh. Recall that Einstein followed Planck's interpretation of U in (4.1) as the average energy of a cavity oscillator and applied the equipartition theorem to an ensemble of charged oscillators in equilibrium with thermal radiation ($U = k_B T$ for the kinetic and potential degrees of freedom of a 1D oscillator). Rayleigh in his 1900 and 1905 papers took a more straightforward approach of proceeding directly to thermal radiation itself. In effect he took U as the average energy (per frequency) of a radiation oscillator. Again one has $U = k_B T$, now for the two polarization degrees of freedom.

Recall our discussion in Section 3.1 showing that radiation (electromagnetic waves) can be regarded as a collection of oscillators—what we called radiation oscillators. This is the same U when we first discussed Wien's displacement law. In other words, the Planck relation (4.1) is now interpreted as a statement about the "radiation density of states" (i.e. the number of states per unit volume per unit frequency),

$$\rho = \frac{N}{V} U \qquad \text{with} \qquad \frac{N}{V} = \frac{8\pi \nu^2}{c^3} \tag{4.5}$$

where V is the volume and N the number of radiation states (oscillators).

Here we present Rayleigh's method of counting radiation wave states (Rayleigh 1900). Consider an electromagnetic wave in a cubic box $V = L^3$

$$\Psi_{n_x n_y n_z}(\mathbf{x}, t) = A \sin k_x x \sin k_y y \sin k_z z e^{i\omega t}.$$

It propagates with a speed which is the product wavelength and frequency $c = \lambda \nu$. In the above expansion, ω is the angular frequency ($2\pi \nu$), and k_i is the i-component of wavevector $\mathbf{k}$, which has a magnitude (wavenumber) equal to 2π divided by the wavelength. Since the boundary condition is vanishing amplitude at the cubic surface (hence the largest the wavelength can be is twice the cubic side $2L$), the possible wavelengths in any one, say x-direction, are $2L/n_x$ where $n_x = 1, 2, 3, 4, \ldots$ This decomposition into normal modes in each direction leads to the wavenumber for 3D space as

$$\sqrt{k_x^2 + k_y^2 + k_z^2} = k = \frac{2\pi}{\lambda} = \frac{2\pi}{2L} n \tag{4.6}$$

with $n = \sqrt{n_x^2 + n_y^2 + n_z^2}$, which is related to the wave frequency as

$$\frac{n}{2L} = \frac{1}{\lambda} = \frac{\nu}{c} \qquad \text{or} \qquad n = \frac{2L}{c} \nu \,. \tag{4.7}$$

Thus each wave state is specified by three positive integers (n_x, n_y, n_z) and the task of counting the number of states is equivalent to counting all the points, labeled by (n_x, n_y, n_z), in the 3D **n**-space. Since we have a dense set of points, this count corresponds to the volume in **n**-space (Fig. 4.1):

$$N dv = \frac{2 \times 4\pi n^2 dn}{8}. \tag{4.8}$$

Here $4\pi n^2$ is the surface area of a sphere[6] with radius n and the factor of 2 accounts for the two polarization states. We have the factor of 8 in the denominator because the counting must be restricted to the positive quadrant of the sphere (n_x, n_y, n_z all being positive). This is the factor missing in Rayleigh's original calculation and subsequently corrected by Jeans. Replacing the variable n by the frequency v as in (4.7), we have

$$\frac{N}{V} dv = \frac{\pi}{V} \left(\frac{2L}{c}\right)^3 v^2 dv. \tag{4.9}$$

Cancelling the volume factor $V = L^3$, we obtain the claimed result for the density of states N/V as given in (4.5). In short, Rayleigh's enumeration of radiation states involves the counting of the number of standing waves in the enclosure.

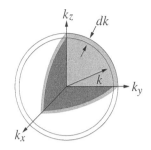

Fig. 4.1 The positive quadrant in the 3D space of wavenumbers. In our notation k is directly related to n as given in Eq. (4.6).

[6]This is a sphere in **n**-space—not to be confused with physical space, which in our case is a cube with side L.

4.2   Radiation entropy and complexion *á la* Einstein

We now discuss how the study of blackbody radiation led Einstein to the proposal of light quanta.

4.2.1   The entropy and complexion of radiation in the Wien limit

Einstein noted that his new result (4.3) is just the Planck distribution

$$\rho(v, T) = \frac{\alpha v^3}{\exp(\beta v/T) - 1} \tag{4.10}$$

taken to the limit of small $\beta v/T$ so that $\exp(\beta v/T) = 1 + \beta v/T + \cdots$:

$$\rho(v, T) \underset{\substack{\text{small} \\ v/T}}{\longrightarrow} \frac{\alpha}{\beta} v^2 T. \tag{4.11}$$

Since the theoretical foundation of the Rayleigh–Jeans law is well-understood classical physics, the new physics must be represented by the opposite limit of large v/T:

$$\rho(v, T) \underset{\substack{\text{large} \\ v/T}}{\longrightarrow} \alpha v^3 \exp(-\beta v/T) \tag{4.12}$$

which is Wien's distribution discussed in the previous chapter. The question was then: what physics did this Wien limit law represent?

Introducing the entropy density per unit radiation frequency

Seeking the answer to this question, Einstein, like Planck, undertook a thermodynamic and statistical study (Einstein 1905a). Here he used an elegant method by first relating the energy density per unit frequency $\rho(v, T)$,

$$\frac{U}{V} = \int \rho(v, T)\, dv \tag{4.13}$$

to the entropy density per unit frequency $\varphi(v, \rho)$,

$$\frac{S}{V} = \int \varphi(v, \rho)\, dv. \tag{4.14}$$

through the equality of

$$\frac{\partial \varphi}{\partial \rho} = \frac{\partial S}{\partial U} = \frac{1}{T}. \tag{4.15}$$

The proof of this relation can be obtained by first showing that the differential $\partial \varphi / \partial \rho$ is independent of frequency v. This follows from the requirement that entropy must be maximized with the condition of constant energy,

$$\delta S - \lambda \delta U = 0, \tag{4.16}$$

where λ is the Lagrange multiplier (cf. Section A.4). Integrating over this relation as in (4.14) and (4.13), we have

$$V \int \frac{\partial \varphi}{\partial \rho} d\rho dv - \lambda V \int d\rho dv = V \int \left(\frac{\partial \varphi}{\partial \rho} - \lambda \right) d\rho dv = 0. \tag{4.17}$$

Since this must be valid for any density distribution, the integrand must also vanish, yielding the condition $\partial \varphi / \partial \rho = \lambda$. Thus the differential must be independent of radiation frequency v. We now consider the infinitesimal change of entropy dS for the system changing from one equilibrium state to another for the infinitesimal input of energy dU:

$$\frac{dU}{T} = dS = V \int \frac{\partial \varphi}{\partial \rho} d\rho dv = \frac{\partial \varphi}{\partial \rho} V \int d\rho dv = \frac{\partial \varphi}{\partial \rho} dU, \tag{4.18}$$

hence the result displayed in (4.15), from which Einstein would deduce the entropy density per unit frequency φ.

The radiation entropy corresponding to the Wien distribution

We first invert the Wien distribution of (4.12) and apply the result of (4.15)

$$\frac{\beta v}{T} = \ln \alpha v^3 - \ln \rho = \beta v \frac{\partial \varphi}{\partial \rho}$$

namely, $d\varphi = \frac{1}{\beta v}(\ln \alpha v^3 - \ln \rho) \, d\rho$ so that

$$\varphi = \int d\varphi = \frac{\ln \alpha v^3}{\beta v} \int d\rho - \frac{1}{\beta v} \int \ln \rho d\rho$$

$$= \frac{\rho}{\beta v} \ln \alpha v^3 - \frac{\rho}{\beta v} \ln \rho + \frac{\rho}{\beta v}. \tag{4.19}$$

To reach the second line we have performed an integration by parts of the last integral in the first line. Thus Einstein found the entropy density corresponding to Wien's limit (4.12) to be

$$\varphi = -\frac{\rho}{\beta v} \left[\ln \frac{\rho}{\alpha v^3} - 1 \right]. \tag{4.20}$$

Multiplying the density by the volume, we have the entropy when the energy density ρ is replace by U/V:

$$S = V\varphi = -\frac{U}{\beta v} \left[\ln \frac{U/V}{\alpha v^3} - 1 \right]. \tag{4.21}$$

It should be noted that, while Planck worked on the entropy of cavity wall oscillators [cf. Eq. (3.32)], here Einstein was working on the entropy corresponding directly to the radiation energy density.[7]

The entropy change due to the volume change

To find the physical significance of this entropy expression, instead of following Planck to make a direct counting of microstates, Einstein took the key step of concentrating on the volume dependence of the entropy change so that he could implement his idea of comparing this result of radiation with the corresponding result for an ideal gas. This would allow him to argue for a particle nature of radiation—the quantum of light. If we make a volume change $V_0 \longrightarrow V$, with energy held fixed, we get from (4.21) the entropy difference of

$$S - S_0 = \frac{U}{\beta v} \ln \frac{V}{V_0}. \tag{4.22}$$

Being in logarithmic form, it is straightforward to extract the complexion through the Boltzmann relation of $S - S_0 = k_B \ln W$. In this way Einstein found the statistical weight W, associated with this volume change,

$$\ln W = \frac{U}{k_B \beta v} \ln \frac{V}{V_0} \quad \text{or} \quad W = \left(\frac{V}{V_0} \right)^{U/(k_B \beta v)}. \tag{4.23}$$

4.2.2 The entropy and complexion of an ideal gas

To interpret radiation as a gas of light quanta, Einstein set out to find the analogous complexion for a volume change of a simple ideal gas. We have the ideal gas law

$$pV = nRT \tag{4.24}$$

[7] Apparently Einstein first thought that his result was very different from that of Planck's. In this 1905 paper no reference was made of Planck's energy quantization result. Only upon further investigation in a year later did Einstein come to the conclusion that their conclusions were compatible. For further comments, see Section 5.1.

where p is the pressure, n the number of moles, and R the gas constant

$$n = \frac{\mathcal{N}}{N_A} \quad \text{and} \quad R = k_B N_A, \tag{4.25}$$

$\mathcal{N}$ being the number of gas molecules and N_A Avogadro's number so that the ideal gas law can be written in an even more transparent form as

$$pV = \mathcal{N} k_B T. \tag{4.26}$$

To find the corresponding entropy, we start with its definition in terms of heat energy change:

$$dS = \frac{dQ}{T} = \frac{p}{T} dV = k_B \mathcal{N} \frac{dV}{V},$$

where we have used the heat input as $dQ = C_V dT + p dV$ at a fixed temperature $dT = 0$, and the ideal gas law of (4.26). Integrating both sides of this equation, we get

$$S - S_0 = k_B \mathcal{N} \ln \frac{V}{V_0}. \tag{4.27}$$

This can be related to the statistical weight of an ideal gas associated with a volume change by the Boltzmann relation $S - S_0 = k_B \ln W$, leading to

$$W = \left(\frac{V}{V_0} \right)^{\mathcal{N}}. \tag{4.28}$$

4.2.3   Radiation as a gas of light quanta

A comparison of (4.23) and (4.28) suggested to Einstein that blackbody radiation in the Wien limit could be viewed as behaving thermodynamically as if it consisted of mutually independent quanta, a gas of photons—by turning $\mathcal{N}$, the number of gas particles, in (4.28) into the number of light particles given by (4.23)

$$\mathcal{N} = \frac{U}{k_B \beta \nu}. \tag{4.29}$$

Recall our previous identification in Eq. (3.29) of $k_B \beta \nu = \epsilon$, the energy quantum, with $k_B \beta = h$ being Planck's constant. This result can then be written as[8]

$$U = \mathcal{N} \epsilon \quad \text{with} \quad \epsilon = h\nu. \tag{4.30}$$

The system is seen as a collection of quanta with their energy being given by Planck's relation. This is how Einstein first argued for a light-quanta description of blackbody radiation in the Wien limit, away from the regime of classical physics. This was such a brilliant idea on the part of Einstein that, by a comparison of the volume dependence, he could reach this conclusion of light quanta without committing himself to any uncertain counting scheme of the blackbody radiation complexion. His revolutionary step is to take this result as suggesting

[8] A word about our notation: In the previous chapter we have used N to denote the number of cavity oscillators, while here $\mathcal{N}$ is the number of quanta in one oscillator. Equation (3.32) gives the average energy of the oscillator as $U = (P/N)\epsilon$, where P/N is the number of energy quanta per oscillator. This is certainly consistent with the present result of $U = \mathcal{N}\epsilon$.

that laws of emission and absorption of radiation must be such as to be consistent with a description of light as consisting of such light quanta. In particular because light could only be emitted and absorbed in integral units of $h\nu$, it was Einstein who interpreted Planck's result as showing the energy of a resonant oscillator itself must also be quantized.

4.2.4   Photons as quanta of radiation

After the trail-blazing work of Einstein and others, present-day readers would naturally identify the light quanta as photons. In the above presentation we had already referred occasionally to Einstein's light quanta as photons. Strictly speaking this would be a misrepresentation of the history. While the 1905 paper certainly showed that radiation energy must be quantized in the Wien limit, it would still be some time before Einstein committed himself to the idea of these quanta as point-like particles. Namely, while Einstein had shown that radiation in the Wien limit with a quantized energy had a thermodynamical behavior similar to an ideal gas under volume change, he was not ready to commit himself to the proposition that these energy quanta, rather than spread out in space, were concentrated at point-like sites. He finally came to the corpuscular interpretation, when in 1909 he strengthened this view through the study of energy fluctuations in radiation. This topic will be discussed in Section 6.1.

In this connection Planck's constant h in the relation $\epsilon = h\nu$ can be viewed as the fundamental physics constant acting as the conversion factor between radiation and particles.

4.3   The photoelectric effect

In Sections 7, 8, and 9 of his 1905 paper, Einstein immediately applied his new quantum idea to several physical situations. Particularly in Section 8, *On the generation of cathode rays by illumination of solid bodies*, he put forth his prediction for the photoelectric effect. This is the phenomenon first discovered in 1887 by Heinrich Hertz (1857–94), in which electrons are emitted from matter (metals and nonmetallic solids, liquids, or gases) as a consequence of their absorption of energy from electromagnetic radiation of very short wavelength, such as visible or ultraviolet light (see Fig. 4.2a). Taking into account the photon energy $h\nu$, Einstein was able to deduce immediately the maximum kinetic energy of the photoelectrons in terms of the incident light frequency:

$$K_{max} = h\nu - W \tag{4.31}$$

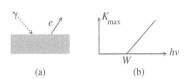

(a)   (b)

Fig. 4.2 (a) Photoelectric effect. (b) Electron kinetic energy spectrum.

where W is the work function representing the threshold energy that an electron must have in order to escape the parent matter. Thus the maximum cathode ray energy spectrum was shown to be independent of the input light intensity, explaining the 1902 experimental discovery of Lenard. It has a linear dependence on the incident light's frequency that can be represented, see Fig. 4.2(b), by a 45° line in the graphic plot of K_{max} vs. $h\nu$, independent of the nature of the substance investigated. Only the number of emitted electrons depends on the intensity. All this was verified through a decade of experimental measurement by others and by Robert A. Millikan (1868–1953), who were able to determine Planck's constant through Eq. (4.31) to an accuracy of 0.5%. It is interesting to note that Millikan was initially convinced that Einstein's light quantum idea had to be wrong, because of the vast body of evidence that had already shown the wave nature of light. He undertook a decade-long painstaking experimental program[9] to test Einstein's theory. His results confirmed Einstein's predictions in every detail, but Millikan was not convinced of Einstein's interpretation, and as late as 1916 he wrote, "Einstein's photoelectric equation ... cannot in my judgment be looked upon at present as resting upon any sort of a satisfactory theoretical foundation", even though "it actually represents very accurately the behavior" of the photoelectric effect (Millikan 1916). Thus, in this case, theory was way ahead of experiment. The whole idea of particles of light was not commonly understood by the physics community until 1924 when the result of Compton scattering was obtained (see Section 7.1).

[9] It required him to build an immaculate laboratory in order to prepare the very clean metal surface of the photoelectrode.

Other applications: Photoluminescence and photoionization

Aside from the application of the light quantum idea to the photoelectric effect, in Section 7 of Einstein's paper, one also finds the explanation of the **Stokes rule**—the observation that the frequency of photoluminescent emission is less than the incident light frequency—by energy conservation. In the last Section 9, Einstein also explained why the incident light must have a frequency, hence energy, greater than the ionization potential of the gas in order for the **photoionization of the gas** to take place. In this way, in this great Einstein paper he proposed the quanta of radiation idea and put forward the various means to test this idea.

4.4   SuppMat: The equipartition theorem

The equipartition theorem follows from the Boltzmann distribution of statistical mechanics. Each degree of freedom corresponds to energy being a quadratic function of position (as in the harmonic oscillator $kx^2/2$) or momentum (as the kinetic energy $p^2/2m$ and rotational kinetic energy $L^2/2I$, etc.):

$$U = \alpha q^2 + \beta p^2$$

where (q, p) are understood to be the generalized position and momentum. We then have, from Boltzmann's principle, the average thermal energy

$$\bar{U} = \frac{\int U \exp\left(-\dfrac{U}{k_B T}\right) dq dp}{\int \exp\left(-\dfrac{U}{k_B T}\right) dq dp}.$$

Concentrating on just one degree,

$$\bar{U} = \frac{\int \alpha q^2 \exp\left(-\dfrac{\alpha q^2}{k_B T}\right) dq}{\int \exp\left(-\dfrac{\alpha q^2}{k_B T}\right) dq} = \frac{1}{2} k_B T$$

where we have used the Gaussian integral results of Section A.2

$$\int \exp(-ax^2)\, dx = \sqrt{\frac{\pi}{a}}, \qquad \text{and} \qquad \int x^2 \exp(-ax^2)\, dx = \frac{1}{2a}\sqrt{\frac{\pi}{a}}.$$

<table>
<tr><td rowspan="4">**5**</td><td></td></tr>
</table>

Quantum theory of specific heat

- Einstein's proposal of light quanta was originally based on the blackbody radiation in the Wien limit. A year later he came to the conclusion that his quantum idea was compatible with Planck's energy quanta. He provided a derivation of Planck's spectrum distribution that was, when compared to Planck's deduction, simpler and less problematic on theoretical ground.
- Einstein argued that the quantum idea should be applicable to thermal properties of matter, as well as to radiation. His theory of specific heat is historically important because it clarified the confused situation that had cast doubt on the kinetic theory of gases and even the molecular structure of matter. This is also the first instance when the quantum idea was shown to be relevant to physical systems well beyond the esoteric case of blackbody radiation.
- We present the Einstein model for the thermal properties of solids, as well its improvement in the form of the Debye model invoking the notion of quanta of sound waves (phonons). With these papers, we can say: the quantum theory of solid state physics has begun.

Soon after proposing the idea of the light quantum, Einstein extended the application of quantum theory in 1906 to the study of specific heat (Einstein 1907a). It cleared up the confused pattern of specific heats of gases, and came up with a new theory of specific heats for solids that explained the temperature dependence observed in some solids. It is important historically as this is the first instance when the quantum idea was shown to be relevant to physical systems well beyond the esoteric case of blackbody radiation.

5.1 The quantum postulate: Einstein vs. Planck

It is commonly thought that Einstein's idea of light quanta came about as an extension of Planck's work on the quantum of energy. But a close read of Einstein's 1905 paper (as discussed in the previous chapter) clearly shows that this is not the case. In the last chapter we have seen how Einstein arrived at his idea of light quanta with $\epsilon = h\nu$ by way of a statistical study of Wien's distribution law of blackbody radiation. This is in contrast to Planck's calculation of the complexion W for the Planck spectral distribution, that is at variance from the then standard statistical mechanical procedure. Planck's work was cited in Einstein's paper (1905a) in two places: 1. (Planck 1900b) for Planck's relation (4.1) between oscillator energy and radiation energy density; and 2. (Planck 1900a) for Planck's distribution as providing the best description of radiation data. On the other hand, Planck's energy quantization result (1900b, 1901) was not mentioned at all. In his discussion leading up to light quanta, Einstein made no use of Planck's spectrum result nor his statistical counting method. Significantly, Einstein did not use Planck's notation h for the quantum of action (as we did in Chapter 4).[1]

At the beginning of Einstein's next paper on the quantum theory (Einstein 1906), after the opening paragraph summarizing his 1905 light quantum proposal, he commented on his attitude towards Planck's quantum result as follows: "At that time it seemed to me that in a certain respect Planck's theory of radiation constituted a counterpart to my work." He then stated his change of opinion after a new analysis of Planck's work, and concluded: "In my opinion the preceding considerations do not by any means refute Planck's theory of radiation; they seem to me rather to demonstrate that, in his radiation theory, Planck introduced a new hypothetical principle into physics—the hypothesis of light quanta." Namely, he and Planck, through different paths, had reached the same conclusion that the energy of the radiation oscillators were quantized as $\epsilon = h\nu$. (Of course Einstein had gone further by interpreting this to mean the quantization of the radiation field itself.) A few months later he proceeded (Einstein 1907a) to give his derivation of Planck's distribution according to Boltzmann's statistics (but without involving any explicit computation of the statistical weight W).

[1] A very helpful historical study can be found in Klein (1977).

5.1.1 Einstein's derivation of Planck's distribution

Consider the states of a radiation oscillator with its quantized energies: 0, $h\nu$, $2h\nu$, $3h\nu$, ..., namely, states having different numbers of energy quanta: $\epsilon_n = n\epsilon$ with $\epsilon = h\nu$. According to Boltzmann, the oscillator would have an average number of quanta given by

$$\langle n \rangle = \frac{\displaystyle\sum_{n=0}^{\infty} n e^{-n\epsilon/k_B T}}{\displaystyle\sum_{n=0}^{\infty} e^{-n\epsilon/k_B T}} = -k_B T \frac{d}{d\epsilon}\left[\ln\left(\sum_{n=0}^{\infty} e^{-n\epsilon/k_B T}\right)\right]. \qquad (5.1)$$

We can easily perform the last sum as it is a simple geometric series,

$$\sum_{n=0}^{\infty} e^{-n\epsilon/k_B T} = 1 + e^{-\epsilon/k_B T} + e^{-2\epsilon/k_B T} + \cdots = \frac{1}{1 - e^{-\epsilon/k_B T}}.$$

Consequently the average number is

$$\langle n \rangle = k_B T \frac{d}{d\epsilon} \ln \left(1 - e^{-\epsilon/k_B T}\right) = k_B T \frac{\frac{1}{k_B T} e^{-\epsilon/k_B T}}{1 - e^{-\epsilon/k_B T}} = \frac{1}{e^{\epsilon/k_B T} - 1}. \tag{5.2}$$

This immediately leads to the Planck's expression (3.42) for the average oscillator energy $U = \langle n \rangle \epsilon$ as

$$U = \frac{h\nu}{e^{h\nu/k_B T} - 1}. \tag{5.3}$$

How did Einstein manage to get this outcome rather than the expected classical EPT result of $U = k_B T$?

The basic classical assumption of Boltzmann's statistics would involve equal probabilities for states having equal phase space volume, i.e. equal energy. Here Einstein assumed equal probability for all energy states which of course have different energies: 0, $h\nu$, $2h\nu$, etc. We shall return to this point in Chapter 7 when we discuss the new quantum statistics.[2]

How can the ultraviolet catastrophe be avoided because of the quantum hypothesis?

In this derivation we see that since the oscillator energy separation between states is $\epsilon = h\nu$, the allowed states of the oscillation modes of very high frequency are widely separated in energy. It takes a great deal of energy to excite such a mode. But the Boltzmann factor $\exp(-\epsilon/k_B T)$ of statistical mechanics tells us that the probability of finding a great deal of energy in any one mode falls off rapidly with the energy. In this way the high-frequency divergence can be avoided. Thus Planck's energy quantization can lead to a radiation distribution that avoids the ultraviolet catastrophe.

Einstein then argued that such a result for the oscillator energy should also be applicable to material oscillators outside the realm of blackbody radiation, such as atoms in gases and solids. With this observation, he proceeded to clarify the great confusion that then existed with respect to the applicability of the equipartition theorem to the study of gases and solids—in fact a confusion that had clouded the whole notion of a statistical basis for thermodynamics.

[2] This 1906 derivation by Einstein differs from Planck's as well as the later one by Bose (1924), to be discussed in Chapter 7. Planck was dealing with cavity-wall oscillators, rather than radiation oscillators. In Einstein's approach, $n h\nu$ is the energy of the nth state of a single radiation oscillator while for Bose it is the energy of n photons.

5.2   Specific heat and the equipartition theorem

In the statistical mechanics of gases and solids we have the basic theorem of equipartition of energy. It was a great triumph in explaining the regularity of measured specific heat capacities. Here we are mainly interested in the heat capacity C defined as the variation of internal energy with respect to temperature (holding the volume fixed)

$$C = \left(\frac{\partial U}{\partial T}\right)_V. \tag{5.4}$$

We shall first review the situation in the pre-quantum era, and will then see how the notion of quantum of energy brings clarity to our understanding of this basic thermal attribute of matter.

5.2.1 The study of heat capacity in the pre-quantum era

As we indicated in Section 4.4, Boltzmann's statistics leads directly to the equipartition theorem (EPT), which was applied to the study of specific heats of substances and produced many successful results—but also several puzzles.

- **Monatomic gas**—This is the simplest case. A monatomic molecule has only three translational degrees of freedom (DOF), $\epsilon = \left(p_x^2 + p_y^2 + p_z^2\right)/2m$. Thus, according to the EPT, the average thermal energy of the gas must be $\langle \epsilon \rangle = 3k_B T/2$. For a mole of the gas, we have

$$U = \frac{3}{2}N_A k_B T = \frac{3}{2}RT, \tag{5.5}$$

  where N_A is the Avogadro number, and $R = N_A k_B$ is the gas constant. This leads to a molar specific heat of

$$C = \frac{\partial U}{\partial T} = \frac{3}{2}R \tag{5.6}$$

  in agreement with experimental measurement.
- **Diatomic gas**—We now have three translational and two rotational DOF[3] (cf. Fig. 5.1), resulting in

$$C = \frac{5}{2}R, \tag{5.7}$$

  also in agreement with observation. What is notable is that this agreement came about only when we had ignored the vibrational degrees of freedom even though there was spectroscopic evidence for the existence the vibrational mode (the presence of absorption lines consistent with this interpretation).
- **Polyatomic gas**—Again, the EPT was successful, if we ignore the vibrational degrees of freedom.
- **Monatomic solid**—Such a solid can be pictured as N_A atoms located on a 3D lattice. An atom at each lattice site can be thought of as a simple harmonic oscillator. A 1D oscillator has two quadratic DOFs (kinetic and potential energies), and a 3D oscillator has six. Thus there are a total of $6N_A$ degrees of freedom for such a solid, hence according to the equipartition theorem, a total thermal energy of $U = 3N_A k_B T$ leading to a specific heat of

$$C = 3R. \tag{5.8}$$

This result, first obtained by Boltzmann in 1876, explained the long-standing observational pattern known as the **Dulong–Petit rule**: various solid substances have a common specific heat value of $3R$.

[3] Rotational kinetic energy corresponds again to a quadratic degree of freedom. See Section 4.4.

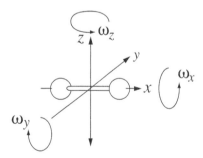

Fig. 5.1 Two rotational degrees of freedom of a diatomic gas molecule, around the y and z axes. It is difficult to excite the third rotational mode because the moment of inertia I around the x axis is very small, thus there is a large rotational energy threshold of $L^2/2I$ for this mode.

[4]In Einstein's 1906 paper on specific heat, he concentrated on the study of solids. There was no discussion on the application of the EPT to gases.

- **The confused situation**—It is puzzling that, to obtain from the EPT correct values for the heat capacity of various substances, we must ignore vibrational DOFs in the case of a gas, yet vibrational energy gives the dominant contribution for a solid. Furthermore, while this rule worked for simple solids, it was also known that the specific heat of some solids (such as diamond) fell below the Dulong–Petit value and were temperature-dependent. This confused state[4] led certain sectors of the physics community to cast doubt on the reality of molecules and atoms.

5.2.2   Einstein's quantum insight

After Einstein's derivation of Planck's distribution law, he wrote (Einstein 1907a):

I believe that we must not content ourselves with this result. For the question arises: If the elementary structures that are to be assumed in the theory of energy exchange between radiation and matter cannot be perceived in terms of the current molecular-kinetic theory, are we then not obliged also to modify the theory for the other periodically oscillating structures considered in the molecular theory of heat? In my opinion the answer is not in doubt. If Planck's radiation theory goes to the root of the matter, then contradictions between the current molecular-kinetic theory and experience must be expected in other areas of the theory of heat as well, which can be resolved along the lines indicated. In my opinion this is actually the case, as I shall now attempt to show.

Einstein realized that the quantum idea could clarify the confused situation in specific heat. He argued that the distribution shown in Eq. (5.3) should be applicable to other oscillators besides those of radiation. To thermally excite a DOF, the thermal energy $k_B T$ must be comparable to or greater than the relevant energy quantum $h\nu$; otherwise the exponential factor in Eq. (5.3) being the dominant one in the denominator would lead to the suppression of this degree of freedom[5]

[5]Just as in the opposite limit of $k_B T \gg h\nu$, we recover the classical result $\langle \epsilon \rangle = k_B T$ from (5.3).

$$\langle \epsilon \rangle \simeq h\nu \exp\left(-\frac{h\nu}{k_B T}\right) \approx 0.$$

Einstein argued that the vibrational DOF could be ignored if $h\nu \gg k_B T$. Indeed one could answer the question concerning the size of the quantum of vibrational energy in gases because an estimate of the quantum of vibration energy $h\nu$ could be obtained through the absorption spectroscopy of gases. The vibrational absorption was known to be in the infrared, a wavelength on the order of a micron corresponding to an energy quantum of $h\nu \simeq 1$ eV, which is much greater than the thermal energy $k_B T \simeq 1/40$ eV at room temperature ($T \simeq 300$ K):

$$\exp\left(-\frac{h\nu}{k_B T}\right) \simeq e^{-40} \ll 1.$$

Thus, Einstein's quantum energy insight explained why the vibrational degrees of freedom of gases can be ignored at room temperature. This not only resolved the riddles in the study of specific heat, but also restored the confidence of the physics community in the validity of the equipartition theorem of statistical mechanics.

5.3 The Einstein solid—a quantum prediction

Since $h\nu/k_B T$ is a temperature-dependent factor, this quantum theory allowed Einstein to make a prediction that for certain solids the Dulong–Petit rule would fail as the quantum distribution (5.3) would lead to a suppression of the heat capacity below the classical value of $C = 3R$ at low temperature.

For this investigation, Einstein constructed a very crude model for the thermal properties of the solid: all lattice sites vibrate at the same frequency ν. Namely, for each solid there is one frequency; hence, in this discussion, ν is taken to be a constant. According to (5.3), the internal molar energy of the solid composed of 1D oscillators would simply be

$$U = N_A \frac{h\nu}{e^{h\nu/k_B T} - 1} \qquad (5.9)$$

and the specific heat can then be calculated as

$$C = \left(\frac{\partial U}{\partial T}\right)_V = N_A \frac{\partial}{\partial T} \frac{h\nu}{e^{h\nu/k_B T} - 1}$$

$$= N_A k_B \frac{(h\nu/k_B T)^2 \exp(h\nu/k_B T)}{\left[e^{h\nu/k_B T} - 1\right]^2}.$$

Generalizing to 3D vibration, we have Einstein's result for the specific heat:

$$C = 3R \left[\mathcal{E}\left(\frac{h\nu}{k_B T}\right)\right], \qquad (5.10)$$

where $\mathcal{E}(x)$ is what is now called the "Einstein function":

$$\mathcal{E}(x) = \frac{x^2 e^x}{(e^x - 1)^2}, \qquad (5.11)$$

where $x = h\nu/k_B T$. Thus there is only one parameter characterizing each solid—the lattice vibration frequency ν. This can also be expressed in terms of the "Einstein temperature" $k_B T_E = h\nu$ so that $x = T_E/T$. We note the following limits of the Einstein function $\mathcal{E}(x)$:

- **High-temperature regime**—When the temperature at which the specific heat is measured is much higher than the Einstein temperature of the solid (i.e. the thermal energy is much higher than the energy quantum, $k_B T \gg h\nu$ so that x is small), the Einstein function can be approximated as

$$\mathcal{E}(x) \simeq \frac{x^2}{(1 + x - 1)^2} = 1. \qquad (5.12)$$

Then according to Eq. (5.10) C has the value of $C = 3R$, showing that the Dulong–Petit rule applies at high temperature $T \gg T_E$.

- **Low-temperature regime**—When the temperature at the measurement is less than the Einstein temperature of the solid (x is large), the Einstein function can be approximated as $\mathcal{E}(x) \simeq x^2 e^{-x}$. We have an exponentially suppressed specific heat

$$C = 3R \left(\frac{h\nu}{k_B T} \right)^2 \exp \left(\frac{-h\nu}{k_B T} \right). \tag{5.13}$$

The experimental result was found to be in qualitative agreement with this prediction. The specific heat drops below the classical value at low temperature because of freezing out the DOFs at the low-energy limit. This is a direct consequence of the energy quantum hypothesis—the presence of a threshold before a DOF can be excited. In fact a fit of the measurement curve by Einstein (Fig. 5.2) showed an Einstein temperature of diamond of $T_E \simeq 1300\,\text{K}$ which is much higher than room temperature.

To recapitulate: Einstein had argued that there was already experimental evidence for the violation of classical physics. Namely, even though classical physics demanded the validity of the Rayleigh–Jeans law, the observed blackbody radiation spectrum could only be fitted by Planck's distribution. To Einstein this was the empirical grounds for the existence of the energy quantum. He then extended this new theory of the radiation oscillator to include oscillators of molecular kinetic theory. Thus Einstein's 1907 paper would be inadequately described as simply an application of the quantum theory to solids and gases. Rather, his paper should be viewed as him arguing for the break from classical physics in the form of a quantum theory. The new theory would bring clarity not only to blackbody radiation, but to a whole range of physical phenomena.

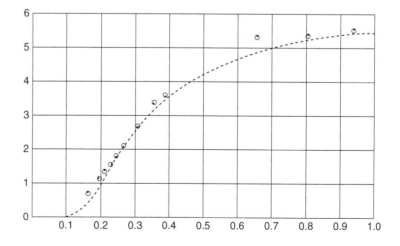

Fig. 5.2 Temperature dependence of diamond's specific heat compared to Einstein's quantum theoretical prediction. The diagram is from Einstein's 1907 paper. The vertical axis is the specific heat in units of calories/mole; the horizontal axis is T/T_E. The theory curve is for $T_E = 1300$ K. The experimental points were the measurements by H.F. Weber (1843–1912), Einstein's professor at ETH.

5.4 The Debye solid and phonons

When more accurate data of specific heats for various solids at very low temperature were obtained, Einstein's theory did not match the measurements in detail: the experimental curve decreases less rapidly than that predicted by Einstein's simple model. This inadequacy was remedied by a more sophisticated theory put forth in 1912 by Peter Debye (1884–1966). In the Einstein model for the thermal properties of a solid, one takes ϵ to be a constant. This means that one has made a very crude approximation of assuming all oscillators (atoms) oscillate with the **same** frequency throughout the solid. That the Einstein model fails at extremely low temperature means it fails in the very low-frequency (hence, long wavelength) regime. A crystalline solid has an intrinsic length-scale—the lattice spacing a. By the long-wavelength region, we mean $\lambda \gg a$; namely, the Einstein model does not provide an adequate description of motions covering many atoms—the correlated motions involving groups of atoms. Debye's theory was able to overcome this shortcoming (Debye 1912).

Debye modeled the solid as an elastic continuum. A disturbance in an elastic medium is simply a sound wave. One way to motivate Debye's approach is as follows: Planck's distribution (5.3) came out of a quantum description of the electromagnetic wave field. The vibrations of lattice sites in a solid can also be thought of as a field of waves—sound waves in this elastic continuum. Debye then adapted (5.3) as a quantum description of this sound field. The quanta of radiation were later termed **photons**; the quanta of sound waves are now called **phonons**.

In this approach by Debye concentrating on the wave aspect of the problem, we can simply take over Rayleigh's calculation of the density of wave states as presented in the previous chapter. The energy quantum of the sound wave can be written as $\epsilon = h\nu = hc_s/\lambda$ where c_s is the speed of sound. Following the treatment as presented in Section 4.1.3 we work with a cubic volume with sides L. The longest wavelength being $\lambda_0 = 2L$ (or the smallest wavenumber of $k_0 = \pi/L$), the general wavelength should be $\lambda = \lambda_0/n$ with $n = 1, 2, 3, 4, \ldots$ Keeping in mind the three spatial components of the wavevector, n is the magnitude of a vector $\mathbf{n}$ with three components

$$n = \sqrt{n_x^2 + n_y^2 + n_z^2} = \frac{2L}{\lambda}. \tag{5.14}$$

which is related to the wave frequency as

$$\frac{n}{2L} = \frac{1}{\lambda} = \frac{\nu}{c} \qquad \text{or} \qquad n = \frac{2L}{c}\nu. \tag{5.15}$$

The counting of states involves the counting of $\left(n_x, n_y, n_z\right)$, i.e. integration over the volume of the positive quadrant in 3D n space. Thus the steps are the same as those taken in Section 4.1.3 except for the following three modifications:

1. The speed of the wave is now c_s instead of c.
2. Because the electromagnetic wave has two polarization states while the sound wave has three (two transverse and one longitudinal), Eq. (4.8) now becomes

$$Nd\nu = \frac{3 \times 4\pi n^2 dn}{8}. \tag{5.16}$$

3. The limits of integration for the frequency integral. For electromagnetic waves, the "medium" being a continuum (the vacuum) with no structure, this allows for a vanishing wavelength $\lambda_{min} = 0$. Equivalently, there is no upper limit to the radiation frequency $\nu_{max} = \infty$. That is not the case for the Debye sound wave in a solid: The wavelength cannot be less than (twice) the lattice spacing $2a$, namely, λ_{min} is of the order (volume/N_A)$^{1/3}$. In fact Debye fixed the upper limit ν_{max} by relating it to the total number of oscillators N_A.

$$N_A = \left(\frac{4\pi}{3}n_{max}^3\right)\frac{1}{8} \quad \text{or} \quad n_{max} = \left(\frac{6N_A}{\pi}\right)^{1/3}. \tag{5.17}$$

The total energy U_N is now given by an integral

$$U_N = 3\int_0^{n_{max}} \frac{4\pi n^2}{8} dn \frac{nh\nu}{e^{nh\nu/k_BT} - 1} \tag{5.18}$$

with n related to the frequency by (5.15). This integral can be evaluated with the usual change of variable:

$$x \equiv \frac{h\nu}{k_BT} \tag{5.19}$$

and the substitution of spatial volume $V = L^3$ so that

$$U_N = \frac{3\pi}{2}\int_0^{x_{max}} V\left(\frac{2k_BT}{hc_s}\right)^3 x^2 dx \frac{k_BTx}{e^x - 1}$$

$$= \frac{12\pi V}{h^3 c_s^3}(k_BT)^4 \int_0^{x_{max}} \frac{x^3 dx}{e^x - 1}, \tag{5.20}$$

where the upper limit can be expressed in terms of the "Debye temperature"

$$h\nu_{max} \equiv k_BT_D, \quad \text{namely,} \quad x_{max} = \frac{T_D}{T}, \tag{5.21}$$

which, after using (5.15) and (5.17), can be written as

$$T_D = \frac{h}{2k_B}\left(\frac{6N_A}{\pi V}\right)^{1/3} c_s. \tag{5.22}$$

The Debye temperature is directly related to c_s, the speed of sound wave propagation in a solid. Recall that $c_s = \sqrt{B/\rho}$ where ρ is the density and B is the bulk modulus, characterizing the elasticity of the solid. Thus a solid with a high T_D has a stiff crystalline structure. Given that

$$\frac{1}{T_D^3} = \frac{4\pi V k_B^3}{3N_A h^3 c_s^3}, \tag{5.23}$$

we can rewrite the total energy in Eq. (5.20) as

$$U_N = 3 \frac{4\pi V k_B^3}{3 N_A h^3 c_s^3} 3 N_A k_B T^4 \int_0^{x_{max}} \frac{x^3 dx}{e^x - 1}$$

$$= 3 N_A k_B T \left(\frac{T}{T_D} \right)^3 \left[3 \int_0^{x_{max}} \frac{x^3}{e^x - 1} dx \right]. \tag{5.24}$$

5.4.1 Specific heat of a Debye solid

From this energy function $U_N(T)$ and we can calculate the predicted specific heat $C = (\partial U / \partial T)_V$.

- **High-temperature regime**: $T \gg T_D$. Thus, $x_{max} \sim 0$, and we can assume that the integration variable x is small and the integrand has the limiting value of $\frac{x^3}{1+x-1} = x^2$ and thus

$$\left[3 \int_0^{x_{max}} \frac{x^3}{e^x - 1} dx \right] \longrightarrow x_{max}^3 = \left(\frac{T}{T_D} \right)^{-3} \quad \text{and} \quad U_N \longrightarrow 3 N_A k_B T,$$

showing the correct classical physics limit of $C = 3R$.
- **Low-temperature regime**: $T \ll T_D$. Thus, $x_{max} \sim$ large, and we have the limit value of

$$\left[3 \int_0^\infty \frac{x^3}{e^x - 1} dx \right] = \frac{\pi^4}{15}$$

and

$$U \longrightarrow 3RT \left(\frac{T}{T_D} \right)^3 \frac{\pi^4}{5} = \frac{3\pi^4 R}{5 T_D^3} T^4. \tag{5.25}$$

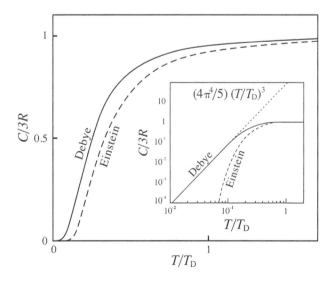

Fig. 5.3 Specific heat at low temperature according to the models of the Einstein solid and Debye solid. The graphs are plotted with $T_E = T_D$. The inset shows the same two curves in a log-log plot. Graph based on Fig. 24.3 in Blundell and Blundell (2009).

This implies a specific heat of

$$C = \frac{12\pi^4 R}{5} \left(\frac{T}{T_D} \right)^3, \tag{5.26}$$

just the T^3 dependence observed in experiments (see Fig. 5.3). By fitting the specific heat curve of different materials in the low-temperature regime, we can deduce their characteristic Debye temperature.

5.4.2 Thermal quanta vs. radiation quanta

In the case of radiation, we recover the classical description of the Rayleigh–Jeans law in the high-temperature regime when the thermal energy is much larger than the quantum of radiation energy ($k_B T \gg h\nu$), and we have Wien's distribution at low temperature, where Einstein had shown the radiation having the property of a gas of photons. For the Debye solid, an analogous situation holds. We have the high-temperature ($k_B T \gg h\nu_{max} = k_B T_D$) classical regime of Dulong–Petit, and at low temperature ($T \ll T_D$) a solid behaves like a collection of phonons.

Waves, particles, and quantum jumps

- The idea of light quanta was developed further by Einstein in 1909. Through a study of radiation energy fluctuation, he proposed that light had the complementary property of wave and particle. This is the first statement ever on wave–particle duality.

- In this discussion, Einstein suggested that the energy quanta were carried by point-like particles—what he termed then as "the point of view of Newtonian emission theory"—what we now call the photon.

- We describe the parallel development in spectroscopy that eventually led to Bohr's quantum model of atomic structure. He postulated that, like radiation, atoms also have quantized energies with transitions characterized by quantum jumps; this led him to the successful explanation of the hydrogen spectrum.

- In three overlapping but nonidentical papers in 1916–17, Einstein used Bohr's quantum jump idea to construct a microscopic theory of radiation–matter interaction. Through what came to be known as Einstein's A and B coefficients, he showed how Planck's spectral distribution followed. The central novelty and lasting feature is the introduction of probability in quantum dynamics.

- In Section 6.4, we present a brief introduction to quantum field theory. The treatment of the harmonic oscillator in the new quantum mechanics is reviewed. A quantized field is a collection of quantum oscillators. We show that the Planck/Einstein quantization result is automatically obtained in this new theoretical framework. This had at last put Einstein's idea of the photon on a firm mathematical foundation.

- The noncommutivity of physical observables in the new quantum theory brings about features that can be identified as creation and annihilation of quantum states. This gives a natural description of the quantum jumps of radiation emission and absorption. In fact they can be extended to the depiction of creation and destruction of material particles as well—a key characteristic of interactions at relativistic energies.

- Finally we explain how the wave–particle duality first discovered by Einstein in the study of radiation energy fluctuation is resolved in quantum field theory.

As we have already mentioned in Chapter 4, it would still be some years before Einstein openly committed himself to a point-like particle interpretation of light quanta. In Section 6.1 we shall discuss Einstein's 1909 study of radiation fluctuation that led him to show, for the first time, that light had not just wave or just particle properties, but a sort of fusion of the two—what came to be known as "wave–particle duality". The next big event in quantum history was the 1913 model for the structure of the atom conceived by Niels Bohr, who applied the Planck/Einstein quantum to the study of the hydrogen spectrum (Bohr 1913). Its spectacular success in effect launched a new era in exploration of the quantum world—what we now call the 'old quantum theory'. Bohr's 'quantum jumps' $E_i - E_f = h\nu$ inspired Einstein in 1916 to propose a detailed study of the radiation mechanism that takes place in a blackbody radiation cavity. He introduced his famous A and B coefficients for the theory of stimulated and spontaneous emissions of radiation. This is the first time that a probability description was invoked in the description of quantum dynamics, and it presaged some of the surprising consequences that would be obtained later in quantum mechanics and quantum field theory. In Section 6.4 we shall present some of the basic elements of quantum field theory to see how it is capable of resolving in one elegant framework the apparent contradictions of waves, particles, and quantum jumps. But Einstein never accepted this new paradigm.

6.1   Wave–particle duality

We have pointed out that Planck did not himself consider the quantum of action as relating directly to any physical entity, and the light quantum proposal of Einstein met considerable resistance from the physics community. This resistance can best be illustrated by the attitude of Robert Millikan, who spent a decade verifying Einstein's prediction for the photoelectric effect. Describing his viewpoint in later years, Millikan wrote this way: "I spent ten years of my life testing that 1905 equation of Einstein's, and contrary to all my expectations, I was compelled in 1915 to assert its unambiguous experimental verification in spite of its unreasonableness since it seemed to violate everything that we knew about the interference of light" (Millikan 1949).

When the idea of the light quantum $\epsilon = h\nu$ was proposed in 1905, there was still the question as what forms the quantum would take. There is the possibility that the energy is distributed throughout space as is the case with waves, or as discontinuous lumps of energy, like particles. By 1909 Einstein was more explicit in proposing that light in certain circumstances was composed of particles (Einstein 1909a,b)—in contradiction to the well-established wave properties of light. Waves cannot have particle properties and particles cannot behave like waves. However, even without the detailed knowledge of quantum electrodynamics, Einstein was able to make some definite statements on the nature of light (wave vs. particle). His argument was based on a study of the energy fluctuations in radiation. Einstein showed that light was neither simply waves nor simply particles, but had the property of being both waves and particles at the same time. The notion of wave–particle duality was born.

6.1.1 Fluctuation theory (Einstein 1904)

In Chapter 2 we discussed Einstein's theory of Brownian motion, which involved the investigation of the fluctuation phenomenon. This is a subject that had long interested Einstein. According to Boltzmann's distribution, the average energy is given by

$$\langle E \rangle = \frac{\int E e^{-\gamma E} \omega(E)\, dE}{\int e^{-\gamma E} \omega(E)\, dE} \tag{6.1}$$

with $\omega(E)$ being the density of states having energy E and $\gamma = (k_{\mathrm{B}} T)^{-1}$. Einstein in 1904 found the following fluctuation relation, after making the differentiation $-\partial/\partial\gamma$ of $\langle E \rangle$ in (6.1):

$$\langle \Delta E^2 \rangle = k_{\mathrm{B}} T^2 \frac{\partial \langle E \rangle}{\partial T}, \tag{6.2}$$

where

$$\langle \Delta E^2 \rangle \equiv \langle (E - \langle E \rangle)^2 \rangle = \langle E^2 \rangle - \langle E \rangle^2 \tag{6.3}$$

is the square deviation from the mean (the variance).

In 1904 Einstein was interested in finding systems with large fluctuations: $\langle \Delta E^2 \rangle \simeq \langle E \rangle^2$, and he studied the volume dependence of such a system. It is plausible to conclude that such an investigation led him to delve into the volume dependence of radiation entropy, which (as we have shown in Chapter 4) was the crucial step in his arriving in 1905 at the idea of light quanta. A study of the fluctuation theory is also instrumental in his finally arriving at the view that light quanta are point-like particles.

6.1.2 Energy fluctuation of radiation (Einstein 1909a)

Consider a small volume $\tilde{v}$, immersed in thermal radiation (see Fig. 6.1) having energy in the frequency interval $(\nu, \nu + d\nu)$ as

$$\langle E \rangle = \tilde{v} \rho(\nu, T)\, d\nu \tag{6.4}$$

(cf. the original definition of radiation energy density ρ given in Section 3.2.3). In his 1909 papers, Einstein used (6.2) to calculate the variance from the various radiation density distributions $\rho(\nu, T)$. For this small volume one obtains

$$\langle \Delta E^2 \rangle = \tilde{v} k_{\mathrm{B}} T^2 d\nu \frac{\partial \rho}{\partial T}. \tag{6.5}$$

This general result holds whether the system is randomly distributed as waves or particles, because it is based on Boltzmann's principle and on the fact that the spectral density at a given frequency depends on temperature only. The fluctuation formulas for the different distribution laws are presented below.

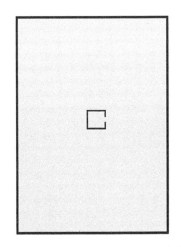

Fig. 6.1 A small volume $\tilde{v}$ immersed in thermal radiation at temperature T.

Radiation in the Rayleigh–Jeans limit fluctuates like waves

For the radiation described by the Rayleigh–Jeans distribution (4.3)

$$\rho_{\text{RJ}} = \frac{8\pi \nu^2}{c^3} k_{\text{B}} T, \tag{6.6}$$

the fluctuation formula (6.5) leads to

$$\left\langle \Delta E^2 \right\rangle_{\text{RJ}} = \tilde{\nu} \frac{8\pi \nu^2}{c^3} k_{\text{B}}^2 T^2 d\nu = \frac{c^3}{8\pi \nu^2} \frac{\langle E \rangle^2}{\tilde{\nu} d\nu}. \tag{6.7}$$

To reach the last expression we have used (6.6) and (6.4) to replace temperature by the average energy. In the following we shall argue that such a variance reflects the wave nature of the system.

Fluctuations of a wave system   A system of randomly mixed waves should display fluctuations. Although the light in an enclosure is distributed uniformly, at a certain point in space and time a light wave of a certain frequency may interfere, constructively or destructively, with another wave of slightly different frequency. This beat phenomenon would cause the energy in this small volume to be larger or smaller than the average value. The result given in (6.7) just reflects a fluctuating wave system. The key feature of wave fluctuation is that, for each radiation oscillator (i.e. degree of freedom, or mode), we have the remarkable result (derived in SuppMat Section 6.5) that the fluctuation in the energy density $\sqrt{\Delta u^2}$ has the same magnitude as the (average) energy density u itself:

$$\sqrt{\Delta u^2} = u. \tag{6.8}$$

This result can be translated into the variance and average energy of the wave system by a consideration of the involved degrees of freedom. The average energy of the system $\langle E \rangle$ requires the summation of all modes, thus a multiplication of the oscillator number in the $(\nu, \nu + d\nu)$ interval $N d\nu$ and the average energy density u for each oscillator:

$$\langle E \rangle = N d\nu u. \tag{6.9}$$

The calculation of the variance $\left\langle \Delta E^2 \right\rangle$ involves a similar sum, i.e. the same multiplication factor, $\left\langle \Delta E^2 \right\rangle = N d\nu \Delta u^2$. The result in (6.8) then implies

$$\left\langle \Delta E^2 \right\rangle = \frac{\langle E \rangle^2}{N d\nu}. \tag{6.10}$$

We have already calculated the wave mode number in Chapter 4 as displayed in (4.5):

$$N d\nu = \frac{8\pi \nu^2}{c^3} \tilde{\nu} d\nu. \tag{6.11}$$

Substituting this expression into (6.10) we obtain a result in agreement with the relation (6.7). This wave fluctuation result is to be expected as the Rayleigh–Jeans law follows from the classical Maxwell wave theory.

Radiation in the Wien limit fluctuates like particles

For radiation described by the Wien distribution, cf. Eq. (4.12),

$$\rho_{\mathrm{W}} = \frac{8\pi h\nu^3}{c^3} e^{-h\nu/k_{\mathrm{B}}T} \tag{6.12}$$

we have $\langle E \rangle = \tilde{\nu}\rho_{\mathrm{W}}d\nu$ and, from formula (6.5), the fluctuation result

$$\left\langle \Delta E^2 \right\rangle_{\mathrm{W}} = \tilde{\nu}h\nu\frac{8\pi h\nu^3}{c^3} e^{-h\nu/k_{\mathrm{B}}T} d\nu = h\nu\langle E \rangle, \tag{6.13}$$

which is clearly different from result expected from fluctuation of system of waves. In fact the fractional fluctuation has the form

$$\frac{\sqrt{\left\langle \Delta E^2 \right\rangle}}{\langle E \rangle} = \sqrt{\frac{h\nu}{\langle E \rangle}}. \tag{6.14}$$

This is exactly the fluctuation that one would expect of a system of particles. We have already discussed such a situation in Chapter 2 on Brownian motion. In particular we have shown that Brownian motion can be modeled as random walks. Equation (2.14) demonstrates that any system of random discrete entities would have a fractional deviation of $N^{-1/2}$ as is the case displayed in (6.14) because, in our case, we have $\langle E \rangle = Nh\nu$.

This result then strengthened Einstein's original proposal that blackbody radiation in the Wien limit behaves statistically like a gas of photons.

Planck distribution: Radiation fluctuates like particles *and* waves

Observationally, radiation is correctly described throughout its frequency range by the Planck spectral law. We now calculate the energy fluctuation from Planck's distribution

$$\rho = \frac{8\pi h}{c^3} \frac{\nu^3}{\exp(h\nu/k_{\mathrm{B}}T) - 1}. \tag{6.15}$$

Remarkably we find the result is simply the sum of two terms, one being the Rayleigh–Jeans terms of (6.7) and the other being the Wien term of (6.13):

$$\left\langle \Delta E^2 \right\rangle_{\mathrm{P}} = \left\langle \Delta E^2 \right\rangle_{\mathrm{RJ}} + \left\langle \Delta E^2 \right\rangle_{\mathrm{W}}. \tag{6.16}$$

This shows that radiation is neither simply waves nor simply particles. This led Einstein to suggest in 1909 that radiation can be viewed as a "fusion" of waves and particles.

Einstein proceeded to the calculation of the pressure fluctuation using explicitly the particle property of a light quantum: a photon has momentum $p = h\nu/c$. Thus, together with the suggestion of light's dual nature, Einstein now stated for the first time his view that quanta were carried by point-like particles.

6.2   Bohr's atom—another great triumph of the quantum postulate

While quantum theory has its origin in the study of blackbody radiation, there was also a parallel development in spectroscopy of the radiation emitted and absorbed by atoms. Bohr's quantum model of the hydrogen atom brought great success in this area. Thus, together with blackbody radiation, they formed the twin foundations of the quantum theory.[1]

[1]For a clear exposition of the 'old quantum theory', we recommend Tomonaga (1962).

6.2.1   Spectroscopy: Balmer and Rydberg

We mentioned in Chapter 3 that, besides blackbody radiation, Gustav Kirchhoff also made major contributions in spectroscopy. But we will start our story with the Swiss high-school mathematics teacher Johann Balmer (1825 – 98). The hydrogen spectrum is particularly simple: it has four lines in the visible range: $H_\alpha = 6563$ Å, $H_\beta = 4861$ Å, $H_\gamma = 4341$ Å, $H_\delta = 4102$ Å (Fig. 6.2). In 1885 Balmer made the remarkable discovery that these wavelengths follow a pattern when written in units of $H = 3645.6$ Å:

$$H_\alpha = \frac{9}{5}H, \quad H_\beta = \frac{16}{12}H, \quad H_\gamma = \frac{25}{21}H, \quad \text{and} \quad H_\delta = \frac{36}{32}H.$$

He then extended this to the relation (the Balmer formula) as

$$\lambda = \frac{n^2}{n^2 - 4}H, \tag{6.17}$$

which covers the original four lines with $n = 3, 4, 5, 6$, and, as it turned out, could also account for the other lines in the ultraviolet region.

   This pattern was later generalized to other hydrogen lines by Johannes Rydberg (1854–1919) in the form of

$$\frac{1}{\lambda} = R\left(\frac{1}{m^2} - \frac{1}{n^2}\right) \tag{6.18}$$

with the **Rydberg constant** $R = 4/H$ and both (m, n) being integers. The case $m = 2$ reduces to the Balmer series (visible), $m = 1$ to the Lyman series (infrared, found in 1906), and $m = 3$ to the Paschen series (ultraviolet, found in 1908).

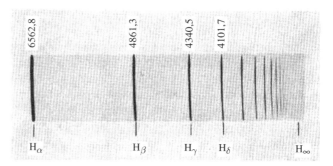

Fig. 6.2 Hydrogen spectral lines $H_\alpha, H_\beta, H_\gamma, H_\delta, \ldots$ Picture from Tomonago (1962).

6.2.2 Atomic structure: Thomson and Rutherford

The discovery of the first subatomic particle, the electron, is traditionally attributed to J.J. Thomson (1856–1940), for his measurement of the charge-to-mass ratio of cathode ray particles in 1897. He proposed a theory of atomic structure that pictures electrons as being embedded in a sphere of uniformly distributed positive charges—a sort of raisins-and-pudding model. The size of the atom had to be put in by hand as there was no way to construct any length-scale from the fundamental constants of charge and mass (e, m) from classical physics. The spectral lines are supposed to result from periodic oscillations of the electrons. However if one identifies the emission lines with the fundamental frequencies, there is no way to get rid of the unwanted higher harmonics.

During the period around 1910, Ernest Rutherford (1871–1937) and his collaborators performed a series of alpha particle scattering experiments. The large scattering angle result led Rutherford to suggest that an atom is mostly empty space, with all the positive charges concentrated in a compact center and electrons circulating around this atomic nucleus. Such a model of the atom still had the deficiencies of no natural path to an atomic size and the presence of higher harmonics. Furthermore, the circulating electrons, according to classic electromagnetism, must necessarily radiate away their energies and spiral into the nucleus. It did not seem to have a way to explain the atom's stability.

6.2.3 Bohr's quantum model and the hydrogen spectrum

Niels Bohr was familiar with Rutherford's atom. In 1913 he found a way to construct an atomic model that overcame the difficulties that Rutherford (and Thomson) had encountered. Moreover, he was able to predict in a simple way the spectrum of the hydrogen atom, with the Rydberg constant expressed in terms of fundamental constants (Bohr 1913). The new input that Bohr had was the quantum of Planck and Einstein.

Planck's constant naturally leads to an atomic scale

We have already mentioned that there is no way to construct an atomic length-scale from the two relevant constants (m, e) of classical mechanics and electromagnetism. With the introduction of Planck's constant, this can be done:

$$l = \frac{h^2}{me^2}. \tag{6.19}$$

One can easily check that this has the dimension of a length.[2] Putting in the values of the electron mass, the charge and Planck's constant (m, e, h), one finds an l of about 20 Å, roughly in the range of the atomic-scale. The Rydberg constant of (6.18) must have the dimension of inverse length, and as we shall see, it is indeed inversely proportional to the length-scale displayed here.

[2] Keeping in mind the Coulomb energy, we see that e^2 has the dimension of (energy·length). The mass m has (momentum^2/energy). Thus the denominator me^2 has (momentum^2·length). With the numerator h^2 being (momentum·length)^2, the ratio $h^2/(me^2)$ has the dimension of a length.

Stationary states and quantum jumps

Bohr reasoned that, since radiation energy is quantized, the atomic energies should similarly form a discrete set. He hypothesized that atoms should be

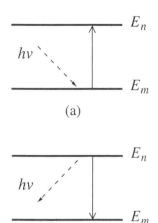

Fig. 6.3 Transitions between atomic states $n \leftrightarrow m$. (a) Absorption of a photon with energy $h\nu$. (b) Emission of a photon.

[3] As recounted by Heilbron (1977).

stable at these quantized values E_n, with $n = 0, 1, 2, 3, \ldots$ Namely, he postulated the existence of a set of stationary states. The absorption and emission of radiation then corresponds to 'jumps' among these quantized states. A state with energy E_m can absorb a photon of frequency ν and makes the transition to a higher energy state E_n, provided energy conservation is respected (the **Bohr frequency rule**):

$$\nu = \frac{E_n - E_m}{h}. \tag{6.20}$$

Such a transition is depicted in Fig. 6.3(a). Significantly, Bohr proposed the revolutionary concept that one must reject any attempt to visualize or to explain the behavior of the electron during the transition of the atom from one stationary state to another. In fact we can interpret Einstein's photoelectric effect as such a transition, if the kinetic energy of the final state electron is ignored. If we accept this possibility, it is then entirely natural to stipulate the inverse process: when an atom makes a downward transition $n \to m$, it should be accompanied by the emission of a photon, as pictured in Fig. 6.3(b). When the frequency rule (6.20) is expressed in terms of wavelength, we have

$$\frac{1}{\lambda} = \frac{\nu}{c} = \frac{E_n}{hc} - \frac{E_m}{hc}. \tag{6.21}$$

Comparing this with the Rydberg formula (6.18), Bohr had a way to connect the atomic energy levels to the Rydberg constant:

$$E_n = -R\frac{hc}{n^2}, \tag{6.22}$$

consistent with the initial assumption that atomic energies are quantized. The energy is negative because it is the binding energy. It is interesting to relate that Bohr was unaware of the Balmer/Rydberg formulas when he started out in his search for an atomic theory. When he was finally told of the Balmer series in 1913, it was a great revelation to him. He later recalled:[3] "As soon as I saw Balmer's formula, the whole thing was clear to me."

Quantization of angular momentum

Bohr then hypothesized that once in these stationary states, the electron's motion was correctly described by classical mechanics. The total energy is the sum of the kinetic and potential energies:

$$E_n = \frac{mv_n^2}{2} - \frac{e^2}{r_n}. \tag{6.23}$$

If for simplicity we take the orbits to be circles, the velocity v_n is related to the centrifugal acceleration v_n^2/r_n, which is fixed by the balance of centrifugal and Coulomb forces $mv_n^2/r_n = e^2/r_n^2$. In this way we find from (6.23) that the total energy is just one-half of the potential energy,

$$E_n = -\frac{e^2}{2r_n}. \tag{6.24}$$

This simple relation makes it clear that quantized energies imply a set of quantized orbits. The higher the energy (i.e. less negative) the larger would be the orbital radius r_n.

What determines the choice of these quantized orbits? Bohr suggested two ways to proceed and he demonstrated that both approaches led to the same conclusion. One can either make the assumption that the classical description will be valid for the description of states with large quantum number n, hence large orbits. (This became known later on as **the correspondence principle**.) Or, the same result was obtained by Bohr with the postulate of angular momentum quantization

$$L_n = n\hbar \equiv n\frac{h}{2\pi}. \tag{6.25}$$

Let us see how Bohr used angular momentum quantization[4] to deduce the quantized orbits and quantized atomic energies. The total energy can be expressed in terms of the orbital angular momentum $E = L^2/2I$. For the presently assumed circular orbits, we have (6.24) with a moment of inertia $I = mr_n^2$:

$$\frac{e^2}{2r_n} = |E_n| = \frac{L_n^2}{2mr_n^2} = n^2\frac{\hbar^2}{2mr_n^2}; \tag{6.26}$$

the last equality follows from (6.25). This fixes the radii of the quantized orbits:

$$r_n = n^2\frac{\hbar^2}{me^2} = n^2 a \tag{6.27}$$

where $a = \hbar^2/(me^2)$ is the **Bohr radius**—just the atomic-scale l of (6.19) divided by $(2\pi)^2$. We can use (6.24) to translate this into the atomic energy

$$E_n = -\frac{e^2}{2a}\frac{1}{n^2}. \tag{6.28}$$

This in turn predicts, through (6.22), the Rydberg constant to be

$$R = -\frac{n^2}{hc}E_n = \frac{e^2/\hbar c}{4\pi}\frac{1}{a} = \frac{\alpha}{4\pi}\frac{1}{a}, \tag{6.29}$$

where we have introduced the shorthand, **fine structure constant** $\alpha = e^2/\hbar c \simeq 1/137$. Putting back all the fundamental constants of (m, e, h), we have

$$R = \frac{2\pi^2 me^4}{ch^3}, \tag{6.30}$$

which was in good agreement with the experimental value of R.

One more bit of interesting history—a sort of icing-on-the-cake (cf. Section 15.6, Longair 2003). One of the first applications made by Bohr of his new theory was to explain the lines in the observed spectrum of the star ς-Puppis. They were thought to be hydrogen lines because of their similarity to the Balmer series. Bohr showed they were really those of the singly ionized helium He^+ which according to the new theory should have exactly the same spectrum as

[4]In Bohr's 1913 paper, he acknowledged that J.W. Nicholson was the first one to discover in 1912 the quantization of angular momentum. We also note that, when the circular electron orbit assumption is relaxed to allow for elliptical trajectories, as first done by Arnold Sommerfeld, the quantum numbers must be extended, besides the principal quantum number n, to include the orbital quantum number $l = 0, 1, \ldots, n-1$.

hydrogen with only the Rydberg constant being four times larger $R_{He}/R_H = 4$—as the factor e^4 in (6.30) has to be replaced by $(Ze^2)^2$ with the atomic number $Z = 2$ for helium. But it was pointed out to him that the experimental value was not exactly 4 but 4.00163. Bohr then realized that the electron mass m in (6.30) should more accurately be the 'reduced mass' $\mu = m_e m_N/(m_e + m_N)$ when the finite nuclear mass m_N was taken into account. Since the helium nucleus is four times larger than the hydrogen nucleus, one then has a ratio of Rydberg constants that is in much closer agreement with observation:

$$\frac{R_{He}}{R_H} = 4\frac{1 + m_e/m_N}{1 + m_e/4m_N} = 4.00160. \tag{6.31}$$

Here is another instance of the importance of high-precision measurements!

6.3　Einstein's *A* and *B* coefficients

During the five-year period prior to 1916, Einstein was preoccupied with the development of general relativity (see Chapters 12–14), which he finalized in 1915–16. In late 1916 he returned his attention to the study of quantum theory (Einstein 1916b,c). Having been inspired by Bohr's papers,[5] he obtained new insights into the microscopic physics concerning the emission and absorption of radiation. In constructing his theory of atomic structure Bohr had used Einstein's quantum idea, which was originally obtained from a thermal statistical study of blackbody radiation. Now Einstein used Bohr's idea of quantum jumps (Fig. 6.3) to construct a microscopic theory of the emission and absorption of radiation by molecular states to show that the resulting radiation distribution is just the Planck spectral law. He found that he could obtain Planck's spectral distribution if, and only if, the quantum jump between two molecular states $m \rightleftarrows n$ involved a monochromatic energy quantum obeying Bohr's frequency condition (6.20). Notably, Einstein's 1916 theory involved the introduction, for the first time, of a probabilistic description of quantum dynamics.

Furthermore, Einstein showed that, if the radiation is pictured as a collection of particles, the energy exchange $\Delta\epsilon = h\nu$ between molecules and radiation would also entail the exchange of momentum. For massless photons, relativity dictates a momentum transfer of $\Delta p = h\nu/c$. In this way he showed that the Planck distribution of radiation energy is precisely compatible with a Maxwell velocity distribution for the molecules. The results Einstein obtained in this investigation, in particular those related to stimulated emission of radiation, laid the foundation for the later invention of the laser and maser. Another aspect of the work was the forerunner of the theory of quantum vacuum fluctuation.

6.3.1　Probability introduced in quantum dynamics

Einstein considered a system in thermal equilibrium, consisting of a gas of particles (called molecules) and electromagnetic radiation (with spectral

[5] When Einstein heard about Bohr's result on astrophysical helium spectrum in a meeting in Vienna in September 1913, he was astonished and said: 'Then the frequency of light does not depend at all on the orbiting frequency of the electron. And this is an *enormous achievement*. The theory of Bohr must be right.' (see p. 137, Moore 1989).

density ρ). Let $\epsilon_1, \epsilon_2, \epsilon_3, \ldots$ be the energies of the molecular states. The relative probability of molecules in the different states is given by Boltzmann's statistics as

$$P_n = g_n e^{-\epsilon_n / k_B T} \tag{6.32}$$

where g_n is the number of states having the same energy ϵ_n (called the degeneracy of the state). The various interactions between radiation with these molecules are considered. The two molecular states with energies $\epsilon_n > \epsilon_m$, as depicted in Fig. 6.3, will be the focus of the following discussion.

Spontaneous emission

Consider first the emission of a photon with the molecule making the $n \to m$ transition as depicted in Fig. 6.3(b). Here Einstein introduced a probabilistic description. He argued that since it is possible for a classical oscillator to radiate without the excitation (i.e. without any perturbation) by an external radiation field, the rate of the probability change (the change of the molecular number) for this spontaneous emission may be written as

$$\left(\frac{dP_n}{dt} \right)_{sp-em} = A_n^m P_n \tag{6.33}$$

where A_n^m is a constant with the lower index denoting the initial state, and the upper index the final state. Einstein noted that this mechanism of spontaneous emission of radiation is generally identical to Rutherford's 1900 statistical description of spontaneous decay of radiative matter. While Einstein could not explain the puzzle of a statistical theory he was the first one to note that it could only be understood in the quantum-theoretical context. Furthermore, Einstein immediately expressed his misgiving that such a probabilistic description seemed to imply an abandonment of strict causality.

Stimulated absorption and emission

In a field of radiation, a molecular oscillator changes its energy because the radiation transfers energy to the oscillator. Depending on the phases of the molecular oscillator and the oscillating electromagnetic field, the transferred work can be positive (absorption) or negative (emission). We call such a processes 'induced' or 'stimulated' because of the presence of the radiation perturbation. We expect the rate of change to be proportional to the radiation density ρ. For the induced absorption, we denote the molecular number change by

$$\left(\frac{dP_m}{dt} \right)_{abs} = B_m^n \rho P_m. \tag{6.34}$$

Similarly for the stimulated emission, we have

$$\left(\frac{dP_n}{dt} \right)_{st-em} = B_n^m \rho P_n. \tag{6.35}$$

The radiation density is fixed to be Planck's distribution

What is the form of the radiation spectral density such that it is compatible with this microscopic description of radiation–matter interaction? To reach equilibrium, the absorption and emission rates must balance out:

$$\left(\frac{dP}{dt}\right)_{\text{abs}} = \left(\frac{dP}{dt}\right)_{\text{st–em}} + \left(\frac{dP}{dt}\right)_{\text{sp–em}} \tag{6.36}$$

or

$$g_m B_m^n e^{-\epsilon_m/k_B T} \rho = g_n e^{-\epsilon_n/k_B T} \left(B_n^m \rho + A_n^m\right). \tag{6.37}$$

We further assume that the energy density ρ goes to infinity as the temperature increases to infinity ($T \to \infty$). The large ρ factor means that we can ignore the A_n^m term in the parentheses; in this way we obtain[6]

$$g_m B_m^n = g_n B_n^m. \tag{6.38}$$

To simplify our writing we shall from now on absorb the degeneracy factor g into the B coefficient. The spectral density that satisfies this dynamic equilibrium condition (6.37) then becomes

$$\rho = \frac{A_n^m}{B_n^m} \frac{1}{e^{(\epsilon_n - \epsilon_m)/k_B T} - 1} \tag{6.39}$$

which is just Planck's law when we apply the Bohr quantum condition (6.20) together with fixing the coefficient ratio to be

$$\frac{A_n^m}{B_n^m} = \alpha \nu^3. \tag{6.40}$$

The constant α can further be determined, for example, by the Rayleigh–Jeans law. Thus

$$\frac{A_n^m}{B_n^m} = \frac{8\pi \nu^2}{c^3} h\nu. \tag{6.41}$$

Recall that we have used the expression for the radiation density of states (4.1) in the derivation of the Rayleigh–Jeans law.

[6] In Einstein's original paper this was justified by the experimental condition that for large temperature ($\nu/T \to 0$) the spectral density $\rho \sim \nu^2 T \to \infty$. It can also be supported by the so-called 'principle of detailed balance'—due to microscopic reversibility in thermal equilibrium.

6.3.2 Stimulated emission and the idea of the laser

Einstein's prediction of stimulated emission became a key element in the invention of the LASER—light amplification by stimulated emission of radiation. Such a device can produce high-intensity collimated coherent electromagnetic waves. In essence a laser is a cavity filled with a "gain medium". We can illustrate the function of this medium by assuming it to be composed of some two-state atoms (e.g. such as the one shown in Fig. 6.3). A positive feedback process, based on stimulated emission, is instituted. The frequency of the input radiation is arranged to match the emission frequency of the medium. This process is amplified by the stimulated emission. If the cavity is enclosed by two mirrors so that light is repeatedly passing back and forth through the gain medium, more and more atoms reside in the excited states (called population inversion) and its intensity can be greatly increased. Because the light originates from a single transition between two fixed levels (unlike an ordinary light source with many different transitions), it is monochromatic with a great deal of coherence. Clearly the invention of the laser required many technical breakthroughs before its realization in the 1950s; nevertheless its basic idea came from the prediction of stimulated emission made by Einstein in 1916.

While stimulated emission can still be understood as a perturbation by an existing field, it would involve a new theoretical framework to understand spontaneous emission. In this case a new photon would have to be created. If it is due to some perturbation, how would the vacuum be the cause? This brings us to the topic of quantum field theory.

6.4 Looking ahead to quantum field theory

At the beginning of this chapter, we discussed the riddle of radiation's wave–particle duality as shown by Einstein's calculation of energy fluctuations (6.16). The Planck's formula for blackbody radiation leads to two terms, one showing the radiation as a system of waves and another as particles. This heightens the apparent contradiction of Einstein's original discovery of thermal radiation (in the Wien limit) behaving thermodynamically like a gas of particles, even though radiation has the familiar wave property of interference, etc. Here we first explain the resolution as provided by the advent of quantum mechanics in 1925–26. The other key property of light quanta that they obey Bose–Einstein statistics will be discussed in Chapter 7.

The new quantum theory is the work of Louis de Broglie, Werner Heisenberg (1901–76), Max Born (1882–70), Pascual Jordan (1902–80), Wolfgang Pauli (1900–58), Erwin Schrödinger, and Paul Dirac (1902–84). In particular electromagnetic radiation is described by a quantized field. This is the subject of quantum electrodynamics. We shall provide, very briefly, some of the basic elements of quantum field theory (QFT).[7] Of course, Einstein never accepted quantum mechanics as a complete theory. His objection to this new quantum theory was mainly in the area of the interpretation of measurement. That will be the topic of our Chapter 8.

[7]A lively and insightful introduction to QFT can be found in Zee (2010).

6.4.1 Oscillators in matrix mechanics

Recall our discussion in Section 3.1 that a radiation field (as a solution to Maxwell's wave equation) can be thought of as a collection of oscillators. Fourier components of waves obey simple harmonic oscillator equations. A quantized radiation field is a collection of quantum oscillators—simple harmonic oscillators as described by quantum mechanics. Quantum field theory is usually presented as the union of quantum mechanics and special relativity. This is so as the Maxwell wave equation satisfies special relativity. When this radiation theory is generalized to other particles, one would work with other relativistic wave equations such as the Dirac equation and Klein–Gordon equation. But the basic features of a quantized field discussed below remain the same.

The essence of quantum mechanics is that physical states are taken to be members of a linear vector space, the Hilbert space, obeying the superposition principle (the basic property of waves), and physical observables are operators represented, for example, by matrices. These observables obey the same dynamical equations as in classical physics, but the kinematics are changed because they, being operators, may no longer be mutually commutative. Thus two operators $\hat{A}$ and $\hat{B}$ may have nonvanishing commutator $\hat{A}\hat{B} - \hat{B}\hat{A} \equiv [\hat{A}, \hat{B}] \neq 0$. As we shall see, this noncommutivity brings about the particle nature of the system. Planck's constant enters the theory through these commutation relations.

Simple harmonic oscillator Hamiltonian in terms of ladder operators

Here is the quantum mechanical description of a simple harmonic oscillator. The total energy (sum of kinetic and potential energies) is represented by the Hamiltonian operator, which can be expressed in terms of the position and momentum operators $(\hat{x}, \hat{p})$. With the angular frequency ω, the Hamiltonian is given by

$$\hat{H} = \frac{\hat{p}^2}{2m} + \frac{1}{2}m\omega^2\hat{x}^2. \tag{6.42}$$

The momentum and position operators are postulated to satisfy the 'canonical' commutation relation

$$[\hat{x}, \hat{p}] = i\hbar. \tag{6.43}$$

We can factorized the oscillator Hamiltonian, in terms of the **ladder operators**:[8]

$$\hat{a}_{\pm} = \frac{1}{\sqrt{2m}}(\mp i\hat{p} + m\omega\hat{x}). \tag{6.44}$$

A simple calculation shows that they have the product relations

$$\hat{a}_+\hat{a}_- = \hat{H} + \frac{i\omega}{2}[\hat{x}, \hat{p}] \quad \text{and} \quad \hat{a}_-\hat{a}_+ = \hat{H} - \frac{i\omega}{2}[\hat{x}, \hat{p}]. \tag{6.45}$$

From the sum and difference of these two expressions, we obtain the Hamiltonian

$$\hat{H} = \frac{1}{2}(\hat{a}_+\hat{a}_- + \hat{a}_-\hat{a}_+) \tag{6.46}$$

and the commutator of the ladder operators, which is just a simple transcription of (6.43),

$$[\hat{a}_{\mp}, \hat{a}_{\pm}] = \pm\hbar\omega. \tag{6.47}$$

[8] Since $\hat{x}$ and $\hat{p}$ are Hermitian operators, these ladder operators are each other's Hermitian conjugates, $\hat{a}_+^\dagger = \hat{a}_-$ and $\hat{a}_-^\dagger = \hat{a}_+$.

An application of this commutation relation to (6.46) leads to

$$\hat{H} = \hat{a}_{\mp}\hat{a}_{\pm} \mp \frac{1}{2}\hbar\omega. \qquad (6.48)$$

Raising and lowering energy levels by $\hat{a}_{\pm}$

Consider the states $\hat{a}_{\pm}|n\rangle$, obtained by applying the ladder operators $\hat{a}_{\pm}$ to an energy eigenstate $|n\rangle$ with $\hat{H}|n\rangle = E_n|n\rangle$. To find the energy of such states, we probe them by the Hamiltonian operator in the form of (6.48):

$$\hat{H}\hat{a}_{\pm}|n\rangle = \left(\hat{a}_{\mp}\hat{a}_{\pm}\hat{a}_{\pm} \mp \frac{1}{2}\hbar\omega\hat{a}_{\pm}\right)|n\rangle. \qquad (6.49)$$

Since (6.47) implies the commutation relation

$$\left[\hat{a}_{\mp}\hat{a}_{\pm}, \hat{a}_{\pm}\right] = \left[\hat{a}_{\mp}, \hat{a}_{\pm}\right]\hat{a}_{\pm} + \hat{a}_{\mp}\left[\hat{a}_{\pm}, \hat{a}_{\pm}\right] = \pm\hbar\omega\hat{a}_{\pm} + 0, \qquad (6.50)$$

the RHS of (6.49), after interchanging the order of $\hat{a}_{\mp}\hat{a}_{\pm}$ and $\hat{a}_{\pm}$ by the commutator (6.50), becomes

$$\hat{H}\hat{a}_{\pm}|n\rangle = \left(\hat{a}_{\pm}\hat{a}_{\mp}\hat{a}_{\pm} \pm \hbar\omega\hat{a}_{\pm} \mp \frac{1}{2}\hbar\omega\hat{a}_{\pm}\right)|n\rangle.$$

We can factor out $\hat{a}_{\pm}$ to the left and use the expression of $\hat{H}$ as given in (6.48) to have

$$\hat{H}\left(\hat{a}_{\pm}|n\rangle\right) = \hat{a}_{\pm}\left(\hat{H} \pm \hbar\omega\right)|n\rangle$$
$$= \hat{a}_{\pm}\left(E_n \pm \hbar\omega\right)|n\rangle = \left(E_n \pm \hbar\omega\right)\left(\hat{a}_{\pm}|n\rangle\right). \qquad (6.51)$$

This calculation shows that the states $\hat{a}_{\pm}|n\rangle$ are also eigenstates of the Hamiltonian with energy values $E_n \pm \hbar\omega$. This explains why $\hat{a}_{+}$ is called the **raising operator** and $\hat{a}_{-}$ the **lowering operator**.

The quantized energy spectrum derived

Just like the classical oscillator case, the energy must be bounded below. We denote this lowest energy state, the ground state, by $|0\rangle$. Since the ground state cannot be lowered further, we must have the condition:

$$\hat{a}_{-}|0\rangle = 0. \qquad (6.52)$$

From this we deduce that the ground state energy E_0 does not vanish:

$$\hat{H}|0\rangle = E_0|0\rangle = \left(\hat{a}_{+}\hat{a}_{-} + \frac{1}{2}\hbar\omega\right)|0\rangle = \frac{1}{2}\hbar\omega|0\rangle. \qquad (6.53)$$

Namely $E_0 = \frac{1}{2}\hbar\omega$, which is often referred to as the **zero-point energy**. On the other hand, all the excited states can be reached by the repeated application of the raising operator[9] to the ground state:

$$(\hat{a}_{+})^n|0\rangle \sim |n\rangle. \qquad (6.54)$$

[9]The proportionality constants will be worked out below when we discuss the normalization of quantum states.

According to (6.51), each application of $\hat{a}_+$ raises the energy by one $\hbar\omega$ unit: we thus derive the energy of a general state $|n\rangle$:

$$\hat{H}|n\rangle = E_n|n\rangle = \left(n + \frac{1}{2}\right)\hbar\omega|n\rangle \qquad (6.55)$$

with $n = 0, 1, 2, \ldots$ and

$$E_n = \left(n + \frac{1}{2}\right)\hbar\omega. \qquad (6.56)$$

This result agrees, up to a constant of $\hbar\omega/2$, with the Planck oscillator energy quantization proposal. Before we move on to quantum field theory we note two technicalities of the quantum theory of oscillators.

The zero-point energy   A comparison of (6.46) and (6.48) with (6.56) shows clearly that the zero-point energy, $E_0 = \frac{1}{2}\hbar\omega$, originates from the noncommutivity of the position with momentum, or equivalently the $\hbar\omega$ factor in the commutator (6.47). Physically one can understand the presence of this ground state energy by the uncertainty principle. Even in the absence of any quanta, an oscillator still has the natural length-scale of $x_0 = \sqrt{\hbar/m\omega}$; thus, the **uncertainly principle**[10] for the position and momentum observables, $\Delta x \Delta p \gtrsim \hbar$, requires a minium momentum of $p_0 = \sqrt{m\hbar\omega}$. This translates into a minium energy of $E_0 = p_0^2/2m = \frac{1}{2}\hbar\omega$—just the zero-point energy.

[10]The uncertainty relation is a direct mathematical consequance of the noncommutivity of observables.

The number operator and the normalization of oscillator states   A simple comparison of (6.55) with (6.48) suggest that we can define a 'number operator' $\hat{n} \equiv \hat{a}_+\hat{a}_-/(\hbar\omega)$ so that $\hat{H} = (\hat{n} + \frac{1}{2})\hbar\omega$ and $\hat{n}|n\rangle = n|n\rangle$. This operator is Hermitian $\hat{n}^\dagger = \hat{a}_-^\dagger\hat{a}_+^\dagger/(\hbar\omega) = \hat{n}$, with real eigenvalues $n = 0, 1, 2, \ldots$. All quantum mechanical states must be normalized (as they have the interpretation of a probability): $\langle n|n\rangle = ||n\rangle|^2 = 1$, and $\langle n-1|n-1\rangle = ||n-1\rangle|^2 = 1$, etc. From these we can find out how the ladder operators act on the number states $\hat{a}_-|n\rangle = c|n-1\rangle$ with the coefficient c determined as follows. Starting with

$$\langle n|\frac{\hat{a}_+\hat{a}_-}{\hbar\omega}|n\rangle = \langle n|\hat{n}|n\rangle = n, \qquad (6.57)$$

we have, using the hermiticity properties $\hat{a}_\pm^\dagger = \hat{a}_\mp$,

$$n\hbar\omega = \langle n|\hat{a}_+\hat{a}_-|n\rangle = |\hat{a}_-|n\rangle|^2 = |c|^2||n-1\rangle|^2 = |c|^2,$$

hence $c = \sqrt{n\hbar\omega}$. Similarly we can work out the effects of $\hat{a}_+$. Thus the effects of the ladder operators are

$$\hat{a}_-|n\rangle = \sqrt{n\hbar\omega}|n-1\rangle \quad \text{and} \quad \hat{a}_+|n\rangle = \sqrt{(n+1)\hbar\omega}|n+1\rangle. \qquad (6.58)$$

6.4.2   Quantum jumps: From emission and absorption of radiation to creation and annihilation of particles

A quantum radiation field is a collection of quantum oscillators. The energy spectrum of the field for each mode is given by the quantized energy as shown in (6.56). Thus the Planck/Einstein quantization result is automatically obtained in the framework of quantum field theory. The first application of the new quantum mechanics to the electromagnetic field was given in the famous three-man paper (*Dreimännerarbeit*) of Born, Heisenberg, and Jordan (1926). This had at last put Einstein's idea of the photon on a firm mathematical foundation. The new quantum mechanics also yields the correct hydrogen spectrum, as shown by Pauli (1926) in matrix mechanics and by Schrödinger (1926) in wave mechanics.

Vacuum energy fluctuation

The new feature of (6.56) is the presence of the zero-point energy. Since the ground state of a field system is identified with the vacuum, quantum field theory predicts a nonvanishing energy for the vacuum state. While the presence of this constant energy term would not affect quantum applications such as the photoelectric effect and specific heat, as we shall see, there are observable effects associated with this nonvanishing vacuum energy. In fact what we have is the fluctuation of the energy in the vacuum state. We have already discussed the zero-point oscillator energy from the viewpoint of the position–momentum uncertainty relation. We also have the uncertainty relation[11] between energy and time, $\Delta E \Delta t \gtrsim \hbar$. This suggest that, for a sufficiently short time interval, energy can fluctuate, even violating energy conservation. The vacuum energy is the (root-mean-square) average of the fluctuation energy.

[11]Time is not a dynamical observable represented by an operator in quantum mechanics. The uncertainty relation follows from the Heisenberg equation of motion with Δt being the characteristic time that a system takes to change.

Emission and absorption of radiation

That the formalism of the quantum oscillator allows one to raise and lower the field energy by units of $\hbar\omega$ can naturally be used to describe the quantum jumps of emission and absorption of radiation. In particular, the amplitude for the emission (i.e. creation) of an energy quantum is directly related to the matrix element:

$$\langle n+1|\,\hat{a}_+\,|n\rangle = \sqrt{(n+1)\,\hbar\omega}. \tag{6.59}$$

The equality follows from (6.58), leading to an emission rate proportional to the factor $(n+1)$.

This is just the result first discovered by Einstein. From the RHS of (6.36) we have the total (induced and spontaneous) emission rate,

$$\left(\frac{dP_n}{dt}\right)_{\text{em}} = \left(\rho\frac{B_n^m}{A_n^m}+1\right)A_n^m P_n = \left(\frac{\rho c^3}{8\pi h\nu^3}+1\right)A_n^m P_n, \tag{6.60}$$

where we have used the relation for Einstein's A and B coefficients (6.41). The language of quantum field theory allows us to express this emission rate in terms of the **number of light quanta** n:

$$\left(\frac{dP}{dt}\right)_{\text{em}} \propto (n+1)P, \tag{6.61}$$

because, according to Eq. (4.1), we have the energy (per radiation oscillator) $U = \rho c^3/8\pi\nu^2$ and $U/h\nu = n$. While the factor of n on the RHS of (6.61) corresponds to the stimulated emission, the factor of 1 in $(n+1)$ reflects the spontaneous emission. Thus in quantum oscillator language, the spontaneous emission term comes from the commutation relation (6.47). It has exactly the same origin as the zero-point energy. This linkage between the vacuum energy and spontaneous emission suggests to us that we can identify spontaneous emission as brought about (i.e. due to the perturbation) by the vacuum energy fluctuation.

Creation and annihilation of particles

In the above we have seen that the raising and lowering ladder operators of the oscillator provide us with a natural description of emission and absorption of radiation in units of the energy quanta. In modern language this is the emission and absorption of photons.

Even with the success of the quantum field theory treatment of radiation, we still have, at this stage, a dichotomy: on one hand we have radiation with its quanta that can be freely created and destroyed; on the other hand, material particles such as electrons and protons were thought to be eternal. Further development of quantum field theory showed that material particles can also be thought of as quanta of various fields, in just the same way that the photon is the quantum of the electromagnetic field. These matter fields are also collections of their oscillators, with their corresponding ladder operators identified as the creation and annihilation operators of these material particles.

This is a major advance in our understanding of particle interactions. Until then the interactions among particles were described by forces that can change the motion of particles. Photons are just like other particles except they have zero rest-mass. While there is no energy threshold for radiation, given enough energy all particles can appear and disappear through interactions. The first successful application of this idea was in the area of nuclear beta decay. The nucleus is composed of protons and neutrons. How is it then possible for one parent nucleus to emit an electron (and a neutrino) while changing into a different daughter nucleus? Enrico Fermi (1901–54) gave the quantum field theoretical answer to this puzzle. He modeled his theory of beta decay on quantum electrodynamics and described the process as the annihilation of a neutron in the parent nucleus followed by the creation of a proton in the final state nucleus along with the creation of the electron (and the neutrino).

Ranges of interactions

In a field theory the interaction between the source particle and test particle is described as the source particle giving rise to a field propagating out from the source and the field then acting locally on the test particle. Since a quantized field can be thought of as a collection of particles, this interaction is depicted as an exchange of particles between the source and test particles. Since the exchanged particle can have a mass, the creation of such an exchange particle (from the vacuum) would involve an energy nonconservation of $\Delta E \simeq mc^2$. But the uncertainty principle $\Delta E \Delta t \geq \hbar$ only allows this to happen for a time interval of $\Delta t \simeq \hbar/mc^2$. This implies a propagation, hence an interaction range, of $R \simeq c\Delta t \simeq \hbar/mc$ (the Compton wavelength of the exchanged particle). Electromagnetic interaction is long range because the photon is massless. Based on such considerations, Hideki Yukawa (1907–81) predicted the existence of a meson, about a couple of hundred times more massive than the electron, as the mediating quantum of nuclear forces which were known to have a finite range of about a fermi ($= 10^{-15}$ m).

We conclude by noting the central point of quantum field theory: The essential reality is a set of fields, subject to the rules of quantum mechanics and special relativity; all else is derived as a consequence of the quantum dynamics of these fields (Weinberg 1977).

6.4.3   Resolving the riddle of wave–particle duality in radiation fluctuation

In this last section we return to the issue of wave–particle duality displayed by the radiation energy fluctuation discussed at the beginning of this chapter. How does quantum field theory resolve the riddle of the radiation fluctuation having two factors (6.16): a wave term plus a particle term?

In quantum field theory a field is taken to be an operator. The above discussion of the radiation field being a collection of quantum oscillators means the replacement of a classical field (a complex number) $Ae^{i\phi_j}$, with appropriate normalization, by an operator $\hat{a}_{j-}e^{i\phi_j} + \hat{a}_{j+}e^{-i\phi_j}$ with $[\hat{a}_{j-}, \hat{a}_{k+}] = \hbar\omega\delta_{jk}$. The calculation of the energy fluctuation of such a wave system follows the same lines as that for classical waves (cf. SuppMat Section 6.5). However the noncommutivity of quantum oscillator operators $\hat{a}_{\pm}$ gives rise to extra terms, as shown in (6.47). The result is that, instead of the classical wave result of (6.8), we now have the mean-square energy density

$$\Delta u^2 = u^2 + u\hbar\omega. \tag{6.62}$$

For the system average we follow the same procedure used in Section 6.1, to obtain $\langle E \rangle = Ndvu$ and $\langle \Delta E^2 \rangle = Ndv\Delta u^2$, and, using the density of states result $N = 8\pi v^3/c^3$ of (4.5), to arrive at the final result of

$$\langle \Delta E^2 \rangle = \frac{\langle E \rangle^2}{Ndv} + \langle E \rangle \hbar\omega = \frac{c^3}{8\pi v^2}\frac{\langle E \rangle^2}{vdv} + \langle E \rangle h v. \tag{6.63}$$

This is just the result (6.16) that Einstein obtained in 1909 from Planck's distribution. Thus these two terms, one wave and one particle, can be explained in a unified framework. Recall that it was based on this result that Einstein first proposed the point-like particles as the quanta of radiation. Alas, as already mentioned above, Einstein never accepted this beautiful resolution of the great wave–particle riddle, as he never accepted the framework of the new quantum mechanics.

The extra particle-like term comes from the commutator (6.47) which is equivalent to $[\hat{x}, \hat{p}] = i\hbar$. Thus it has the same origin as the zero-point energy and the energy quantization feature of the quantized wave system. This elegant resolution of the wave–particle duality was discovered by Pascual Jordan (Born, Heisenberg, and Jordan 1926).[12] Somehow this result is not well-known generally; the full story of, and a careful re-derivation of, Jordan's contribution was given by Duncan and Janssen (2008).

[12]There is ample historical evidence showing that Jordan was alone responsible for this section of the *Dreimännerarbeit*.

Quantum field theory can account for another fundamental feature of a system of many particles: its quantum statistics property. As we shall discuss in the next chapter, photons obey Bose–Einstein statistics and the new quantum mechanics requires its state to be symmetric under the exchange of any two photons. It turns out that the commutation relation that is being discussed in this section [cf. Eq. (6.47)] is just the elegant mathematical device needed to bring about this required symmetry.[13] This quantum statistical property leads directly to the Planck distribution for a thermal photon system. Planck's distribution yields a fluctuation showing the wave–particle duality. Thus quantum field theory gives a completely self-consistent description of the electromagnetic radiation. In this theory one can see the effects of waves and particles simultaneously.

[13]For a fermionic system, the particle creation and annihilation operators are postulated to obey anticommutation relations $\hat{a}_{j-}\hat{a}_{k+} + \hat{a}_{k+}\hat{a}_{j-} \equiv \{\hat{a}_{j-}, \hat{a}_{k+}\} = \hbar\omega\delta_{jk}$ so that a multi-fermion system is antisymmetric under the interchange of two identical fermions.

6.5   SuppMat: Fluctuations of a wave system

Here is a calculation of the fluctuations of randomly superposed waves. This presentation follows that given by Longair (2003, p. 369). The energy density is proportional to the field squared $|F|^2$. For the case of electromagnetic waves, F can be the electric or magnetic field. We assume that all polarization vectors are pointing in the same direction, reducing the problem to a scalar field case, and all waves have the same amplitude A. [Cf. Eq. (3.4) in Section 3.1] In this way, we have the energy density as

$$|F|^2 = A^2 \left(\sum_{j=1}^{N} e^{i\phi_j}\right)^* \left(\sum_{k=1}^{N} e^{i\phi_k}\right) = A^2 \left(N + \sum_{j\neq k} e^{i(\phi_k - \phi_j)}\right). \qquad (6.64)$$

When the phases of the waves are random, the second term in the parentheses being just sines and cosines, averages out to zero:

$$u = \langle|F|^2\rangle = NA^2. \qquad (6.65)$$

Namely, the total average energy density of a set of incoherent waves is simply the sum of the energy density of each mode.

To calculate the mean-squared energy, we need to calculate the square of the energy density (6.64). The result is

$$\left||F|^2\right|^2 = A^4 \left|\left(N + \sum_{j\neq k} e^{i(\phi_k - \phi_j)}\right)\right|^2$$

$$= A^4 \left(N^2 + 2N\sum_{j\neq k} e^{i(\phi_k - \phi_j)} + \sum_{j\neq k} e^{-i(\phi_k - \phi_j)}\sum_{l\neq m} e^{i(\phi_m - \phi_l)}\right).$$

Again the second term (with coefficient $2N$), as well as most of the terms in the double sum, average out to zero. The terms in the double sum that survive

are those with matching indices $j = l$ and $k = m$; there are thus N^2 such terms.
The result then is

$$\left\langle ||F|^2|^2 \right\rangle = A^4 \left(N^2 + 0 + N^2 \right) = 2N^2 A^4. \tag{6.66}$$

We have the variance of the fluctuating wave energy:

$$\Delta u^2 = \left\langle ||F|^2|^2 \right\rangle - \left\langle |F|^2 \right\rangle^2 = N^2 A^4 = u^2. \tag{6.67}$$

This is the claimed result for wave fluctuations as displayed in (6.8).

Bose–Einstein statistics and condensation

- The physics community's acceptance of the photon idea did not come about until the discovery and analysis of Compton scattering in 1923. We present a short introduction to this subject. The remainder of this chapter is devoted to the statistical analysis of photons as identical bosons.

- Just about all of Einstein's principal contributions to quantum theory are statistical in nature: his first paper on the photon, his work on specific heats, on stimulated and spontaneous radiative processes, and now on quantum statistics. This last effort was prompted by a paper that S. Bose sent to him in 1924.

- Bose was dissatisfied with the logical foundation of Planck's radiation theory. He presented a derivation of Planck's distribution using the particle approach from the very beginning. We present Bose's derivation in detail so as to understand the implicit assumptions he made in this pioneering work.

- In the meantime de Broglie put forth his idea that matter, under certain circumstance, could behave like waves. This inspired Einstein to extend Bose's analysis of radiation to systems of matter particles. Here he made the discovery of the astounding possibility of Bose–Einstein condensation (BEC).

- The papers of de Broglie and Einstein directly influenced Schrödinger in his creation of the Schrödinger equation. This prompted Pais to bestow onto Einstein the title of "godfather of wave mechanics".

- The ultimate understanding of Planck's spectral distribution came about in modern quantum mechanics with its notion of indistinguishable particles. A multiparticle system must be described by a wavefunction that is either symmetric or antisymmetric under the interchange of two identical particles. The spin-statistics theorem instructs us that particles with half-integer spin obey Fermi–Dirac statistics and particles with integer spin (like photons) obey Bose–Einstein statistics.

- Section 7.4 is devoted to some basics of Bose–Einstein condensation. In particular we show that this phenomenon of a macroscopic number of particles "condensing" into the momentum space ground state can only take place when the particles' wavefunctions start to overlap. The production of BEC and a demonstration of its macroscopic quantum behavior in the laboratory setting was achieved in the 1990s. We briefly describe this success.

- In SuppMat Section 7.5, we discuss radiation pressure resulting from photon collisions with the enclosure. In SuppMat Section 7.6, we show why, in the context of modern quantum statistics, Planck found the right answer using the statistical weight he wrote down in 1900. In SuppMat Section 7.7, we discuss the role of particle indistinguishability, making it possible for BEC to take place.

As we have already recounted previously, there was persistent resistance to Einstein's 1905 photon proposal. This lasted till 1923 when Arthur H. Compton (1892–1962) performed X-ray scattering off a graphite target and provided the analysis showing that a light quantum has not only energy but also momentum. This brought about the general acceptance of the photon idea. Arnold Sommerfeld (1868–1951), one of the leading lights in physics, had this to say about the result of Compton scattering (Sommerfeld 1924): "It is probably the most important discovery which could have been made in the current state of physics." With this general acceptance, the investigation of radiation made further progress with the first correct statistical analysis of radiation as a specific case of Bose–Einstein quantum statistics. These will be the main topics of this chapter.

7.1 The photon and the Compton effect

Compton carried out experiments with X-rays scattering on a graphite target. In the classical theory, these incoming electromagnetic waves cause electrons in the carbon atoms to oscillate with the same frequency as the incident waves and re-emit the final state waves with the same frequency. In the particle picture of light, a photon carries momentum as well as energy. The energy E and momentum p are related[1] by $\epsilon = pc$, which is the $m = 0$ case of the general relativistic energy and momentum relation $\epsilon^2 = (pc)^2 + (mc^2)^2$. For a quantized photon energy $\epsilon = h\nu$, one has the simple relation between photon momentum and wavelength

$$p = \frac{h\nu}{c} = \frac{h}{\lambda}. \tag{7.1}$$

The particle description of this scattering of light by electrons leads to a distinctive result.[2] From momentum–energy conservation, one expects the final state photon to have a smaller momentum than the incident photon, hence a longer wavelength. The exact relation between this shift of wavelength and scattering angle can easily be worked out (see Fig. 7.1). One starts with the energy conservation relation:

$$\epsilon + mc^2 = \epsilon' + \sqrt{(p_e c)^2 + (mc^2)^2}$$

[1] This is compatible with a classical radiation field with field energy density (u) given by $u = (E^2 + B^2)/2$ and field momentum density given by the Poynting vector $\mathbf{S} = \mathbf{E} \times \mathbf{B}/c$, with their magnitudes related by $u = cS$, because, in an electromagnetic wave, the electric and magnetic field strengths are equal, $E = B$.

[2] This analysis was independently worked out by Compton (1923) and Debye (1923).

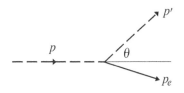

Fig. 7.1 The momentum diagram of Compton scattering: A photon with momentum **p** scatters off an electron to produce another photon (**p′**) with the recoil electron having momentum **p**$_e$.

which, after using $(\epsilon - \epsilon') = (p - p')c$, can be written as

$$\left[(p - p')c + mc^2\right]^2 = \left(p_e c\right)^2 + (mc^2)^2. \tag{7.2}$$

The corresponding momentum conservation relation can similarly be written down, with the electron recoil momentum being $\mathbf{p}_e = \mathbf{p} - \mathbf{p}'$. This implies an equation among the magnitudes of the various momenta,

$$p_e^2 = p^2 + p'^2 - 2pp' \cos \theta. \tag{7.3}$$

Eliminate p_e^2 from Eqs. (7.2) and (7.3), we immediately obtain the correlation between wavelength shift and scattering angle as

$$\lambda' - \lambda = \lambda_c(1 - \cos \theta). \tag{7.4}$$

We have used the definition of the **Compton wavelength** $\lambda_c = h/mc$, which is a very small length $\lambda_c = 0.0024$ nm even for an electron mass. In order to see such a tiny wavelength shift, it helps to work with an electromagnetic wave having short wavelength in the first place. This explains why the effect was discovered in the scattering experiment involving X-rays. Still, it was remarkable that Compton was able, in the first experiment making measurements at scattering angles of $45°$, $90°$, and $135°$, to determine the Compton wavelength λ_c to an accuracy that was less than one percent off the modern value. Compton concluded his paper (Compton 1923) this way: "The experimental support of the theory indicates very convincingly a radiation quantum carries with it directed momentum as well as energy."

We have in Section 3.4.3 derived the result for radiation pressure and radiation energy density $p = u/3$ by way of thermodynamic arguments. It may be easier for a modern-day reader to understand this result from the viewpoint of momentum changes as suffered by photons after collisions with the enclosure wall (cf. SuppMat Section 7.5). In the next few sections we shall see how the idea of the photon would finally lead to a more consistent and deeper understanding of blackbody radiation.

7.2   Towards Bose–Einstein statistics

The Compton scattering result was obtained in 1923. This had finally established the reality of photons for the general community of physicists. In 1924 Einstein received a letter from Satyendranath Bose (1894–1974) of Kolkata asking his opinion of an enclosed paper,[3] in which Bose used the particle properties of a photon (having the energy–momentum relation of $\epsilon = pc$ with $p^2 = p_x^2 + p_y^2 + p_z^2$) to obtain another derivation of Planck's spectral distribution of blackbody radiation. He stated that he was motivated by the observation that Planck's derivation was not logically self-consistent. Planck arrived at a non-classical physics result of energy quantization while using a relation between radiation density $\rho(\nu, T)$ and average energy U of an oscillator,

$$\rho = \frac{8\pi \nu^2}{c^3} U, \tag{7.5}$$

[3] The paper was written in English and sumitted to *Philosophical Magazine* in 1923. After its rejection by that journal, Bose sent it to Einstein who translated it into German and arranged its publication in *Zeitschrift der Physik* (Bose 1924). A re-translation back into English can be found in the *American Journal of Physics* (Bose 1976).

deduced by a classical electromagnetic calculation (cf. Section 3.2.1). In Section 4.1.3 this result was obtained by a counting of wave states. Bose set out with a particle approach from the very beginning.

After deriving this density of states, Bose proceeded to deduce Planck's distribution. Bose's approach is to adhere closely to Boltzmann's procedure. In contrast to Planck's method (cf. Sections 3.3.1 and 3.3.2), his statistical analysis involves a counting of phase space cells. But to have the calculation ending in Planck's distribution, rather than Boltzmann's, clearly he had to deviate from the latter's procedure. In order to see all this, we shall first provide a brief review of Boltzmann's statistical program.

7.2.1 Boltzmann statistics

Boltzmann's analysis proceeded by identifying particles with cells in position–momentum phase space. He also used the device of discrete energy packets ϵ, but would set $\epsilon = 0$ at the end of a calculation in order to recover a continuous energy. One regards the P energy packets as forming $P + 1$ energy levels, with the lowest level having zero energy $\epsilon_0 = 0$, the second level having energy $\epsilon_1 = \epsilon$, so on up to the highest level $\epsilon_P = P\epsilon$. The N particles (cells) can have various energies; such a configuration (label it as σ) can be described by a set of cell numbers $(N_0, N_1, N_2, \ldots, N_P)$. N_0 is the number of cells at the ground level ϵ_0, N_1 is the number at the next level ϵ_1, etc. Namely, we have the total number of particles and total energy as given by

$$N = \sum_{j=0}^{P} N_j, \qquad U = \sum_{j=0}^{P} \epsilon_j N_j = \sum_{j=0}^{P} j\epsilon N_j. \tag{7.6}$$

The probability complexion for such a macrostate configuration σ is calculated by counting the number of ways N cells can have different energies (i.e. N is divided into different sets of cells). We note that the number of ways we can select out N_0 cells from the total N is $N!/[N_0!(N - N_0)!]$, then the number of ways to select out N_1 cells from the remaining $N - N_0$ is $(N - N_0)!/[N_1!(N - N_0 - N_1)!]$, etc. Thus the total number of ways one can divide up N oscillators into a distribution of $(N_0, N_1, N_2, \ldots, N_P)$ is the product[4]

$$W_\sigma = \frac{N!}{N_0!(N - N_0)!} \times \frac{(N - N_0)!}{N_1!(N - N_0 - N_1)!} \times \cdots$$

$$= \frac{N!}{N_0!N_1!\ldots N_P!}, \tag{7.7}$$

leading to the entropy

$$S = k_B \ln W_\sigma = k_B \left(\ln N! - \sum_j \ln N_j! \right)$$

$$\approx k_B \left(N \ln N - \sum_j N_j \ln N_j \right),$$

[4] In the so-called "correct Boltzmann counting" procedure, one would insert the ad hoc factor of $1/N!$ in order to make the resultant entropy an extensive thermodynamical function. This is ultimately justified by quantum mechanics with its concept of identical particles. We do not make this insertion here because this feature is irrelevant for our present discussion.

where we have approximated the factorial logarithms by Stirling's formula (cf. Appendix A, Section A.3). One then maximizes the entropy, under the constraint of the two conditions given in (7.6), to find the configuration that corresponds to the equilibrium state:

$$0 = \delta \sum_j \left(N_j \ln N_j + AN_j + B\epsilon_j N_j \right)$$
$$= \sum_j \delta N_j \left(\ln N_j + 1 + A + B\epsilon_j \right)$$

where we have used the Lagrangian multipliers A and B to incorporate the two constraint conditions (cf. Appendix A, Section A.4). For an independent variation δN_j, the coefficient must vanish

$$\ln N_j = -\alpha - \beta \epsilon_j + \text{constant} \tag{7.8}$$

giving the primitive form of the Boltzmann distribution, $N_j \propto \exp(-\alpha - \beta \epsilon_j)$. One then has to appeal to analysis of other systems such as an ideal gas to fix the parameters of α and $\beta = 1/k_B T$, etc.

7.2.2   Bose's counting of photon states

Bose obtained the relation (7.5) by a counting of the photon states (instead of wave states as done by Rayleigh). His approach was to count the cells in the particle's position–momentum phase space, which is quantized in units of Planck's constant h^3. First, we will show that quantum theory naturally tells us that the phase space volume for an oscillator is quantized in this way.

As we have already shown in Eq. (3.55) in SuppMat Section 3.5.1, the energy equation of a 1D harmonic oscillator with frequency ν traces out an elliptical curve in the 2D phase space $dqdp$, with an area equal to the ratio of the oscillator energy to its frequency U/ν. In quantum theory, the oscillator energy (above the zero-point energy) is quantized, $U_n = nh\nu$. Thus, each oscillator state occupies an elliptical area of h. Clearly for 3D oscillators with phase space $d^3\mathbf{q}d^3\mathbf{p}$, the 6D volume has a volume of h^3—each oscillator state occupies a phase space volume of h^3.

Density of photon states

Recall the relation of radiation energy density ρ in terms of the number of radiation oscillators N and their respective energy U, see Eqs. (3.13) and (4.5); we have $\rho d\nu = (UN/V)d\nu$. Bose counted the number of photon states N by counting the number of cells occupied in the 6D phase space:

$$N d\nu = 2 \frac{d^3\mathbf{q}d^3\mathbf{p}}{h^3} = \frac{2V}{h^3} 4\pi p^2 dp \tag{7.9}$$

[5] Previously this factor of 2 had always been introduced as the two polarizations of the electromagnetic wave. But here Bose in 1924 was inserting it when working with photons. The whole idea that a particle can have intrinsic spin was not proposed until 1925 (for the electron). Nevertheless, Bose introduced this factor of 2 in his derivation with the comment: "it seems required"!

where the factor of 2 corresponds to the two polarization states of a photon.[5] The momentum variable p is then replaced by the frequency ν through the relation of $p = \epsilon/c = (h/c)\nu$ so that

$$N d\nu = \frac{8\pi V}{h^3} \left(\frac{h}{c} \right)^3 \nu^2 d\nu \quad \text{or} \quad \rho = \frac{N}{V} U = \frac{8\pi \nu^2}{c^3} U. \tag{7.10}$$

This is Planck's relation (7.5).

Bose's derivation of the Planck distribution

Just like Planck, Bose then proceeded to a statistical calculation of the complexion that would lead to Planck's spectral distribution. Recall that all Einstein's previous discussions of blackbody radiation had avoided any explicit statistical analysis. In fact he had mildly criticized Planck's statistical approach as being without foundation (cf. Section 3.4.1). But he was supportive of Bose's new analysis.

Bose assumed that there are N^s quanta, distributed among Z^s phase space cells (i.e. potential states that a photon can occupy) at frequency ν^s (the superscript s is a label of the state, having frequency interval $d\nu$). These cells can be at different energy levels. There are w_r^s cells each holding r quanta, having energy $\epsilon_r^s = rh\nu^s$: thus w_0^s cells at ground state $\epsilon_0^s = 0$, w_1^s cells at $\epsilon_1^s = h\nu^s$, and w_2^s cells at $\epsilon_2^s = 2h\nu^s$, etc. In total there are N photons in the frequency interval $d\nu$. Their relations can be expressed as

$$N = \sum_s N^s, \qquad U = \sum_s N^s h\nu^s, \qquad (7.11)$$

$$N^s = \sum_r r w_r^s, \qquad Z^s = \sum_r w_r^s.$$

The number of microstates, according to Bose, should "simply" be the product of the number of ways (for each ν^s) that Z^s cells can be partitioned into a distribution of $(w_0, w_1, w_2, \ldots,)$, much like (7.7):

$$W = \prod_s \frac{Z^s!}{w_0^s! w_1^s! w_2^s! \ldots}. \qquad (7.12)$$

Just as in Section 7.2.1, one then maximizes the logarithm of this statistical weight $\delta \ln W = 0$ with

$$\ln W = \sum_s Z^s \ln Z^s - \sum_{s,r} w_r^s \ln w_r^s$$

holding the total number of cells Z^s and total energy U fixed. Using the method of Lagrangian multipliers β and λ^s (see Appendix A.3), he obtains the condition[6]

$$\sum_s \sum_r \delta w_r^s \left(\ln w_r^s + 1 + \lambda^s + r\beta h\nu^s \right) = 0 \qquad (7.13)$$

which implies the solution

$$\ln w_r^s = -r\beta h\nu^s + \text{constant}$$

or $w_r^s = A^s e^{-r\beta h\nu^s}$ with the coefficient A^s related to $Z^s = \sum_r w_r^s$ by:

$$w_r^s = Z^s \left(1 - e^{-\beta h\nu^s} \right) e^{-r\beta h\nu^s}. \qquad (7.14)$$

Bose proceeds to calculate the photon number distribution by[7]

$$N^s = \sum_r r w_r^s = Z^s \left(1 - e^{-\beta h\nu^s} \right) \sum_r r \left(e^{-\beta h\nu^s} \right)^r$$

$$= Z^s \frac{e^{-\beta h\nu^s}}{1 - e^{-\beta h\nu^s}} = \frac{Z^s}{e^{\beta h\nu^s} - 1}. \qquad (7.15)$$

[6]From (7.11), we insert into $\delta \ln W = 0$ the factors $\lambda^s \delta Z^s = \lambda^s \sum_r \delta w_r^s = 0$ and $\beta \delta U = \beta \sum_s \delta N^s h\nu^s = \beta \sum_{s,r} r \delta w_r^s h\nu^s = 0$.

[7]Here one uses the result of $\sum_r r x^r = x/(1 - x)^2$, which can be gotten from the familiar geometric series $\sum_r x^r = 1/(1 - x)$ by a simple differentiation with respect to the variable x.

Using his result (7.10) for the density of states $Z^s = 8\pi \nu^{s2} V/c^3$, Bose deduces the total radiation energy in the interval

$$\frac{U}{V} d\nu = \sum_s \frac{N^s h\nu^s}{V} d\nu = \frac{8\pi \nu^2}{c^3} \frac{h\nu}{e^{\beta h\nu} - 1} d\nu. \qquad (7.16)$$

The Lagrangian multiplier can be fixed as $\beta = 1/k_B T$ by noting that the entropy is $S = k_B \ln W$, and differentiating with respect to energy must be identified with the absolute temperature $T = \partial U/\partial S$. In this way Bose arrives at the Planck's distribution:

$$\rho(\nu, T) = \frac{8\pi h}{c^3} \frac{\nu^3}{e^{h\nu/k_B T} - 1}. \qquad (7.17)$$

Implicit assumptions in Bose's derivation

Bose's statistical counting method, compared to Planck's, certainly looks more like the traditional Boltzmann procedure. Clearly he had gone beyond the established approach so that he was able to arrive at Planck's distribution instead of the usual Boltzmann result. Apparently Bose did not realize in what fundamental ways his derivation departed from the usual classical statistical mechanical method. He made no comments about these assumptions in his paper. And in later years he remarked that he did not realize "what he did was all that new"!

Instead of counting photons directly, he divided the phase space into cells, and asked how many photons were in a cell (rather than which photons were in a cell) and thus implicitly assumed indistinguishability of photons. He assumed statistical independence of cells. Thus there is no statistical independence of particles. In his calculation he imposed the condition of conservation of phase space cells (i.e. $Z^s = \sum_r w_r^s = $ constant). Since each cell has an indefinite number of photons, he had implicitly assumed photon number nonconservation (i.e. $N = \sum_s N^s \neq$ constant). Interestingly, this constant Z^s condition is actually irrelevant for his result, as we see in the above calculation that the corresponding Lagrangian multiplier λ^s drops out in the result of the cell distribution (7.14). Furthermore, he had also assumed that (in modern language) the photon has intrinsic spin—again without much of a comment.

7.2.3   Einstein's elaboration of Bose's counting statistics

Einstein arranged the publication of Bose's paper and he also sent in a related contribution (Einstein 1924), extending Bose's case of photon statistics to the general case of noninteracting particles (i.e. atoms and molecules). In the meantime (1923–24) Louis de Broglie made his suggestion that matter, under certainly circumstances, could have wave-like behavior. Einstein was very enthusiastic about this idea. His paper (Einstein 1925) was the first of anyone who actually referred to de Broglie's new suggestion of matter waves. He justifies his application of Bose's photon counting method to matter particles by saying that if particles can be waves, they should obey similar statistics to photons.

Density of nonrelativistic particle states

To count the nonrelativistic gas particle states, Einstein followed Bose by calculating the corresponding phase space volume in units of h^3 as in (7.9):

$$N dv = \frac{d^3 \mathbf{q} d^3 \mathbf{p}}{h^3} = \frac{V}{h^3} 4\pi p^2 dp. \tag{7.18}$$

Instead of photons, we have nonrelativistic particles with the momentum p related to the kinetic energy ϵ by $p = \sqrt{2m\epsilon}$ so that

$$N dv = \frac{2\pi V}{h^3} (2m)^{3/2} \epsilon^{1/2} d\epsilon = \mathcal{Z}(\epsilon) d\epsilon. \tag{7.19}$$

The particle density (number per unit energy) $\mathcal{Z}(\epsilon)$ is given by

$$\mathcal{Z}(\epsilon) = 2V \sqrt{\frac{\epsilon}{\pi}} \left(\frac{2\pi m}{h^2} \right)^{3/2}. \tag{7.20}$$

Particles obeying Bose–Einstein statistics are called bosons.

Distribution of identical bosons

Einstein improved upon Bose's method in his derivation of the distribution of gas particles. Recall our comment that it is irrelevant to impose the condition $Z^s = \sum_r w_r^s = \text{constant}$. Instead of working with the statistical weight of (7.12), Einstein wrote down the complexion in a form more similar to Planck's Eq. (3.38):

$$W = \prod_s \frac{(N^s + Z^s - 1)!}{N^s! (Z^s - 1)!} \simeq \prod_s \frac{(N^s + Z^s)!}{N^s! Z^s!}. \tag{7.21}$$

This is a counting of the ways that one can distribute N^s identical particles into Z^s cells.[8] Einstein then maximizes $\ln W$ under the constraint of holding the particle number $N = \Sigma N^s$ and the energy $U = \Sigma N^s \epsilon^s$ fixed—by using two Lagrangian multipliers, which can eventually be identified with $1/k_B T$ and the chemical potential μ:

[8] Recall in Eq. (3.34) that Planck was counting the ways of distributing P quanta into N oscillators.

$$N^s = \frac{1}{e^{(\epsilon^s - \mu)/k_B T} - 1}. \tag{7.22}$$

This came to be known as Bose–Einstein statistics. Because photon number is not conserved (that is, photons can be freely emitted and absorbed), there is not the requirement of $N = \Sigma N^s$, and the Lagrangian multiplier μ, identified with the chemical potential, is absent (or, $\mu = 0$):

$$N^s = \frac{1}{e^{\epsilon^s/k_B T} - 1}. \tag{7.23}$$

This is the correct distribution for photons as discussed in Section 7.2.2.

Using the expression for the density of states obtained in (7.20), the total number density is

$$N = \int \frac{\mathcal{Z}(\epsilon) d\epsilon}{e^{(\epsilon^s - \mu)/k_B T} - 1}$$

$$= \frac{2V}{\sqrt{\pi}} \left(\frac{2\pi m}{h^2} \right)^{3/2} \int \frac{\epsilon^{1/2} d\epsilon}{e^{(\epsilon^s - \mu)/k_B T} - 1}. \tag{7.24}$$

The discovery of Bose–Einstein condensation

For the case of the ground state (with $\epsilon^s = 0$) we have from (7.22):

$$N_0 = \frac{1}{e^{-\mu/k_B T} - 1}.$$

The requirement that $N_0 > 0$ (i.e. $e^{-\mu/k_B T} > 1$) implies that the chemical potential μ must in general be nonpositive. Einstein discovered that when the temperature drops below a certain temperature (called the condensation temperature T_c) when the chemical potential approaches zero (from below) the above expression for the ground state population becomes macroscopic in size:

$$N_0 = \frac{k_B T}{-\mu} \ggg 1. \tag{7.25}$$

Einstein commented on this remarkable result:

I maintain that, in this case, a number of molecules steadily growing with increasing density goes over in the first quantum state (which has zero kinetic energy) while the remaining molecules distribute themselves according to the parameter value of $\mu = 0$ ... A separation is effected; one part condenses, the rest remains a 'saturated ideal gas'.

Einstein had discovered a purely statistically induced phase transition, which we now call "Bose–Einstein condensation". In Section 7.4 we shall provide a more detailed discussion of this condensation phenomenon.

7.3   Quantum mechanics and identical particles

Modern quantum mechanics came into being in 1925–26. In this new theory physical states are identified with vectors and physical observables with operators of the Hilbert space. These vectors and operators may appear rather differently depending on what basis vectors one chooses to represent this geometric space. This is reflected in the two separate discoveries of quantum mechanics. In the spring of 1925 Werner Heisenberg, following his study of dispersion relations, had proposed a rather abstract version of a quantum theory, which came to be known as *matrix mechanics*. At the end of 1925 Erwin Schrödinger wrote down his wave equation thus initiated *wave mechanics*. The hydrogen spectrum was obtained in both matrix and wave mechanics. Soon, in 1926, P.A.M. Dirac, and independently Heisenberg and Schrödinger, had shown that wave and matrix mechanics are equivalent to each other: they were just two different representations of the same theory. In the following discussion we shall mostly use the more accessible language of wave mechanics.

7.3.1   Wave mechanics: de Broglie–Einstein–Schrödinger

In 1924 Louis de Broglie proposed in his doctoral thesis that, associated with every matter particle (with momentum p) there was also a wave with wavelength $\lambda = h/p$ (de Broglie 1924). Recall that Einstein had introduced the idea of light quanta—particle-like properties of radiation; he stated his general motivation as wanting to have a more symmetric description of matter and radiation. In fact de Broglie explicitly acknowledges that his matter wave idea was inspired by this particle–wave dualism[9] as stated in Einstein's 1905 photon paper. Thus we should not be surprised that de Broglie's matter wave idea received Einstein's enthusiastic support. The question: What would be the equation that governs the behavior of this matter wave? It is also of historical interest to note that Erwin Schrödinger first became aware of de Broglie's idea from reading Einstein's 1924–25 papers. This led directly to the creation of his wave equation, which immediately came to be known as the Schrödinger equation, at the end of 1925.

[9]Recall our discussion in Section 6.1 of Einstein's 1909 investigation of radiation fluctuation leading him to the notion of particle–wave duality. In a paper submitted at the beginning of 1925 Einstein showed that a relation, just like Eq. (6.16), holds as well for his quantum gas of material particles as discussed in Section 7.2.3.

7.3.2 Identical particles are truly identical in quantum mechanics

The concept of identical particles in quantum mechanics is qualitatively different from any analogous notion in classical physics. In classical physics ultimately no particles can be identical. Two electrons with identical charge and mass can still be distinguished because we can in principle follow their individual trajectories and tell apart which is which particle. In quantum mechanics, it is impossible to have precise particle trajectories. In classical physics we can also in principle distinguish two particles by putting labels on them (e.g. paint them different colors) without interfering with their motion. In quantum mechanics, on the other hand, it is not possible to keep track of particle trajectories or to put extra labels involving an incompatible observable.[10] Thus identical particles are truly identical in quantum mechanics. The interchange of any two identical particles leaves no observable consequence—no change in the measurement probability. The wavefunction, being the probability amplitude, must therefore be either symmetric or antisymmetric with respect to such an exchange of identical particles. As we commented at the end Section 6.4, quantum field theory can account for symmetry properties of the identical particle nature of a field's quanta. The commutation relation discussed in Eq. (6.47) is just the elegant mathematical device needed to bring about this required symmetry.

[10]Because observables are represented by operators. Two observables are said to be incompatible if their respective operators do not commute, $\hat{A}\hat{B} - \hat{B}\hat{A} \equiv [\hat{A}, \hat{B}] \neq 0$. This leads to an uncertainty relation of their observable values, $\Delta A \Delta B \geq \hbar$. The 'extra label' that one would wish to place on a particle must be an incompatible one as a particle has already been labeled by a complete set of compatible observables.

7.3.3 Spin and statistics

In the meantime the quantum mechanical concept of particle spin had emerged.[11] It was then proposed that there is a direct relation between the particle spin and the symmetry property of a wavefunction, and hence the statistical properties of such identical particles. A system of particles having integer spin (e.g. photons with spin 1) must have a symmetric wavefunction and obey Bose–Einstein statistics; these particles are called bosons. A system of particles with half-integer spin (e.g. electrons with spin 1/2) must have an antisymmetric wavefunction, and obey Fermi–Dirac statistics. They are called fermions. The spin-statistics theorem was proven by Wolfgang Pauli and others

[11]Electron spin was first proposed by Uhlenbeck and Goudsmit (1925).

in the framework of quantum field theory (based on quantum mechanics and special relativity) (Pauli 1940).

7.3.4 The physical implications of symmetrization

The physical implications of the concept of indistinguishable particles are remarkable. The antisymmetric property of the fermionic wavefunction means that two identical fermions cannot be in the same state. This explains Pauli's exclusion principle—crucial, among other consequences, in the explication of the structure of multi-electron atoms. The totally symmetric wavefunction of a boson system also leads to highly counter-intuitive results. Just consider the calculation of statistical weight. Boltzmann's statistics (7.7) would yield results that are in accord with our intuition. For instance, compare the two cases of distributing 10 (distinguishable) particles into two cells: in one case 10 particles are in one cell and none in the other cell, and in the other case each cell has five particles. The ratio of weights for these two cases is

$$\frac{10!}{10!0!} : \frac{10!}{5!5!} = 1 : 252.$$

This is to be contrasted with the quantum distribution of 10 identical bosons yielding the rather counter-intuitive result of the statistical weight for each case being unity—hence the ratio 1 : 1 for the above-considered situation. In each case there is only one totally symmetric wavefunction. When Einstein first worked out the Bose–Einstein (BE) counting, he commented:

The BE counting "expresses indirectly certain hypothesis on the mutual influences of the molecules which for the time being is of a quite mysterious nature."

While we now know that this is just the correlation induced by the requirement of a totally symmetric wavefunction,[12] on a deeper physical level this mutual influence is still no less mysterious today.

[12]For a discussion of the role of particle indistinguishability making BE condensation possible, see SuppMat Section 7.7.

The final resolution of the counting schemes of Bose (1924), Einstein (1905), and Planck (1900)

We have seen how quantum Bose–Einstein statistics naturally explains how Bose's implicit assumptions are all justified. It also justifies Einstein's original classical statistical mechanical argument of Wien radiation being a gas of photons. Even though Einstein avoided making an explicit calculation of the statistical weight, since he used the analogy of a classical ideal gas, implicitly he had assumed the Boltzmann statistics of (7.7). However we can justify it now because in the Wien limit ($\epsilon^s \gg k_B T$) the average photon number (7.23) is vanishing small: $N^s = e^{-\epsilon^s/k_B T} \simeq 0$. Thus in this limit, the statistical weight of Boltzmann counting (7.7) is indistinguishable from Bose–Einstein ($W = 1$). The Planck spectral distribution is of course understood as the consequence of BE statistics. Nevertheless it is useful to work out the way of seeing how Planck's statistical analysis can lead to the correct result. This calculation can be found in the SuppMat Section 7.6.

7.4 Bose–Einstein condensation

Let the energies of the first excited and the (lowest energy) ground states be ϵ_1 and $\epsilon_0 = 0$ so that they have the energy gap $\Delta\epsilon = \epsilon_1$. If the available thermal energy is not much bigger than this energy gap $k_B T \lesssim \Delta\epsilon$, it would not surprise us to find many molecules in their ground state. On the other hand, Bose–Einstein condensation (BEC) is the phenomenon that, at a temperature below some critical value T_c, a macroscopic number of molecules would stay in ("condense into") the ground state, even though the available thermal energy $k_B T$ is much bigger than the energy gap, $k_B T \gg \Delta\epsilon$. What Einstein had shown was that the chemical potential can be extremely small ($|\mu| \ll \epsilon_1$) in the low-temperature regime $T < T_c$. For example, liquid helium has an energy gap of $\epsilon_1 \simeq 10^{-18}$ eV. But below a temperature of $\lesssim O(1 \text{ K})$, i.e. a thermal energy of $\lesssim O(10^{-4} \text{ eV})$, the chemical potential of helium[13] has such a small value $\mu \simeq -10^{-26}$ eV that the Bose–Einstein distribution (7.22) would imply

$$N_1 = \frac{1}{e^{(\epsilon_1 - 0)/k_B T} - 1} \simeq \frac{k_B T}{\epsilon_1}. \tag{7.26}$$

Comparing this to the ground state occupation number given in (7.25), one finds that most of the molecules would condense into the ground state,

$$\frac{N_0}{N_1} = \frac{\epsilon_1}{-\mu} \gg 1. \tag{7.27}$$

One must bear in mind this is condensation in momentum space (rather than the everyday condensation in configuration space).

Bose–Einstein condensation as a macroscopic quantum state That a macroscopic number of molecules are in one quantum state would lead to quantum mechanical behavior on the macroscopic scale. This was first pointed out in 1928 by Fritz London (1900–54). London suggested that superfluid helium was an example of a Bose–Einstein condensate. His related work on superconductivity also decidedly influenced the later development of BCS theory in which electron pairs (the Cooper pairs) form the Bose–Einstein condensate.[14]

7.4.1 Condensate occupancy calculated

To calculate the condensation temperature, below which a macroscopic fraction of the particles are in the ground state, we must know how to add up the occupancy for every state. Only then can we compare the occupancy in the ground state with those in the excited states. In general the total particle number N is given by (7.24) where the discrete sum can be replaced by integration using the density of states $\mathcal{Z}(\epsilon)$. This replacement, while applicable for the excited states, is not valid for the ground state, as $\mathcal{Z}(\epsilon = 0) = 0$. We will simply separate out the ground state with its occupancy labeled N_0:

$$N = N_0 + N_{\text{ex}}$$

$$= \frac{1}{e^{-\mu/k_B T} - 1} + \frac{2}{\sqrt{\pi}} \left(\frac{2\pi m}{h^2}\right)^{3/2} V \int_0^\infty \frac{\sqrt{\epsilon}\, d\epsilon}{e^{(\epsilon - \mu)/k_B T} - 1}. \tag{7.28}$$

[13]Helium exhibits superfluid behavior below the critical temperature of 2.17 K. We can estimate the size of its chemical potential by the relation (7.25), $-\mu = k_B T/N_0$, with the approximation $N_0 \simeq N$ because a significant fraction of all molecules would have condensed into the ground state. We then obtain a value $-\mu = \left(10^{-4} \text{ eV} \cdot \text{K}^{-1}\right) \times (1 \text{ K}) \times 10^{-22} = 10^{-26}$ eV.

[14]London later substantiated his original suggestion by showing that the phase change of superfluid helium had properties consistent with a BEC transition (London 1938). The BCS theory is named after its originators: John Bardeen, Leon Cooper, and Robert Schrieffer.

As we have already discussed, the chemical potential is extremely small in the region $T < T_c$. Nevertheless, μ must be kept in N_0 because $\epsilon_0 = 0$, while setting $\mu = 0$ in N_{ex},

$$N = -\frac{k_B T}{\mu} + \frac{2}{\sqrt{\pi}} \left(\frac{2\pi m}{h^2}\right)^{3/2} V \int_0^\infty \frac{\sqrt{\epsilon} d\epsilon}{e^{\epsilon/k_B T} - 1}. \tag{7.29}$$

Let us carry out the calculation for N_{ex} with a change of integration variable $x = \epsilon/kT$:

$$N_{ex} = \left(\frac{2}{\sqrt{\pi}} \int_0^\infty \frac{\sqrt{x} dx}{e^x - 1}\right) \left(\frac{2\pi m k_B T}{h^2}\right)^{3/2} V = 2.612 \frac{V}{v_Q} \tag{7.30}$$

[15]The quantum volume is the cube of the quantum length $v_Q = l_Q^3$ with $l_Q = h/(2\pi m k_B T)^{1/2}$, also called the "de Broglie thermal wavelength" because, except for a $O(1)$ factor of $\sqrt{\pi}$, it is the de Broglie wavelength associated with the thermal momentum $p = (2m k_B T)^{1/2}$.

where $v_Q = \left(\frac{h^2}{2\pi m k_B T}\right)^{3/2}$ is the particle's quantum volume.[15] Rewriting this in terms of the particle's physical volume $v = V/N$, we have

$$\frac{N_{ex}}{N} = 2.612 \frac{v}{v_Q}. \tag{7.31}$$

In other words, the fraction of particles in the excited states is directly proportional to how small a particle's quantum volume has become with respect to its physical volume.

7.4.2   The condensation temperature

Because of the absence of a significant number of molecules in the ground state when $T > T_c$, we can define the condensation temperature T_c by $N_{ex}(T_c) \equiv N$. Namely, we have the relation

$$N_{ex} = 2.612 \left(\frac{2\pi m k_B T_c}{h^2}\right)^{3/2} V = N \tag{7.32}$$

or

$$T_c = \frac{h^2}{2\pi m k_B} \left(\frac{N}{2.612 V}\right)^{2/3}. \tag{7.33}$$

Furthermore, taking the ratio of Eq. (7.30) and Eq. (7.32), we have

$$\frac{N_{ex}}{N} = \left(\frac{T}{T_c}\right)^{3/2}, \tag{7.34}$$

which is plotted in Fig. 7.2.

T_c for noninteracting helium   Let us use (7.33) to calculate the condensation temperature for the He system: one mole $N = N_A = 6 \times 10^{23}$ of helium with molar volume $V = 27.6$ cm^3. This yields $T_c(\text{He}) = 3.1$ K, which is not too far off from the experimental value of 2.17 K, considering that we have completely ignored mutual interactions, which are rather complicated collisions involving many particles.

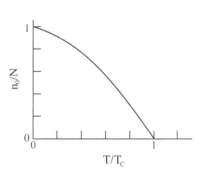

Fig. 7.2 Fractional ground state as a function of temperature.

Behavior of the chemical potential when $T \lesssim T_c$

From (7.34) we can also see the behavior of the chemical potential

$$N = N_0 + N_{\text{ex}} = -\frac{k_B T}{\mu} + N\left(\frac{T}{T_c}\right)^{3/2} \quad \text{or} \quad \frac{k_B T/N}{\mu} = \left(\frac{T}{T_c}\right)^{3/2} - 1. \tag{7.35}$$

Thus

$$\mu(T) = \frac{\frac{k_B T}{N}}{\left(\frac{T}{T_c}\right)^{3/2} - 1} = \begin{cases} -\dfrac{k_B T}{N} & \text{for } T \to 0 \\[2ex] -\dfrac{k_B T_c}{N}\dfrac{2T_c}{3\delta T} & \text{for } T \to (T_c - \delta T). \end{cases} \tag{7.36}$$

Namely, the chemical potential, which started out being extremely small in magnitude at low temperature, suddenly increases at the condensation temperature (as shown by the factor $T_c/\delta T$). It is this behavior that brings about the phase transition at T_c.

Overlap of wavefunctions when $T \lesssim T_c$

Instead of μ, it is useful to have a more direct way to understand the meaning of the condensation temperature. From Eq. (7.31) $N_{\text{ex}}/N = 2.6\left(a/l_Q\right)^3$ where a and l_Q are the interatomic separation $\left(v = a^3\right)$ and the quantum length $\left(v_Q = l_Q^3\right)$, respectively. Thus at $T = T_c$ when $N_{\text{ex}} = N$, we have $a \simeq l_Q$. Namely, the condensation temperature is the low temperature when the thermal de Broglie wavelength $\sim 1/\sqrt{T}$, hence the effective quantum size of atoms, becomes so large that atomic wavefunctions begin to overlap. As the temperature falls below T_c the overlap is enhanced, more particles condense into the ground state, and N_0 becomes ever increasing.

7.4.3 Laboratory observation of Bose–Einstein condensation

The prediction of BEC is for noninteracting bosons. Thus a dilute gas is much closer to theoretical considerations, rather than the dense helium case. The helium density has the value

$$\left(\frac{N}{V}\right)_{\text{He}} = \frac{6 \times 10^{23}}{27.6} \text{ cm}^{-3} = 2 \times 10^{22} \text{ cm}^{-3}. \tag{7.37}$$

In a modern experiment, using a laser and magnetic cooling techniques, experimenters have achieved the confinement of $O\left(10^4\right)$ atoms (e.g. rubidium 86) in a volume $V = 10^{-9}$ cm^3, thus a density of

$$\left(\frac{N}{V}\right)_{\text{gas}} = \frac{10^4}{10^{-9}} \text{ cm}^{-3} = 10^{13} \text{ cm}^{-3} \tag{7.38}$$

which is a billion times smaller than the helium case. This implies a decrease in condensation temperature, through (7.33),

$$T_c^{\text{gas}} = \frac{m_{\text{He}}}{m_{\text{Rb}}} \left[\frac{(N/V)_{\text{gas}}}{(N/V)_{\text{He}}} \right]^{2/3} T_c^{\text{He}}$$

$$= \frac{4}{86} \left[\frac{10^{13}}{10^{22}} \right]^{2/3} T_c^{\text{He}} \simeq 10^{-7} \text{ K}. \tag{7.39}$$

Given this extremely low condensation temperature the experimental study of BEC of a low-density gas is very difficult. One can use laser cooling and at the same time trap the atoms in a magneto-optical trap. Further manipulation of the trapped atoms by evaporation cooling enables experimentalists to reach the critical temperature when the BEC takes place. The positions of the atoms can be recorded using laser beams. In Fig. 7.3 we display the result of one such experiment. With this technique, the signature of the BEC is the appearance of a sharp peak. The atoms in the condensed phase are in the ground state (of the momentum space) and expand only slowly once released from the trap. The atoms in the excited states move relatively rapidly out of their steady state positions.

The success of producing dilute-gas BEC in the laboratory setting at Boulder, MIT, and elsewhere came about some 70 years after its first theoretical proposal by Einstein. The 2001 Nobel Prize for Physics was awarded to Eric Cornell (1961–), Wolfgang Ketterle (1957–), and Carl Wieman (1951–) for their experimental work in this area.

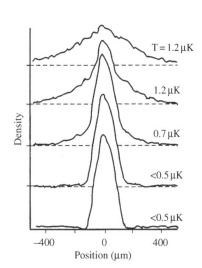

Fig. 7.3 Observation of BEC in a trapped atomic gas with $T_c = 1.7$ μK when a central peak appears representing the probability $|\psi|^2$. As the temperature is lowered, the thermal cloud is depleted, and atoms accumulate in the Bose condensate: the number of particles on the wing-sides (excited state particles) diminishes while the central peak (ground state particles) rises. Reproduction of Fig. 7 from Stenger *et al.* (1998).

7.5   SuppMat: Radiation pressure due to a gas of photons

The pressure ($\mathcal{P}$) is the force per unit area A and the force is the rate of momentum change $\Delta p / \Delta t$:

$$\mathcal{P} = \frac{\text{force}}{\text{area}} = \frac{1}{\Delta A \Delta t} \Delta p. \tag{7.40}$$

We denote the photon momentum by $\mathbf{p}$ (not to be confused with the pressure denoted by capital script $\mathcal{P}$). A given photon with energy and momentum of $(\epsilon, \mathbf{p})$ with $\epsilon = cp$, colliding with the wall, imparts a momentum of $\Delta p = 2p_z$, where p_z is the photon momentum component in the direction perpendicular to the wall (call it $\hat{z}$). Let $n(q)$ be the photon density function—the number of photons per unit spatial volume and per unit momentum space volume, in the momentum interval $(p, p + dp)$. The total pressure is the sum of the momenta that all the photons deposit onto the wall; we need to integrate over the momentum and configuration spaces (see Fig. 7.4). The configuration space is the volume of the parallelepiped with base area ΔA and perpendicular height $c\Delta t \cos\theta$, with θ being the angle between the photon momentum and the normal to the area (hence, $p_z = p \cos\theta$). All photons in this volume $[c\Delta t \cos\theta \Delta A]$ would collide with the wall in the interval of Δt. The sum of (7.40) is then

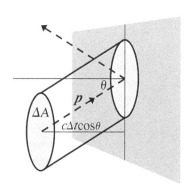

Fig. 7.4 Momentum change of a photon bouncing off a wall.

$$\mathcal{P} = \frac{1}{\Delta A \Delta t} \int n(p)\, \Delta p d^3 \mathbf{p}[c\Delta t \cos\theta\, \Delta A]$$

$$= \int_{g_z > 0} n(p)\, 2p_z d^3 \mathbf{p} \left[c\frac{p_z}{p} \right] = \frac{2}{3}\int_{p_z > 0} [cp] n(p)\, d^3 \mathbf{p}. \qquad (7.41)$$

To reach the last equality on the RHS, we have used the fact that radiation is isotropic $p_x^2 = p_y^2 = p_z^2$ and thus the momentum magnitude squared is $p^2 = p_x^2 + p_y^2 + p_z^2 = 3p_z^2$. Furthermore, the density function $n(p)$ depends only on $p = |\mathbf{p}|$ so that the integrand must be an even function of p; we can extend the integration to the full range of $(-\infty, +\infty)$ and thus remove the factor of 2:

$$\mathcal{P} = \frac{1}{3}\int_{-\infty}^{+\infty} [cp] n(p)\, d^3 \mathbf{p} = \frac{1}{3}\int_{-\infty}^{+\infty} [\epsilon] n(p)\, d^3 \mathbf{p} = \frac{1}{3} u. \qquad (7.42)$$

Namely, the integral is the radiation density—the sum of all photon energies (ϵ) per unit spatial volume—as n is the number per unit spatial and momentum volumes.

7.6   SuppMat: Planck's original analysis in view of Bose–Einstein statistics

We have already explained how Planck's distribution can be derived as the $\mu = 0$ case of the Bose–Einstein statistics (7.21). In the context of later development, it is perhaps still useful to take a closer look at Planck's original statistical weight written down in Eq. (3.34),

$$W_N = \frac{(P+N)!}{P!N!}, \qquad (7.43)$$

to see how it leads at the correct result despite of its unorthodox analysis. Let us recall that P is the total number of quanta in a system of N oscillators. Planck converts this W_N into the entropy of the oscillator $S = S_N/N = (k_B \ln W_N)/N = k_B \ln W$, in term of the statistical weight of a single oscillator $W = (W_N)^{-N}$. Planck does not perform any maximization of the entropy, subject to the energy constraint, but simply makes the substitution $U = (P/N)\epsilon$ and, after a differentiation of $\partial S/\partial U = 1/T$, obtains the distribution

$$U = \frac{\epsilon}{e^{\epsilon/k_B T} - 1}. \qquad (7.44)$$

We now wish to interpret Planck's analysis in view of Bose–Einstein statistics.

Accord to Bose–Einstein statistics, one starts with the statistical weight of (7.21),

$$W_N = \prod_s \frac{(n^s + 1)!}{n^s!}. \qquad (7.45)$$

For our purpose, we have the density of states (degeneracy) $Z^s = 1$, and we have changed the notation for the photon number from N^s to n^s (so as not to have it confused with the oscillator number discussed above). To avoid the maximization of entropy (with Lagrangian multiplier, etc.), we simply impose the energy condition $U = \Sigma_s n^s \epsilon^s$ in the form of $U = \langle n \rangle \epsilon$, where $\langle n \rangle$ is the average number of photons in a state having energy $\epsilon^s = \epsilon = h\nu$. With this average value, the statistical weight of (7.45) is simplified to

$$W_N = \left[\frac{(\langle n \rangle + 1)!}{\langle n \rangle !} \right]^{\langle n \rangle}. \tag{7.46}$$

Taking its logarithm and using $\langle n \rangle = U/\epsilon$, we have

$$S = k_B \ln W = k_B (\ln W_N)/\langle n \rangle$$

$$= k_B [(\langle n \rangle + 1) \ln(\langle n \rangle + 1) - \langle n \rangle \ln \langle n \rangle]$$

$$= k_B \left[\left(\frac{U}{\epsilon} + 1 \right) \ln \left(\frac{U}{\epsilon} + 1 \right) - \frac{U}{\epsilon} \ln \frac{U}{\epsilon} \right],$$

which is just the entropy expression (3.40) obtained by Planck, leading to Planck's distribution of (7.44) or equivalently, the average photon number:

$$\frac{U}{\epsilon} = \langle n \rangle = \frac{1}{e^{\epsilon/k_B T} - 1}. \tag{7.47}$$

7.7   SuppMat: The role of particle indistinguishability in Bose–Einstein condensation

[16] Here we follow Schroeder (2000, p. 321).

It will be illuminating to see how Bose–Einstein condensation follows from the indistinguishability of particles.[16] To see such an effect, we shall contrast two cases—in one all particles are somehow distinguishable, while in the other case, they are not.

N distinguishable particles

We shall discuss this case using two different approaches.

The approach of considering one-particle systems separately   This is the most straightforward approach. The partition function

$$Z_1 = \sum_\epsilon e^{-\epsilon/kT} \tag{7.48}$$

is essentially the number of accessible single-particle states. Namely, because higher energy states are exponentially suppressed, it counts all states with energy on the order of kT. Each state is roughly equally probable, hence there is about equal chance $1/Z_1$ for a particle to occupy any one of these states. This situation is not changed when we consider the whole system of N (independent) particles. The ground state being one of these many states, the fraction of particles in the ground state, when compared to the particles in all the excited state, is negligible. There is no BEC.

The approach of considering all *N* particles as a system   The above approach, while straightforward in explaining the absence of BEC for distinguishable particles, does not really highlight the crucial role played by the particles' distinguishability. Here is an alternative approach of considering the N particles as a whole. For an excited state of this system with energy U, the associated probability[17] is $W(U) = e^{-U/kT}$. While the ground state $U_0 \equiv 0$ has a probability of $W(0) = 1$, a typical excited state with $U = NkT$ has a probability of $W(U) = e^{-N}$. Thus it appears that the ground state probability is overwhelmingly large. This conclusion is incorrect because we have not taken into account the fact that while an individual excited state has small probability, there are an enormous number (ν) of such excited states. This multiplicity ν can be calculated by remembering that each one of the distinguishable particles can be in any one of these Z_1 single-particle states. Thus $\nu = (Z_1)^N$, or $N_{ex} = \nu e^{-N} = (Z_1/e)^N$, which is a large number as long as $Z_1 > e$. [*Comment*: While Z_1 can be large, the likely situation is still $Z_1 \ll N$.] Hence we have the situation that vastly more particles are in the excited state, confirming the above argument that there is no BEC.

N indistinguishable particles

The probability of an N-particle system being in each of the excited states is still e^{-N}. However, the number of excited states (of the system as a whole) ν for the case of N indistinguishable particles is much less than that for the distinguishable case. Now ν is the number of ways one can distribute N indistinguishable particles among the various single-particle states (Z_1):

$$\nu = \binom{N + Z_1 - 1}{N} = \frac{(N + Z_1)!}{N! \, Z_1!} \simeq \frac{(N + Z_1)^{N+Z_1}}{N^N \, Z_1^{Z_1}}, \qquad (7.49)$$

where we have used Stirling's approximation of $X! \simeq X^X$. For $N \gg Z_1$ we have

$$\nu \simeq \frac{N^{N+Z_1}}{N^N \, Z_1^{Z_1}} = \left(\frac{N}{Z_1}\right)^{Z_1}. \qquad (7.50)$$

While this is still fairly large, but the product $\nu e^{-N} \lesssim 1$, and we expect that it is now possible for a significant fraction of the particles to be in the ground state. This indicates why BEC becomes possible in a system of identical bosons.

[17] Since all the probabilities discussed here have a common partition function in their denominator, we shall not bother to display it.

8

Local reality and the Einstein–Bohr debate

- According to the orthodox interpretation of quantum mechanics (Niels Bohr being its leading voice), the attributes of a physical object (position, momentum, spin, etc.) can be assigned only when they have been measured. Einstein advocated, as more reasonable, the local realist viewpoint that a physical object has definite attributes whether they have been measured or not.

- The orthodox view that measurement actually produces an object's property implies that the measurement of one part of an entangled quantum state would instantaneously produce the value of another part, no matter how far the two parts have been separated. Einstein, Podolsky, and Rosen devised a thought experiment in order to shine a light on this "spooky action-at-a-distance" feature of the orthodoxy; its discussion and debate have illuminated some of the fundamental issues related to the meaning of quantum mechanics.

- Such discussions led later to Bell's theorem showing that these seemingly philosophical questions could lead to observable results. The experimental vindication of the orthodox interpretation has sharpened our appreciation of the nonlocal features of quantum mechanics. Nevertheless, the counter-intuitive picture of objective reality as offered by quantum mechanics still troubles many, leaving one to wonder whether quantum mechanics is ultimately a complete theory.

8.1 Quantum mechanical basics—superposition and probability

Recall our discussion of wave–particle duality in Section 6.1. Physical objects are found to be neither simply waves nor simply particles, but to have wave and particle attributes simultaneously—two seemingly contradictory properties at the same time. In the new quantum mechanics (QM) they are represented by quantum states, which are taken to be vectors in a linear algebra space, called the Hilbert space. These vectors can be added and obey equations that are linear—hence display the property of waves. These waves are interpreted as probability waves.

In less abstract language, a central quantity in quantum mechanics is the wavefunction, for example, the position wavefunction $\psi(x)$. It satisfies the Schrödinger wave equation, which is a linear differential equation; any linear combination of its solutions is still a solution. The wavefunction $\psi(x)$ has the interpretation that when we make a measurement of a particle's position, $|\psi(x)|^2\, dx$ is the probability of finding the particle in the interval $(x, x + dx)$. Namely, there is the possibility of a quantum mechanical position state in which the particle is in more than two, in fact an infinite number of, positions simultaneously.

Before a measurement is made, the particle can be in all these positions simultaneously. A measurement of the position would yield one particular value, say x_A. To make a measurement of a particle's position finding any particular value, according to the orthodox interpretation, is a random process, only subject to the likelihood as predicted by the probability distribution given by the wavefunction. Many physicists were ill at ease with the probability feature being built right into the foundation of the theory. Einstein famously objected: "God does not play dice!"

8.2 The Copenhagen interpretation

If a measurement of the position finds the particle to be at x_A, then immediately afterwards one should find the particle at x_A as well. Namely, due to the measurement, the particle "jumps" from the state being simultaneously in all positions to a state with definite position at x_A. According to the interpretation of the wavefunction as given above, the measurement causes the "collapse" of the wavefunction. Thus there are two fundamentally distinctive categories of physical processes in quantum mechanics:

1. **Smooth evolution of the wavefunction:** The Schrödinger equation completely determines the behavior of the wavefunction. There is nothing random about this description.
2. **Quantum mechanical measurement:** A measurement to obtain a particular result, according to quantum mechanics, is a random process. The theory only predicts the probability of getting any particular outcome. A measurement, which involves the interaction between the micro and macro realms of physics, collapses the wavefunction. This collapse of the wavefunction is not described by the Schrödinger equation; it is necessarily a non-local process as the wavefunction changes its value everywhere instantaneously.

8.2.1 The Copenhagen vs. the local realist interpretations

The first category of processes is noncontroversial, while questions related to measurement (the second category) bring out the strangeness of the QM theory.

One asks the question: "If the measurement finds the particle at x_A, where **was** the particle **just before** the measurement?"

- **Local realists:** "A particle has an objective reality; it has a set of attributes whether they have been measured or not. Therefore, it was at x_A just before the measurement finding it at x_A." According to Einstein, Schrödinger, *et al.,* this is the answer a reasonable theory would have given.
- **Orthodox quantum mechanical interpretation:** Bohr, Heisenberg, Born, and Jordan, *et al.* (the Copenhagen School) would have replied: "The particle was not really anywhere." Not only it is impossible to know, but it's **not even meaningful** to ask such a question (like asking the marital status of a table). The framework of every theory determines the relevant issue; in quantum mechanics, such a question should not have even been asked! Thus, a Copenhagen theorist would say: "The measurement compels the particle to assume a position. Observations not only disturb what's measured, they produce it."

To local realists, like Einstein, a particle must have objective reality (mass, spin, position, etc.) *independent* of whether these properties are being measured or not. (The moon is there whether you look at it or not.) Thus quantum mechanics must be an incomplete theory—the particle is at x_A, yet the theory cannot tell us it is so.

8.3   EPR paradox: Entanglement and nonlocality

From 1928 onward, Einstein had engaged Niels Bohr (the leading proponent of the Copenhagen school) in a series of debates (some public, but mostly private) as to the meaning of measurements in quantum mechanics. To sharpen his argument, to bring out the strangeness of the theory more clearly, Einstein, with his collaborators Boris Podolsky and Nathan Rosen, published in 1935 a paper in which a thought experiment was discussed in order to bring out clearly the underlying nonlocal nature of quantum mechanics.[1] The influence of this paper has grown over the years as subsequent developments showed that the question it raised was of fundamental importance to the meaning of quantum mechanics.

[1]Bohr's rejoinder can be found in Bohr (1935).

Local reality in physics   Let us recall briefly the history of the locality concept in physics. One aspect of Newton's theory of gravitation that he himself found unsatisfactory is the invocation of the "action-at-a-distance" force. Somehow the source particle can act instantaneously on the test particle some distance away. The same situation holds for Coulomb's law. This was later remedied with the introduction of the Faraday–Maxwell field. In a field theory such an interaction is pictured as a two-step process: the source particle brings about a field everywhere (with the field emanating from the source and propagating outward at a finite speed). The field then acts on the test particle **locally**. Thus, through field theory, locality was restored back to physics. Einstein now points out that quantum mechanics brought about a new form of nonlocality.

The EPR thought experiment We shall present the Einstein–Podolsky–Rosen (EPR) "paradox" as simplified and sharpened by David Bohm (1917–92). Consider the decay of a spin-zero particle, for definiteness take it be a neutral pion (a spin-zero elementary particle), into an electron and positron pair (both have spin one-half):

$$e^+ \longleftarrow \pi^0 \longrightarrow e^-.$$

The decay products will speed away from each other in opposite directions. To have angular momentum conservation, the spins of the daughter pair must be opposite each other in order for their sum to be a spin-zero state. In quantum mechanics such a final state is a superposition of two states: in one the electron spin is up, in another it's down:[2]

$$\psi_0 = \frac{1}{\sqrt{2}} \left(\psi_\uparrow^{(-)} \psi_\downarrow^{(+)} - \psi_\downarrow^{(-)} \psi_\uparrow^{(+)} \right) \tag{8.1}$$

where $\psi^{(-)}$ is the wavefunction of the electron e^-, and $\psi^{(+)}$ is the wavefunction of the positron e^+. The subscripts indicate their respective spin orientation (in some definite direction, say the z direction). Since these two terms have coefficients with equal magnitude, there is equal probability for either outcome to take place. These two terms, the two (product) wavefunctions $\psi_\uparrow^{(-)} \psi_\downarrow^{(+)}$ and $\psi_\downarrow^{(-)} \psi_\uparrow^{(+)}$, superpose to make up one quantum state as the final state of this decay process.

The entangled states

We now measure the spin orientations of the $e^+ e^-$ pair. Let us concentrate on the electron spin. There is a 50% chance of finding the electron's spin being up and 50% chance down. But once the electron spin is measured, say finding it to be spin up, we are 100% sure that the positron spin must be down—this is so no matter how far away the positron has traveled: to the other side of the lab bench, or to the other side of the galaxy. One can perform such a measurement repeatedly and the spin orientations of these widely separated particle pairs are always 100% correlated. Such a correlation[3] of the spin states is described as being "entangled". In quantum mechanics, these two particles are in one entangled quantum state; any change will affect both particles together instantaneously.

Entanglement as viewed by local realists To a local realist this entanglement by itself does necessarily represent a deep puzzle. We encounter this sort of total correlation often in our daily experience. Let's say Chris and Alex have two coins, one gold and the other silver. They also have two boxes, each holding one coin so that Chris carries one and Alex the other. After they departed from each other, Chris opens his box finding a silver coin. It does not matter how far he had traveled, he knows immediately that Alex has the gold one.

[2]The minus sign, which is irrelevant for our discussion, reflects the relative phases of Clebsch–Gordan coefficients in the addition of two spin-1/2 states to form a spin-zero angular momentum state.

[3]Mathematically, it has the feature that the probability of finding any particular combinations of electron and positron spin states is not a simple product of probabilities of finding electron and positron spin states separately.

Thus to a local realist, the perfect correlation of the e^+e^- spins simply means they were correlated **before** they were measured. Namely, since the beginning, the silver coin had been in the box Chris took and the gold coin in the box Alex had.

Copenhagen interpretation of entanglement   However the interpretation given by the Copenhagen school is very different. This orthodox view would say that, before the measurement, the electron and positron were not in definite spin states. Their quantum state is not just one or the other, but a superposition of these two possibilities: the electron is in both spin up and down states, while the positron also does not have definite spin, but always has its spin pointing in the opposite direction to the electron spin. Because the electron and positron are entangled, according to the orthodox interpretation, a measurement of the electron spin **compels** the positron, no matter how far away it had traveled, to jump into some definite spin state (opposite to that of the measured electron). Einstein found such an instantaneous effect so strange that he called it "spooky action-at-a-distance". This comes about because of the claim that the states do not have any definite attributes until they have been measured. It is the measurement here that compels the positron to jump into its spin state over there!

Local realist hidden-variable theories   To the local realists, the two particles always had some definite spin orientation, yet quantum mechanics can only predict it with some probability. This just means that quantum mechanics is an incomplete theory. The suggestion is made that there is a set of yet unknown variables; their specification in a more complete theory would then lead to definitive predictions. Such local realist theories are often referred to as "hidden-variable theories".

Hidden-variable theories can account for the quantum mechanical result in simple situations   In simple situations, for example, measuring the spin components in the same orientation or in two perpendicular orientations, hidden-variable theories can account for the quantum mechanical result Putting this in more quantitative terms, both hidden-variable theories and QM will find the average (as denoted by $\langle \ldots \rangle$) product value of electron and positron spins (in units of $\hbar/2$) in any particular direction, whether in the z direction or x direction, to be

$$\langle S_z^{(-)} S_z^{(+)} \rangle = \langle S_x^{(-)} S_x^{(+)} \rangle = -1. \tag{8.2}$$

The precise quantum mechanical calculation is presented in SuppMat Section 8.4, see Eq. (8.23). The same result can be understood in the framework of the local realist interpretation: as any pair of electrons and positrons is produced, its members have opposite spin immediately after their birth. This clearly holds as well when we average over all the measured values.

In the above we discussed only one spin orientation at a time. Now consider measuring spins in two independent directions, say the z and x directions. A quantum mechanical calculation [Eq. (8.24) below] shows that

$$\langle S_z^{(-)} S_x^{(+)} \rangle = 0. \qquad (8.3)$$

It is not difficult for hidden-variable theories to reproduce this result. For example, one can have a theory that allows independent spin orientations in the two perpendicular directions to have electron spin up and positron spin down, randomly (namely, equally likely). They all average out to zero.

8.3.1 The post-EPR era and Bell's inequality

Many (most?) working physicists took an agonistic viewpoint. Since these issues concern the interpretation of the situation before measurements, one can adopt a "shut-up-and-calculate" attitude. One just uses the Schrödinger equation to compute quantities of practical interest and ignores the "philosophical puzzles". Then came the surprise when John S. Bell (1928–90) published a paper (Bell 1964) showing that what were assumed for the situations "before the measurement" (e.g. whether $e^+ e^-$ spins were already correlated) actually have experimentally observable consequences (Bell's inequality). Since the 1980s a whole series of experimental results have demonstrated that the local realist viewpoint[4] is not supported by observation. The orthodox way of interpreting entanglement has gained ground. It is interesting to note that currently the researchers whose work has more bearing on these "philosophical issues" are actually the ones pursuing the very practical ends of constructing "quantum computers", for which QM entanglement is of paramount importance. One way or another, Einstein's thoughts on the deep meanings of quantum mechanics still exerts an influence on current investigations.

[4] Here we ignore the "many-worlds interpretation" of quantum mechanics, advocated by Hugh Everett. Some would argue this interpretation as the ultimate "realist" theory.

Bell's inequality derived

Basically, what Bell did was to extend the above discussion of spin values S_z and S_x to more than two directions. In such richer systems, one can deduce relations that can distinguish the local realist interpretation from that of QM, independent of the assumed forms of any hidden-variable theory.

Let us again consider the spin measurement of an electron and positron produced by a parent system having zero angular momentum. We can measure the spin in any direction perpendicular to the $e^+ e^-$ pair's motion (call it the $\hat{\mathbf{y}}$ direction); we broaden our consideration from just the $\hat{\mathbf{z}}$ and $\hat{\mathbf{x}}$ directions to three directions $(\hat{\mathbf{a}}, \hat{\mathbf{b}}, \hat{\mathbf{c}})$ in the x-z plane. According to the local realists, the particles must have definite spin values at all times: the electron's spin in the $\hat{\mathbf{a}}$ direction can take on value of $S_a^{(-)} = \pm 1$ (in units of $\hbar/2$), similarly $S_b^{(-)} = \pm 1$ and $S_c^{(-)} = \pm 1$. For notational simplicity we will write the electron spin values as $S_a^{(-)} \equiv E(a, \lambda) = \pm 1$, $S_b^{(-)} \equiv E(b, \lambda) = \pm 1$, and $S_c^{(-)} \equiv E(c, \lambda) = \pm 1$, respectively. We have explicitly displayed their dependence on the hidden variable λ. For positron spins, $S_a^{(+)}$, $S_b^{(+)}$, and $S_c^{(+)}$, we write $P(a, \lambda) = \pm 1$, $P(b, \lambda) = \pm 1$, and $P(c, \lambda) = \pm 1$, respectively. Since they must form a spin-zero system, we must have $E(a, \lambda) = -P(a, \lambda)$, $E(b, \lambda) = -P(b, \lambda)$, and $E(c, \lambda) = -P(c, \lambda)$.

Instead of presenting Bell's original derivation, we shall present one only involving simple arithmetic (d' Espagnat 1979). The electron spin in three directions $(\hat{\mathbf{a}}, \hat{\mathbf{b}}, \hat{\mathbf{c}})$ has $2^3 = 8$ possible configurations. Thus in the local

realist's approach we can think[5] of the following possible electron and positron spin configurations as soon as the particles are produced:

	E_a	E_b	E_c	$\longleftrightarrow$	P_a	P_b	P_c
N_1	+	+	+		−	−	−
N_2	+	+	−		−	−	+
N_3	+	−	+		−	+	−
N_4	+	−	−		−	+	+
N_5	−	+	+		+	−	−
N_6	−	+	−		+	−	+
N_7	−	−	+		+	+	−
N_8	−	−	−		+	+	+

$$(8.4)$$

N_i is the number of events having the spin configuration in the *i*th row. Thus there are N_1 events with $S_a^{(-)} = E(a, \lambda) = +1$, $E(b, \lambda) = +1$, and $E(c, \lambda) = +1$, etc. (The exact values of N_i are to be determined, hopefully, in some hidden-variable theory.) Because the electron and positron spins must be anti-aligned (in order to have zero total angular momentum), the positron spin configuration in the second group of columns must be exactly opposite to those in the electron column (first group): thus $P_a = -E_a$ and $P_b = -E_b$, etc.

The probability of having the *i*th row configuration is $p_i = N_i / \Sigma N$ with ΣN being the total number of events. Thus, according to the local realists (lr) approach, the average value of a spin product $\left\langle S_a^{(-)} S_b^{(+)} \right\rangle_{\text{lr}} \equiv \langle a, b \rangle$ is

$$\langle a, b \rangle = \sum_i (p_i) \, [E(a) \, P(b)]_i = \sum_i N_i \, [E(a) \, P(b)]_i \, / \Sigma N. \qquad (8.5)$$

From the table in (8.4) we see that the values of the spin products are $[E(a) \, P(b)]_{1,2,7,8} = -1$ and $[E(a) \, P(b)]_{3,4,5,6} = +1$ (with the row number being indicated by the subscript). This allows us to write out weighted sums such as (8.5) explicitly,

$$\langle a, b \rangle = (-N_1 - N_2 + N_3 + N_4 + N_5 + N_6 - N_7 - N_8) \, / \Sigma N.$$

Similarly, we can calculate the average product value for spins in the *a* and *c* direction:

$$\langle a, c \rangle = (-N_1 + N_2 - N_3 + N_4 + N_5 - N_6 + N_7 - N_8) \, / \Sigma N;$$

thus

$$\langle a, b \rangle - \langle a, c \rangle = 2 \, (-N_2 + N_3 + N_6 - N_7) \, / \Sigma N. \qquad (8.6)$$

We also have

$$\langle b, c \rangle = (-N_1 + N_2 + N_3 - N_4 - N_5 + N_6 + N_7 - N_8) \, / \Sigma N,$$

$$1 = (+N_1 + N_2 + N_3 + N_4 + N_5 + N_6 + N_7 + N_8) \, / \Sigma N,$$

or

$$1 + \langle b, c \rangle = 2 \, (N_2 + N_3 + N_6 + N_7) \, / \Sigma N. \qquad (8.7)$$

A comparison of (8.6) and (8.7) leads to **Bell's inequality**,

$$|\langle a, b \rangle - \langle a, c \rangle| \le [1 + \langle b, c \rangle], \tag{8.8}$$

or, in the explicit notation of spin products, the average values according to the local realists (lr) must obey

$$\left|\left\langle S_a^{(-)} S_b^{(+)} \right\rangle_{\text{lr}} - \left\langle S_a^{(-)} S_c^{(+)} \right\rangle_{\text{lr}}\right| \le \left[1 + \left\langle S_b^{(-)} S_c^{(+)} \right\rangle_{\text{lr}}\right]. \tag{8.9}$$

This result is independent of any assumption of the N_i values. The significance of Bell's inequality is that any realist hidden-variable theory must satisfy such a relation.

The quantum mechanical result

How does Bell's inequality compare to quantum mechanics result? The quantum mechanical (QM) result for the correlation of spin components in two general directions (with angle θ_{ab} between them) is calculated in (8.27) below:

$$\left\langle S_a^{(-)} S_b^{(+)} \right\rangle_{\text{QM}} = -\cos \theta_{ab}. \tag{8.10}$$

The two sides of Bell's inequality (8.9) have the quantum mechanical values

$$\text{LHS} = |-\cos \theta_{ab} + \cos \theta_{ac}| \tag{8.11}$$

and

$$\text{RHS} = 1 - \cos \theta_{bc}. \tag{8.12}$$

To see that Bell's inequality is incompatible with this QM result, consider, for example, the case of $(\hat{\mathbf{a}}, \hat{\mathbf{b}}, \hat{\mathbf{c}})$ with $\hat{\mathbf{a}}$ and $\hat{\mathbf{b}}$ being perpendicular, $\theta_{ab} = \pi/2$, and $\hat{\mathbf{c}}$ being 45° from $\hat{\mathbf{a}}$ and $\hat{\mathbf{b}}$: $\theta_{ac} = \pi/4$ and $\theta_{bc} = \pi/4$. Thus, the LHS (8.11) would be $|-\cos \frac{\pi}{2} + \cos \frac{\pi}{4}| = 1/\sqrt{2} \simeq 0.7$; and the RHS (8.12) would be $1 - \cos \frac{\pi}{4} \simeq 0.3$, which is clearly **not** greater than the LHS, as required by Bell's inequality. That is, in this situation, no matter what choice one makes of the N_i values, the hidden-variable theory will not be able to mimic the QM prediction.

8.3.2 Local reality vs. quantum mechanics—the experimental outcome

John Clauser (1942–) and his collaborators were the first ones to carry out, in 1972, an experimental test of Bell's inequality and found that the spooky prediction of QM do occur (Clauser and Shimony 1978). Alain Aspect (1947–) and his collaborator performed experiments (Aspect *et al.* 1981) and were able to show more convincingly that such entanglement connections also take effect instantaneously (at a speed faster than light speed), with results in agreement with QM. Thus the nonlocal feature, what Einstein termed the "spooky action-at-a-distance" effect, does seem to be a fundamental part of nature. Quantum entanglement seems to say that if you have a system composed of more than one particle, the individual particles are actually not individual. It leaves us with a rather strange picture of reality as it seems one is not allowed in principle to assign objective attributes, independent of actual measurements.

John Bell, when referring to the implications of Aspect's experiment, speaks for many when he said,

For me it's a dilemma. I think it's a deep dilemma, and the resolution of it will not be trivial; it will require a substantial change in the way we look at things.

An aside: Quantum computer

Nowadays, physicists do not regard the peculiarities of quantum systems as a problem, but rather an opportunity. A proper appreciation of the profound counter-intuitive properties of quantum multiparticle systems and the nature of entanglement allows the possibility of using this peculiar behavior for potential applications such as quantum computing. A quantum computer is a device that makes direct use of QM phenomena, such as superposition and entanglement, to perform operations on data. Quantum computing will be a revolutionary new form of computation.

Conventional computers manipulate **bits,** each of which can take on values of 0 or 1. Thus two bits can be in four states: 00, 01, 10, 11, and n bits encompass 2^n states. But a classical computer can only be in any one of these states sequentially.

Quantum computers manipulate quantum bits, called **qubits,** which are quantum states, i.e. a **superposition** of the classical states. Namely, a quantum computer can be in many of the classical states simultaneously. For example, using an electron spin (with spin up or down), a qubit can be in $|\downarrow\rangle \equiv |0\rangle$ and $|\uparrow\rangle \equiv |1\rangle$ states, as well as in the state of $a\,|0\rangle + b\,|1\rangle$; and a two-qubit system in $a\,|00\rangle + b\,|10\rangle + c\,|01\rangle + d\,|11\rangle$, etc. The complex numbers a, b, c, d are the relative phases and amplitudes within the superposition. Thus, while a classical computer acts on binary numbers stored in the input register to output another number, a quantum computer acts on the **whole superposition** in qubits of its input register, thus achieving enormous parallelism.

* * *

The 2005 reprint of Pais' Einstein biography (Pais 1982) includes a new Foreword by Roger Penrose (1931–). This short essay is, in this author's opinion, a particularly insightful appraisal of Einstein's scientific achievement. Relevant to our discussion of Einstein's view of quantum mechanics, Penrose has this to say:

*It must be said that some of Einstein's objections to quantum theory have not really stood the test of time—most notably it was "unreasonable" that the theory should possess strange **non-local** aspects (puzzling features Einstein correctly pointed out). Yet, his most fundamental criticism does, I believe, remain valid. This objection is that the theory seems not to present us with any fully **objective** picture of physical reality. Here, I would myself side with Einstein (and with certain other key figures in the development of the theory, notably Schrödinger and Dirac) in the belief that quantum theory is not yet complete.*

8.4 SuppMat: Quantum mechanical calculation of spin correlations

In quantum mechanics, a state $|\psi\rangle$ is a vector[6] in the Hilbert space. It can be expanded in term of a complete set of basis vectors $\{|i\rangle\}$:

$$|\psi\rangle = \sum_i \psi_i |i\rangle. \tag{8.13}$$

The basis vectors are usually taken to be the eigenvectors of some operator A (representing some observable): $A |i\rangle = a_i |i\rangle$, where a_i is a number (the eigenvalue). This means that if the system is in the state $|i\rangle$, a measurement of the observable A is certain to obtain the result of a_i. The coefficient of expansion $\psi_i = \langle \psi | i \rangle$ is interpreted as the probability amplitude. A measurement of A of the system in the general state of $|\psi\rangle$ will result in obtaining one, say a_j, of the possible eigenvalues $\{a_i\}$, with probability $p_j = |\psi_j|^2$. The familiar wavefunction $\psi(x)$ is simply the coefficient of expansion in the representation space having position eigenstates $\{|x\rangle\}$ as basis vectors. In this case Eq. (8.13) becomes

$$|\psi\rangle = \int dx \psi(x) |x\rangle. \tag{8.14}$$

The orthonormality condition of the basis vectors $\langle i | j \rangle = \delta_{ij}$ means we have $\langle j |A| i \rangle = \delta_{ij} a_i$. Thus the average value of an observable A:

$$\langle A \rangle = \sum_i p_i a_i = \sum_i |\psi_i|^2 a_i, \tag{8.15}$$

can be obtained efficiently by taking the expectation value of an operator (i.e. sandwich the operator between the bra and ket vectors of the state):

$$\langle A \rangle = \sum_{i,j} \psi_j^* \psi_i \delta_{ij} a_i = \sum_{i,j} \psi_j^* \psi_i \langle j |A| i \rangle = \langle \psi |A| \psi \rangle. \tag{8.16}$$

To reach the last expression we have used the expansion of Eq. (8.13).

8.4.1 Quantum mechanical calculation of spin average values

Spin states

Here we shall mostly deal with spin eigenstates $|s, m\rangle$, which are labeled by eigenvalues of the total spin $S^2 = S_x^2 + S_y^2 + S_z^2$ and one spin component, say S_z, respectively:

$$S^2 |s, m\rangle = s(s+1) \hbar^2 |s, m\rangle, \qquad S_z |s, m\rangle = m\hbar |s, m\rangle. \tag{8.17}$$

In our notation of setting $\frac{1}{2}\hbar \equiv 1$, we shall simply label the spin state by suppressing the s value and concentrate on S_z with $m = \pm\frac{1}{2}$:

$$S_z |S_z\pm\rangle = \pm |S_z\pm\rangle. \tag{8.18}$$

In the representation space spanned by the basis vectors $|S_z\pm\rangle$

$$|S_z+\rangle \equiv |\uparrow\rangle \doteq \begin{pmatrix} 1 \\ 0 \end{pmatrix}, \qquad |S_z-\rangle \equiv |\downarrow\rangle \doteq \begin{pmatrix} 0 \\ 1 \end{pmatrix}, \tag{8.19}$$

[6] A reader who is not familiar with Dirac notation may simply think of the "ket" vector $|\psi\rangle$ as a column vector, and the "bra" vector $\langle\phi|$ as a row vector. The inner product is represented by a bracket $\langle\phi| \psi\rangle$ which is the scalar resulting from the multiplication of a row and a column vector. Similarly, an operator is represented by a matrix, and the expectation value $\langle\psi |A| \psi\rangle$ as the multiplication of a row vector and a matrix, then with another column vector.

the spin operators $S_{x,y,z}$ are represented by the Pauli matrices $\sigma_{x,y,z}$

$$S_z \doteq \sigma_z = \begin{pmatrix} 1 & 0 \\ 0 & -1 \end{pmatrix}, \qquad S_x \doteq \sigma_x = \begin{pmatrix} 0 & 1 \\ 1 & 0 \end{pmatrix}, \qquad (8.20)$$

as can be checked by

$$S_z \, |\uparrow\rangle = |\uparrow\rangle \qquad \text{and} \qquad S_z \, |\downarrow\rangle = - \, |\downarrow\rangle . \qquad (8.21)$$

Spinless state resulting from adding two spin $\frac{1}{2}$ states

Adding the electron and positron spin operators $\mathbf{S}^{(-)} + \mathbf{S}^{(+)} = \mathbf{S}$ we can have total spin $S = 1$ or $S = 0$. Concentrating on the $S = 0$ state with its z component $S_z = S_z^{(-)} + S_z^{(+)}$ and spin value $m^{(-)} + m^{(+)} = M_s = 0$, the total spin state is labeled as $|S = 0, M_s = 0\rangle \equiv |0, 0\rangle$. This total spin-zero state is related to the individual electron/positron S_z eigenstates $\left|S_z^{(-)}+\right\rangle \left|S_z^{(+)}-\right\rangle \equiv \left|\uparrow^{(-)}\downarrow^{(+)}\right\rangle$, etc. as

$$|0, 0\rangle = \frac{1}{\sqrt{2}} \left(\left|\uparrow^{(-)}\downarrow^{(+)}\right\rangle - \left|\uparrow^{(-)}\downarrow^{(+)}\right\rangle \right) . \qquad (8.22)$$

This is an example of the expansion discussed in (8.13). Namely, the final state is a superposition of the electron/positron states with the z component of electron spin up and positron spin down and vice versa, with respective expansion coefficients $\pm 1/\sqrt{2}$. This is the same relation as (8.1) but expressed in terms of Dirac notation. Thus the probability of finding the state with the electron spin up and positron spin down is $1/2$ and the probability of finding the state with the electron spin down and positron spin up is also $1/2$.

8.4.2   Spin correlation in one direction

From this we can check that the quantum mechanical formalism yields the average value

$$\left\langle S_z^{(-)} S_z^{(+)} \right\rangle_{\text{QM}} = \left\langle 0, 0 \left| S_z^{(-)} S_z^{(+)} \right| 0, 0 \right\rangle = -1, \qquad (8.23)$$

showing that the electron–positron spins must point in the opposite directions. Because of (8.22), this involves calculating the type of terms such as

$$S_z^{(-)} S_z^{(+)} \left|\uparrow^{(-)}\downarrow^{(+)}\right\rangle = \left(S_z^{(-)} \left|\uparrow^{(-)}\right\rangle\right) \left(S_z^{(+)} \left|\downarrow^{(+)}\right\rangle\right) = - \left|\uparrow^{(-)}\downarrow^{(+)}\right\rangle ,$$

leading to $S_z^{(-)} S_z^{(+)} |0, 0\rangle = - |0, 0\rangle$ and the claimed result of (8.23) because of the normalization condition $\langle 0, 0| \, 0, 0 \rangle = 1$. This calculation expresses the fact that $S_z^{(-)}$ and $S_z^{(+)}$ are in different spin spaces so we can have $S_z^{(-)}$ act directly on $\left|\uparrow^{(-)}\right\rangle$ and $S_x^{(+)}$ act directly on $\left|\downarrow^{(+)}\right\rangle$ as in (8.21).

8.4.3   Spin correlation in two directions

Two directions that are perpendicular

We also expect

$$\langle S_z^{(-)}S_x^{(+)}\rangle_{QM} = \langle 0,0| \, S_z^{(-)}S_x^{(+)} \, |0,0\rangle = 0. \tag{8.24}$$

Namely, spin values in perpendicular directions are uncorrelated. In contrast to (8.21), the spin operator S_x flips the spin in the z direction:

$$S_x\,|\uparrow\rangle = |\downarrow\rangle \qquad \text{and} \qquad S_x\,|\downarrow\rangle = |\uparrow\rangle, \tag{8.25}$$

because, as the representations in (8.19) and (8.20) show,

$$\begin{pmatrix} 0 & 1 \\ 1 & 0 \end{pmatrix}\begin{pmatrix} 1 \\ 0 \end{pmatrix} = \begin{pmatrix} 0 \\ 1 \end{pmatrix}, \quad \begin{pmatrix} 0 & 1 \\ 1 & 0 \end{pmatrix}\begin{pmatrix} 0 \\ 1 \end{pmatrix} = \begin{pmatrix} 1 \\ 0 \end{pmatrix}.$$

A simple exercise shows that $\langle 0,0\,|S_x^{(-)}S_x^{(+)}|\,0,0\rangle = -1$. We next calculate the action of $S_z^{(-)}S_x^{(+)}$ on the state $|0,0\rangle$ of (8.22):

$$S_z^{(-)}S_x^{(+)}\,|0,0\rangle = S_z^{(-)}S_x^{(+)}\left(|\uparrow^{(-)}\downarrow^{(+)}\rangle - |\downarrow^{(-)}\uparrow^{(+)}\rangle\right)/\sqrt{2}$$

$$= \left(|\uparrow^{(-)}\uparrow^{(+)}\rangle + |\downarrow^{(-)}\downarrow^{(+)}\rangle\right)/\sqrt{2}. \tag{8.26}$$

When multiplied with the bra vector $\langle 0,0|$ of (8.22), the orthogonality conditions[7] such as $\langle\uparrow^{(-)}|\,\downarrow^{(-)}\rangle = \langle\downarrow^{(+)}|\,\uparrow^{(+)}\rangle = 0$ lead to $\langle 0,0|\,S_z^{(-)}S_x^{(+)}\,|0,0\rangle = 0$.

Two general directions

Here we calculate the quantum mechanical expectation value of a product of spins in two general directions, calling them $\hat{\mathbf{a}}$ and $\hat{\mathbf{b}}$. We are free to choose $\hat{\mathbf{a}} = \hat{\mathbf{z}}$ and $\hat{\mathbf{b}}$ in the x-z plane: $\hat{\mathbf{b}} = \cos\theta\hat{\mathbf{z}} + \sin\theta\hat{\mathbf{x}}$. Thus $S_a^{(-)} = S_z^{(-)}$ and $S_b^{(+)} = \cos\theta S_z^{(+)} + \sin\theta S_x^{(+)}$:

$$\left\langle 0,0\,\left|S_a^{(-)}S_b^{(+)}\right|\,0,0\right\rangle = \cos\theta\,\langle 0,0\,|S_z^{(-)}S_z^{(+)}|\,0,0\rangle + \sin\theta\,\langle 0,0|\,S_z^{(-)}S_x^{(+)}\,|0,0\rangle.$$

Knowing the results of (8.23) and (8.24), we immediately have

$$\left\langle 0,0\,\left|S_a^{(-)}S_b^{(+)}\right|\,0,0\right\rangle = \cos\theta(-1) + \sin\theta(0) = -\cos\theta. \tag{8.27}$$

[7] An even simpler way is to note that $|\uparrow^{(-)}\uparrow^{(+)}\rangle = |S=1, M_s=+1\rangle \equiv |1,+1\rangle$ and $|\downarrow^{(-)}\downarrow^{(+)}\rangle = |1,-1\rangle$; then the orthogonality condition is $\langle 0,0|\,1,+1\rangle = \langle 0,0|\,1,-1\rangle = 0$.

SPECIAL RELATIVITY

Part
III

Prelude to special relativity

- Relativity is explained as a coordinate symmetry. Special relativity means that physics is the same in all inertial frames; general relativity, in all coordinate frames.

- The equations of Newtonian mechanics are covariant under Galilean coordinate transformations. This is called Galilean (or Newtonian) relativity. Its absolute time leads to the familiar velocity addition rule: $\mathbf{u}' = \mathbf{u} - \mathbf{v}$ with $\mathbf{v}$ being the relative velocity between two inertial frames. Since Maxwell's equations predict an electromagnetic wave propagating with speed c making no reference to any coordinate frames, these equations do not have Galilean symmetry.

- It was universally accepted in the nineteenth century that Maxwell's equations were valid only in the rest-frame of the aether—the purported elastic medium for light propagation. The outstanding issue in physics was termed 'electrodynamics of a moving body'—the problem of matter moving in the aether-frame and its manifestation in electromagnetic phenomena.

- We discuss the observational result of the aberration of starlight and Fizeau's experimental measurement of the light speed in a moving medium, which was found to be in agreement with Fresnel's formula constructed under the hypothesis that the aether was partially entrained by a moving body.

- By the 1890s Lorentz presented a theory of aether–matter interaction which could account for stellar aberration, Fizeau's experiment, and Fresnel's drag. The key element in his theory is the introduction of the 'local time', which he took to be a mathematical construct that facilitated his proof: Maxwell's equations are unchanged, up to first order in v/c, under a Galilean transformation when augmented by this effective 'local time'. We present the calculational details of this 'theorem of corresponding states' of Lorentz in a SuppMat section.

- The Michelson–Morley experiment was capable of detecting an $O(v^2/c^2)$ deviation of light speed from the value c. Their famous null result forced Lorentz to introduce the idea of length contraction. The subsequent modification of the Galilean position transformation and local time led him to the Lorentz transformation, under which the

Maxwell equations remain unchanged (to all orders) for a body moving in the aether frame.

- Einstein was strongly motivated by the aesthetics of symmetry principles. He was seeking a physical description that would have the widest validity—in this case, the invariance principle of relativity. He motivated this in the opening paragraph of his 1905 paper by the magnet–conductor thought experiment.

- But if the relativity principle was valid for Maxwell's equations, this would apparently be in conflict with the constancy of light speed implied by these equations. After years of rumination on this subject, Einstein realized that the classical concept of time had no experimental foundation, and (when signal transmission is not instantaneous) the notion of simultaneity, and hence the measured time, must be coordinate-dependent. With this new insight he was able to work out the entire theory of special relativity in just five weeks.

- In Section 9.4.3, we give a summary of the influence that prior investigators in physics and philosophy had on Einstein in his formulation of the special theory of relativity.

9.1   Relativity as a coordinate symmetry

Relativity means that physically it is impossible to detect absolute motion. We are all familiar with the experience of sitting in an airplane, and not able to "feel" the speed of the plane when it moves with a constant velocity; or, in a train and observing another passing train on the next track, we may find it difficult to tell which train is actually in motion. This can be interpreted as saying that no physical measurement can detect the absolute motion of an inertial frame of reference. Hence the basic notion of relativity—only relative motion is measurable.

One can describe relativity as a "coordinate symmetry"—the laws of physics are unchanged under a coordinate transformation. Special relativity is the symmetry when we restrict our coordinate frames to inertial frames of reference; in general relativity we allow all coordinate frames including accelerated frames as well. Because the laws of physics are the same in different coordinate frames, there is no physical effect we can use to tell us which coordinate frame we are in. Hence absolute motion is not detectable.

9.1.1   Inertial frames of reference and Newtonian relativity

What is an inertial frame of reference? What is the transformation that takes us from one inertial frame to another?

Inertial frames of reference are the coordinate systems in which, according to Newton's first law of motion, a particle will, if no external force acts on it, continue its state of motion with constant velocity (including the state of rest). Galileo Galilei (1564–1642) and Isaac Newton (1643–1727) taught us that a physical description would be the simplest when given in these coordinate systems. The first law of motion provides us with the definition of an inertial system; its implicit message that such coordinate systems exist is its physical content. There are infinite sets of such frames, differing by their relative orientation, displacement, and relative motion with constant velocities. For simplicity we shall ignore the transformations of rotation and displacement of the coordinate origin, and concentrate on the relations among the rectilinear moving coordinates—the **boost** transformation. Consider a coordinate system $\{r_i'\}$ moving with a constant velocity v_i with respect to another coordinate frame $\{r_i\}$. The relation between these two inertial frames is given by the Galilean transformation (Fig. 9.1)

$$r_i' = r_i - v_i t. \tag{9.1}$$

Since time is assumed to be absolute $t' = t$, and the velocity of a particle measured in these two frames is $u_i = dr_i/dt$ and $u_i' = dr_i'/dt'$, respectively, we have the familiar **velocity addition rule**:

$$u_i' = u_i - v_i. \tag{9.2}$$

Because the relative velocity v_i is constant, a particle's acceleration remains the same with respect to these two frames $a_i' = a_i$. When we multiply this by the invariant mass $ma_i' = ma_i$, this means that force is also unchanged, $F_i' = F_i$. In this way we see, for example, that Newton's law of gravitation has the same equation-form in these two frames:

$$\mathbf{F} = G_N \frac{m_A m_B}{r_{AB}^3} (\mathbf{r}_A - \mathbf{r}_B) \quad \text{and} \quad \mathbf{F}' = G_N \frac{m_A m_B}{r_{AB}'^3} (\mathbf{r}_A' - \mathbf{r}_B')$$

where $r_{AB} = |\mathbf{r}_A - \mathbf{r}_B|$, etc. That the equations of Newtonian mechanics and gravitational theory are form-invariant (i.e. they are "covariant") under coordinate transformation among inertial frames is called **Newtonian relativity**.[1]

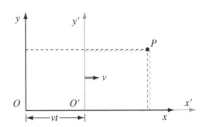

Fig. 9.1 The point P is located at (x, y) in the S system and at (x', y') in the S' system, which is moving with velocity v in the x direction. The coordinate transformations are $x' = x - vt$ and $y' = y$.

[1] It is also called *Galilean relativity* as it is the symmetry under Galilean transformation, first described qualitatively by Galileo.

9.2   Maxwell's equations

James Clerk Maxwell (1831–79) formulated a consistent theory that unified electricity, magnetism, and optics. The simplest way to see that Maxwell's equations are not covariant under Galilean transformation is through a discussion of electromagnetic waves. Writing down Maxwell's equations in the Heaviside–Lorentz unit system,[2] we have the magnetic Gauss' law and Faraday's law:

$$\nabla \cdot \mathbf{B} = 0, \qquad \nabla \times \mathbf{E} + \frac{1}{c} \partial_t \mathbf{B} = 0, \tag{9.3}$$

then the Coulomb–Gauss law and Ampere's law (with displacement current).

$$\nabla \cdot \mathbf{E} = \rho, \qquad \nabla \times \mathbf{B} - \frac{1}{c} \partial_t \mathbf{E} = \frac{1}{c} \mathbf{j}, \tag{9.4}$$

where ρ and $\mathbf{j}$ are, respectively, the charge density and current density.

[2] See Appendix A, Eq. (A.35) for a discussion.

9.2.1   The electromagnetic wave equation

According to Faraday's law, a changing magnetic field induces an electric field, while a changing electric field, which is the displacement current, induces, according to Ampere's law, a magnetic field. The mutual induction can bring about a sustained propagation of the electromagnetic field (a wave) in free space. We can easily demonstrate mathematically that Maxwell's equations in free space ($\rho = \mathbf{j} = 0$) imply a wave equation. We do this by first taking the time derivative of Faraday's equation

$$\nabla \times \partial_t \mathbf{E} + \frac{1}{c}\partial_t^2 \mathbf{B} = 0. \tag{9.5}$$

After applying Ampere's law $\partial_t \mathbf{E} = c\nabla \times \mathbf{B}$ to the first term and using the vector identity

$$(\nabla \times \nabla \times \mathbf{B})_i = \varepsilon_{ijk}\nabla_j(\nabla \times \mathbf{B})_k = \varepsilon_{ijk}\varepsilon_{klm}\nabla_j\nabla_l B_m = -\nabla^2 B_i, \tag{9.6}$$

where we have summed over repeated indices, and used the vector identity [cf. Eq. (A.11)] $\varepsilon_{ijk}\varepsilon_{klm} = \delta_{il}\delta_{jm} - \delta_{im}\delta_{jl}$ and the no-monopole law $\nabla_m B_m = 0$ of (9.3), we can then turn Eq. (9.5) into the wave equation for the magnetic field[3] in free space:

[3]It is obvious that we can derive exactly the same wave equation for the electric field.

$$\nabla^2 B_i - \frac{1}{c^2}\partial_t^2 B_i = 0. \tag{9.7}$$

The solution is that the field, which depends on position $\mathbf{r}$ and time t, is some function depending only on their combination $\mathbf{k} \cdot \mathbf{r} - \omega t$, with the magnitude k multiplied by the angular frequency ω equal to c. This solution is a wave propagating with a constant speed c—the same speed in any inertial frames of reference. Symbolically we can write this as an equality of the light speed c' measured in coordinate frame S' in relative motion with respect to another frame S in which the light speed is c:

$$c' = c. \tag{9.8}$$

The constancy of light velocity clearly violates the velocity addition rule of (9.2), which is a direct consequence of the Galilean transformation. This suggests that Maxwell's equations do not have Galilean symmetry.[4]

[4]One expects the violation to be $O(v/c)$. See Section 9.5.1 for some details.

The wave theory of light triumphed in the nineteenth century

In the seventeenth century, two rival theories were proposed to explain the propagation of light: the corpuscular theory of Newton and the wave theory of Christiaan Huygens (1629–95). However, by the mid-nineteenth century the wave theory became dominant because only this theory could explain the observed interference phenomena. Furthermore, after Maxwell's prediction of the electromagnetic wave with a propagation speed just matching that of the measured light speed, any doubt that light propagated as a wave was removed.

9.2.2   Aether as the medium for electromagnetic wave propagation

Because the only experience one had with waves in that era was with mechanical waves (the sound wave being the prototype), it was universally assumed that light waves would also require an elastic medium for their propagation. Thus the existence of a luminiferous aether was generally believed. Since light travels everywhere with high speed, one had to assume (again based on one's experience with mechanical waves) that this medium must be extremely stiff, yet it could permeate into all available spaces, large or small.

In the corpuscular theory of light one would expect its velocity (like the speeds of bullets) to be constant with respect to the source that emits the light particles. The conventional velocity addition rule (9.2) should apply. In wave theory the light propagation velocity should be independent of the motion of the source, but constant relative to its medium. The constancy of light speed predicted by Maxwell's wave theory seemed to imply that this theory of electromagnetism applies only in the rest-frame of aether.

"The electrodynamics of a moving body" was then an outstanding problem for nineteenth-century physicists, all subscribers of the wave theory. How should electromagnetism be described for sources and observers moving with respect to the aether-frame? In particular the solar system was pictured as filled with aether and the earth was viewed to be traveling in this aether-frame (experiencing an "aether wind"). Thus electromagnetic phenomena, for example the propagation of light on the moving earth, were expected to show deviations from Maxwell's description. In this debate, an issue that had been raised was whether the aether would be dragged along by the moving object (e.g. the earth). If this drag was complete, then even a moving observer would not detect any deviation from Maxwell's predictions. But a whole range of possibilities, from full to partial drag of aether, had been suggested.

9.3   Experiments and theories prior to special relativity

Here we will provide a summary of some of the relevant experimental and theoretical results that were likely known to Einstein when he formulated his new theory in 1905.

9.3.1   Stellar aberration and Fizeau's experiment

The effects of objects moving in the aether-frame that were under active discussion were stellar aberration and Fizeau's experiment. Einstein explicitly cited them as having had an influence on his thinking when formulating his theory of special relativity. Stellar aberration is the effect of earth's velocity (moving around the sun) on the apparent direction of light coming from a star. Fizeau's experiment involved the measurement of light velocity in moving water. Both phenomena were interpreted as having a bearing on the issue of whether the aether was being dragged along by a moving body.

Stellar aberration and the velocity addition rule

The phenomenon of stellar aberration was first reported in 1725 and later explained by James Bradley (1693–1762). When a star is observed, its apparent position is displaced from its true position because the velocity of light is finite and the resultant velocity is the vector sum of this light velocity and the velocity of the observer. As the earth moves around the sun, the star appears to trace out a circle.

For simplicity of discussion we will assume that the original direction of the starlight is vertically downward and this velocity addition results in the tilting of the incoming light direction.[5] The velocity addition can most easily be carried out in the rest-frame of the earth. In this system the star moves in a direction opposite to the original orbital direction of the earth. The tilting angle (Fig. 9.2) is clearly given by

$$\tan \alpha = \frac{v}{c}, \tag{9.9}$$

which is independent of the distance between the star and the observer.[6] Since the earth's orbital speed around the sun was known to be around 30 km/s, one predicts the aberration angle $\alpha = 20.5$ arc second, or an angular diameter of $2\alpha = 41''$ in agreement with observation.

[5] Recall our experience with this tilting effect when moving in the rain. Rain drops are seen to follow, instead of a vertically downward direction, oblique trajectories.

[6] One is not to confuse aberration with the phenomenon of stellar parallax, which involves the observation of relatively nearby astronomical objects. The distances for objects in our astronomical neighborhood can be deduced from such a change of position of the observer with respect to distant stars. Even for the nearest stars, the parallax angle is much smaller than the aberration angle.

Fresnel's drag coefficient and Fizeau's experiment

The explanation of stellar aberration posed no problem for the corpuscular theory of light. One simply applies the velocity addition rule of (9.2) with $u = c$. For the wave theory with its propagation medium of aether, there is the question whether the aether was dragged by or entrained within an object (e.g. the earth, the telescope lens, etc.). Stella aberration would seem to imply that the aether was dragged completely so that the light velocity remained the same in a moving medium. Still now one needs to understand the bending of a light ray in a moving medium. Augustin-Jean Fresnel (1788–1827) proposed a wave theory in which the aether was unaffected by the motion of earth, but partially entrained by another moving (v) medium, with an index of refraction n, so that the velocity of light in that medium V, instead of c/n, becomes

$$V = \frac{c}{n} + v\left(1 - \frac{1}{n^2}\right). \tag{9.10}$$

Fig. 9.2 Stellar aberration. The tilted light ray passes through the telescope (a) in the rest-frame of the sun, with the earth E in motion; (b) in the rest-frame of earth, with the star S in motion. (c) In a 12-month duration, the star is seen to trace out a circle with an angular diameter of 2α.

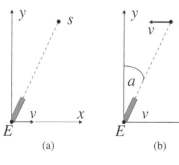
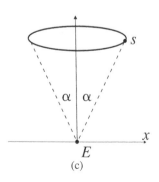

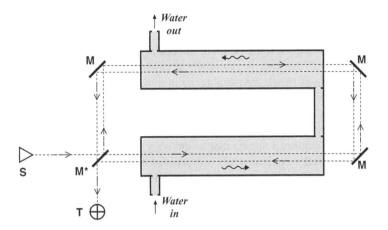

Fig. 9.3 Schematic view of Fizeau's experiment. Water flows through a ⊐ shaped vessel; the light emitted from S is split, at mirror M^*, into two beams: one always propagates along the direction of water flow, the other always against it. Both light rays are detected at T.

In other words, the velocity addition rule (9.2) was still to apply,[7] but the aether is partially dragged with a drag coefficient of $(1 - 1/n^2)$. Stellar aberration could be accommodated because air's index of refraction is very close to unity, and there is no modification of the light velocity.

While we may be less interested in the original justification of this formula by Fresnel, involving the behavior of aether inside a moving body, the important historical sequence of events was that this proposal was verified by Hippolyte Fizeau (1819–96) in an 1851 experiment. Fizeau arranged his apparatus so that the effect of moving water on light's propagation speed could be measured in an interferometer (Fig. 9.3). Remarkably, Fresnel's formula was verified. After Maxwell's theory become the dominant framework to understand electromagnetic waves, the question was then how could Fresnel's formula be derived in this theoretical framework?

[7] As we will see in Section 10.6.1, this actually was an early signal for the modification of the velocity addition rule of (9.2).

9.3.2 Lorentz's corresponding states and local time

One can summarize the above discussion of experiments and theories concerning the effect of a moving body on light propagation as follows: to first order in v/c, no deviation from Maxwell's prediction had been observed. The question was how could Fresnel's formula (which explained both stellar aberration and Fizeau's experiment) be justified by the aether/matter interaction in the framework of Maxwell's theory of electromagnetism.

Lorentz's program (1895)

Hendrik Lorentz (1853–1928) was one of the leading thinkers on electromagnetism[8] and electromagnetic (EM) interactions of matter (called electrons). To undertake this difficult theoretical investigation, he had to make a series of ad hoc assumptions in order to derive the electromagnetic fields resulting from the interaction of matter and aether in relative motion. One then checks that these "transformed fields" (namely, the fields as seen by an observer in motion with respect to the aether) still satisfied, at least up to $O(v/c)$, the same Maxwell equations so that the new wave equation in the

[8] Among Lorentz's contributions on electromagnetism is his proof that there are not four fields $(\mathbf{E}, \mathbf{B}, \mathbf{D}, \mathbf{H})$: electric, magnetic, displacement and magnetic strength, but two microscopic fields $(\mathbf{E}, \mathbf{B})$, which is of great historical importance.

moving frame would have a light speed not deviating from the value of c by more than this order.

Lorentz characterized his program as seeking "the theorem of corresponding states": The theorem is essentially a calculational tool that sets up the correspondence between the phenomena in moving systems and those in stationary systems by introducing transformed coordinates and fields. To every state (solution of Maxwell's equations) of an electromagnetic system in the aether-frame with coordinates $\mathbf{r}$, there exists a corresponding state (the transformed field mentioned above) that would obey a set of equations identical in form to Maxwell's equations when expressed in terms of the coordinates $\mathbf{r}'$ of a system moving with respect to the aether-frame.[9] If the form-invariance of the equations is not exact, the noncompliant terms should be small, at most of order v^2/c^2.

By the 1880s Lorentz made the remarkable discovery that his program could be accomplished more efficiently if one followed a prescription of introducing a mathematical quantity that he called the "local time" (it being a position-dependent quantity):

$$t' = t - \frac{v}{c^2}x. \tag{9.11}$$

Lorentz was still working in the framework of an absolute time. His local time was just a convenient mathematical device that summarizes a series of effects. Although it worked in the formalism like the kinematical time, it had nothing to do with the clock rate in a coordinate frame. But the part of the mathematics used in checking Lorentz's corresponding states as solutions of Maxwell-like equations in the moving frame is, formally speaking, the same as checking the covariance of Maxwell's equations under the coordinate transformation of observers in relative motion (say, in the x direction):

$$x' = (x - vt), \quad y' = y, \quad z' = z, \quad \text{and} \quad t' = t - \frac{v}{c^2}x. \tag{9.12}$$

We carry out this calculation in SuppMat Section 9.5.2, where we find that the transformed fields (i.e. the corresponding states in Lorentz's investigation)

$$\mathbf{E}' = \mathbf{E} + \frac{1}{c}\mathbf{v} \times \mathbf{B} \quad \text{and} \quad \mathbf{B}' = \mathbf{B} + \frac{1}{c}\mathbf{v} \times \mathbf{E}, \tag{9.13}$$

did obey Maxwell's equations, if terms of order $O(v^2/c^2)$ are dropped. This means that the light speeds in these two frames in relative motion differ at most by the order of v^2/c^2. This is an extremely small difference even for the speed of earth's motion in the solar system. In this way, Lorentz succeeded in explaining, in one theoretical framework, all the experimental and theoretical developments concerning issues related to light propagation. We shall refer to these results as (Lorentz 1895), by the publication date when Lorentz's work was presented in detail. In this paper, one finds another innovation by Lorentz: his assumption that the force on a charge q is given by a first-order equation

$$\mathbf{f} = e\left(\mathbf{E} + \frac{1}{c}\mathbf{v} \times \mathbf{B}\right), \tag{9.14}$$

which we now call the **Lorentz force law**.

[9] The coordinates $\mathbf{r}$ and $\mathbf{r}'$ are related by the usual Galilean transformation.

Lorentz's explanation of stellar aberration and Fresnel's formula

The explanation of stellar aberration and Fresnel's formula in (Lorentz 1895) actually would not involve the corresponding states, i.e. the transformed electromagnetic fields of (9.13). The spacetime transformation of (9.12) was all that is needed. Of course Lorentz never took his local time t' as physical time, but a convenient summary of a set of EM effects. However, in terms of the calculational steps, there is no difference; we would simply take Lorentz's local time to be the "transformed time coordinate". As we shall see, the local time was the key element in Lorentz's explanation.

Fresnel's drag coefficient derived   For simplicity we shall work in one spatial dimension of x. An electromagnetic wave can always be represented by a function of space and time $F(kx - \omega t)$ with k being the wavenumber and ω the angular frequency. The propagation speed is given by the ratio ω/k. The product $\phi = kx - \omega t$, which counts the peaks and troughs of the wave, is its phase. This counting could not be affected by the motion of the medium or observer; it must be an invariant:

$$\phi = kx - \omega t = k'x' - \omega't'. \tag{9.15}$$

Substituting in the coordinate transformation of (9.12), we have

$$k'x' - \omega't' = k'(x - vt) - \omega'\left(t - \frac{v}{c^2}x\right)$$
$$= k'\left(1 + \frac{v}{c^2}\frac{\omega'}{k'}\right)x - \omega'\left(1 + v\frac{k'}{\omega'}\right)t.$$

Phase invariance (9.15) implies

$$k = k'\left(1 + \frac{v}{c^2}\frac{\omega'}{k'}\right), \qquad \omega = \omega'\left(1 + v\frac{k'}{\omega'}\right), \tag{9.16}$$

or, taking their ratio,

$$\frac{\omega}{k} = \frac{\omega'}{k'}\frac{1 + v\frac{k'}{\omega'}}{1 + \frac{v}{c^2}\frac{\omega'}{k'}}. \tag{9.17}$$

Let the wave speed in the S' system be $\omega'/k' = c/n$; this relation, to first order in the velocity v, immediately leads to the propagation speed $\omega/k = V$ as given by Fresnel's formula (9.10). We recall that in the aether theory Fresnel's drag also explained both stellar aberration and Fizeau's experiment.

Aberration of light   We will not try to derive the aberration of light from Fizeau's drag coefficient, but rather to show that they both sprang from the same physics: Lorentz's local time. Since we will be concerned with directional change, the invariant phase of the waves in 3D space will be written out:

$$\phi = \mathbf{k} \cdot \mathbf{r} - \omega t = \mathbf{k}' \cdot \mathbf{r}' - \omega't'. \tag{9.18}$$

In particular from Fig. 9.4(a), we have a plane wave propagating in the x'–y' plane

$$\mathbf{k}' \cdot \mathbf{r}' - \omega't' = k'\cos\theta'x' + k'\sin\theta'y' - \omega't'.$$

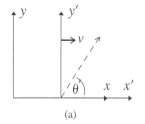

(a)

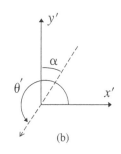

(b)

Fig. 9.4 (a) The light ray direction as defined in the S' system, which has a relative velocity v with respect to the S system. (b) The relation between the two separately defined tilting angles α and θ' is displayed.

Following essentially the same steps as the derivation of Fresnel's formula given above, we have

$$\mathbf{k}' \cdot \mathbf{r}' - \omega't' = k'\left(\cos\theta' + \frac{v}{c^2}\frac{\omega'}{k'}\right)x + k'\sin\theta'y - \omega'\left(1 + v\cos\theta'\frac{k'}{\omega'}\right)t$$

and obtain from (9.18)

$$k\cos\theta = k'\left(\cos\theta' + \frac{v}{c}\right) \quad \text{and} \quad k\sin\theta = k'\sin\theta' \tag{9.19}$$

as well as

$$\omega = \omega'\left(1 + \frac{v}{c}\cos\theta'\right). \tag{9.20}$$

Here we have used the approximation $\omega/k = \omega'/k' = c$ as we have already shown that, using the local time, Lorentz had demonstrated that Maxwell's equations, hence its wave equation, are unchanged if terms of order v^2/c^2 or higher are ignored. By taking the ratio of the two equations in (9.19) we obtain the change of propagation directions

$$\tan\theta = \frac{\sin\theta}{\cos\theta} = \frac{\sin\theta'}{\cos\theta' + \frac{v}{c}}, \tag{9.21}$$

which can be inverted ($v \to -v$) to read

$$\tan\theta' = \frac{\sin\theta}{\cos\theta - \frac{v}{c}}.$$

Thus for a vertically incident light ray $\theta = 3\pi/2$, we have $\tan\theta' = c/v$, or in terms of the tilting angle α defined in the previous section, cf. Fig. 9.4(b),

$$\tan\alpha = \cot\theta' = \frac{v}{c},$$

which is just the required result of (9.9).

We also note that this same derivation also yields the Doppler frequency shift effect as given in (9.20).

In short, at this point it seemed that Lorentz's 1895 theory could account for all the physics results about the electrodynamics of a moving body. But, in the meantime, experimental exploration continued its forward march, from the $O(v/c)$ to the $O(v^2/c^2)$ effects.

9.3.3   The Michelson–Morley experiment

Albert Michelson (1852–1931) in collaboration with E.W. Morley (1838–1923) succeeded in performing in 1887 an experiment showing that the speed of light has no measurable dependence on coordinate frames. From this set of observations, one could conclude that the deviation must be **less** than the order of v^2/c^2. The interferometer set-up involves two light rays with one traveling along the direction of earth's motion around the sun, and the other being in the perpendicular direction (Fig. 9.5).

Using the velocity addition rule of (9.2), the light velocity would be different when traveling along a trajectory aligned or anti-aligned with the aether wind

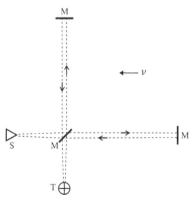

Fig. 9.5 A schematic version of the Michelson–Morley interferometer.

(i.e. the motion of the observer in the aether-frame), with the respective speeds being $c - v$ and $c + v$. The time taken by a light pulse to cover a round trip over a length l in the longitudinal direction would then be

$$\Delta t_{\shortparallel} = \frac{l}{c - v} + \frac{l}{c + v} = \gamma^2 \frac{2l}{c}, \tag{9.22}$$

where we have used the shorthand (commonly referred to as the "Lorentz factor")

$$\gamma \equiv \sqrt{\frac{1}{1 - \left(\frac{v}{c}\right)^2}}. \tag{9.23}$$

On the other hand, one can work out the round trip time $\Delta t_\perp$ to cover the same length perpendicular to the aether wind by solving the equivalent problem in one coordinate frame where the light travels along the diagonal length of a rectangular triangle with two perpendicular lengths being l and $v \triangle t_\perp / 2$ (Fig. 9.6). The Pythagorean relation yields $(c \triangle t_\perp/2)^2 = l^2 + (v \triangle t_\perp/2)^2$, which has the solution of

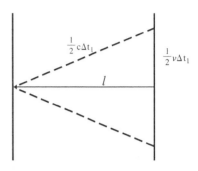

Fig. 9.6 Light traverses a length l as viewed by an observer in motion perpendicular to the light propagation direction.

$$\Delta t_\perp = \gamma \frac{2l}{c}. \tag{9.24}$$

Thus we expect the time intervals for light traveling in the parallel and in the perpendicular directions to differ by a factor of γ. Namely, their difference would be tiny, being $O(v^2/c^2)$,

$$\Delta t_{\shortparallel} - \Delta t_\perp = \{\gamma^2 - \gamma\} \frac{2l}{c} \simeq \left\{ \left[1 + \left(\frac{v}{c}\right)^2 \right] - \left[1 + \frac{1}{2}\left(\frac{v}{c}\right)^2 \right] \right\} \frac{2l}{c} = \left(\frac{v}{c}\right)^2 \frac{l}{c},$$

where we have used the expansion of $(1 - \epsilon)^{-\frac{1}{2}} = 1 + \epsilon/2 + \cdots$ etc. for both the γ^2 and the γ factors, and have discarded the higher order terms in v/c.

The interferometer of Michelson and Morley was capable of detecting such a small time difference by observing the interference pattern when these light beams were brought together after having travelled in their respective directions. Furthermore, it was predicted that a rotation of the interferometer by 90° should bring about a fringe shift of this v^2/c^2 order.

We can summarize the situation as follows: Newtonian relativity holds for laws of mechanics. But one expected it to fail in optical and electromagnetic phenomena. Lorentz constructed a theory of matter/aether interaction so that the relativity principle still holds to the order of v/c. But his theory predicted that it should fail at second order, leading to an effect that the Michelson–Morley experiment should be able to detect. Their famous null result was the strongest experimental evidence that light propagates at the same speed in all inertial frames of reference. Related to this, Fresnel's partially dragged aether could not accommodate this Michelson–Morley result.

9.3.4 Length contraction and the Lorentz transformation

In order to explain the null result of Michelson and Morley, Lorentz, and independently G.F. Fitzgerald (1851–1901) although only in qualitative terms, proposed in 1892 that the length of any object (under some yet to be understood molecular forces) would contract along the direction of relative motion with respect to the aether-frame, by a factor of γ,

$$l_{\shortparallel} = \frac{l'_{\shortparallel}}{\gamma} \tag{9.25}$$

where l' is the length measured in the aether rest-frame. The travel time in the longitudinal direction (9.22) would then be reduced by this factor, leading to the equality $\Delta t_{\shortparallel} = \Delta t_{\perp}$ and the no-fringe shift result of Michelson and Morley (Lorentz 1895).

To implement this length contraction in his theorem of corresponding states, Lorentz (starting in the late 1890s) inserted a factor γ in the longitudinal coordinate relation in Eq. (9.12):

$$x' = \gamma(x - vt), \qquad y' = y, \quad z' = z. \tag{9.26}$$

But for self-consistency[10] another factor γ was also required for the local time expression:

$$t' = \gamma\left(t - \frac{v}{c^2}x\right). \tag{9.27}$$

As discussed in SuppMat Section 9.5, other γ factors must also be inserted into the field transformation of (9.13):

$$\mathbf{E}'_{\shortparallel} = \mathbf{E}_{\shortparallel}, \quad \text{and} \quad \mathbf{E}'_{\perp} = \gamma\left[\mathbf{E}_{\perp} + \frac{1}{c}(\mathbf{v} \times \mathbf{B})_{\perp}\right] \tag{9.28}$$

$$\mathbf{B}'_{\shortparallel} = \mathbf{B}_{\shortparallel}, \quad \text{and} \quad \mathbf{B}'_{\perp} = \gamma\left[\mathbf{B}_{\perp} - \frac{1}{c}(\mathbf{v} \times \mathbf{E})_{\perp}\right].$$

Collectively the transformations displayed in (9.26), (9.27), and (9.28) are known as the **Lorentz transformation**. Lorentz then obtained the astonishing result (finally published in 1904) that the covariance of Maxwell's equations is exact—not just to the next order v^2/c^2, but to all orders!

Still, it should be emphasized that Lorentz always viewed results such as Eq. (9.28) as dynamical consequences of the matter–aether interaction. These results and the related Lorentz transformation are valid only for, and carry no implication beyond, electromagnetism.

9.3.5   Poincaré and special relativity

Any discussion of the history of special relativity would be incomplete without mention of the contribution and insight of Henri Poincaré (1854–1912). He was among the first to stress the fundamental importance of the principle of relativity—beyond any particular theories used to explain the nonobservability of the aether medium. He was the first one to give a physical interpretation to Lorentz's local time. In this connection he stressed the coordinate-dependent nature of time, and suggested the possibility of clock synchronization by light signal exchanges. In this way he showed the possibility of interpreting the local time as the time measured in a moving frame. He completed the proof of Lorentz covariance of Maxwell's equations by extending it to the case with sources, and named the transformation rule the "Lorentz transformation".

[10]The demonstration of this requirement proceeds as follows. Assume the modification has the general form of $ct' = Act - B\frac{v}{c}x$, where (A, B) are even functions of the constant velocity v. The result of $A = B = \gamma$ follows immediately when we require that the inverse transformation be obtained when the relative velocity is reversed $v \to -v$.

However, he never gave up on the reality of aether. His formulation of the relativity principle still distinguished observers at rest in the aether-frame and those in motion. Over all, he never formulated a coherent theory that completely discarded the notion of aether and absolute time.

9.4 Reconstructing Einstein's motivation

Einstein thought hard about Maxwell's theory of electromagnetism for many years[11] before he published his special relativity paper in 1905. It was a self-contained, and complete, theory on the new kinematics for all physics. Since there was no prior publication by Einstein on this subject, it has always been a difficult task for historians to reconstruct his motivation. Nevertheless, one can get a general idea from some remarks made by Einstein in later years, and by a careful reading of the 1905 paper itself. Our presentation will be brief; for readers interested in more thorough accounts, I recommend (Norton 2004, Pais 1982, and Stachel 1995).

Looking over Einstein's oeuvre, it is quite clear that his motivation for a new investigation was seldom related to any desire to account for this or that experimental puzzle. His approach often started with fundamental principles that would have the widest validity in the field, and had always been motivated by a keen sense of aesthetics in physics.[12]

We recall his introductory remarks in his light quantum paper (Einstein 1905d) as reported in Chapter 4: He was dissatisfied with the asymmetric ways matter and radiation were described: one as composed of particles and the other as waves. He was seeking a more unified theory. Here he was concerned that electromagnetism seemed to single out a particular frame of reference—the aether-frame. He was looking for a more symmetric description, valid for all frames. The principle of relativity would give us the same physics in all inertial coordinate systems. And later on, in his general theory of relativity, he would be seeking the same description in all frames.

9.4.1 The magnet and conductor thought experiment

Einstein began his 1905 paper by the consideration of a simple electromagnetic situation: one magnet and one conductor in relative motion. In the framework of aether it gave rise to very asymmetric descriptions, depending how one ascribes the motion with respect to the aether.

Case I The magnet is at rest (i.e. the rest-frame of the magnet coincides with the aether-frame; call it the *S* coordinate system) giving rise to a magnetic field **B**. The conductor (having free charges q) is in motion, with velocity **v** (Fig. 9.7). A charge moving with the conductor feels the Lorentz force (9.14) per unit charge as:

$$\frac{\mathbf{f}}{e} = \frac{1}{c}\mathbf{v} \times \mathbf{B}. \tag{9.29}$$

[11]When Einstein was asked by R.S. Shankland in the early 1950s how long he had worked on special relativity, he replied that he had started at age 16 and worked for 10 years (Shankland, 1963).

[12]As an aside, I mention that Einstein admired Albert Michelson greatly. As the highest form of praise, he called Michelson "a great artist in physics". Shankland (1963), reporting on his discussion of the Michelson–Morley experiment with Einstein, made the following comments: "Despite having had little mathematics or theoretical training, yet Michelson carried out such a successful experimental program that probed the fundamental issues in physics . . . This he [Einstein] feels was in large measure due to Michelson's artistic sense and approach to science, especially his feeling for symmetry and form. Einstein smiled with pleasure as he recalled Michelson's artistic nature—here there was a kindred bond."

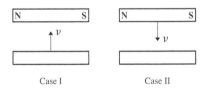

Case I   Case II

Fig. 9.7 A magnet and a conductor in relative motion.

Case II The conductor is at rest (call it the S' coordinate system). The magnet is in motion, giving rise to a changing magnetic field (Fig. 9.7). This changing magnetic field, according to Faraday's law, will induce an electric field everywhere:

$$\nabla' \times \mathbf{E} = -\frac{1}{c}\partial'_t\mathbf{B}. \tag{9.30}$$

[13]For the relative velocity $(-v)$ in the x direction, the derivative relation $\partial'_t = \partial_t + v\partial_x$ follows from the position transformation $x' = x + vt$ just as it is shown in Eq. (9.43). Also, without a nontrivial local time, the spatial derivatives are simply unchanged, $\nabla' = \nabla$.

Translating this into the aether frame with[13] $\partial'_t = \partial_t + \mathbf{v}\cdot\nabla$, the RHS of the above equation becomes $-(\mathbf{v}\cdot\nabla)\mathbf{B}/c$ as we have $\partial_t\mathbf{B} = 0$ because the magnet is at rest in the S frame. Using a vector identity, similar to (9.6), we can expand this RHS further as

$$(\mathbf{v}\cdot\nabla)\mathbf{B} = \mathbf{v}(\nabla\cdot\mathbf{B}) - \nabla\times(\mathbf{v}\times\mathbf{B}) = -\nabla\times(\mathbf{v}\times\mathbf{B}).$$

After noting the Galilean derivative relation $\nabla' = \nabla$, we have from (9.30)

$$\nabla\times\mathbf{E} = \frac{1}{c}\nabla\times(\mathbf{v}\times\mathbf{B}).$$

[14]Formally the equality $\mathbf{E} = \frac{1}{c}\mathbf{v}\times\mathbf{B}$ may be marred by the presence of a possible gradient factor ∇g. However in order to detect the electric force, one has to use a conductor in the form of a loop. Any possible gradient term will not contribute to the loop line-integral of the electromotive force: $\oint(\nabla g)\cdot d\mathbf{l} = \oint dg = 0$.

This allows us to make the identification for the induced electric field.[14] This produces an electric force $\mathbf{f} = e\mathbf{E}$. This result is identical to that in (9.29) of Case I:

$$\mathbf{E} = \frac{1}{c}\mathbf{v}\times\mathbf{B} = \frac{\mathbf{f}}{e}. \tag{9.31}$$

Since Case I and Case II have the same relative motion leading to the same electric force, Einstein pointed out that we should naturally ascribe the physical consequence to the relative motion of the magnet and the conductor. Yet an aether theory would depict them as having very different origins, depending whether the descriptions were given in the aether-frame or not. Situations such as this had compelled him to postulate the principle of relativity—not just for mechanics but for all physics, including electromagnetism. Given the difficulties of implementing Galilean symmetry in Maxwell's equations, what one has to do is to change, not the principle of relativity, but the relations among inertial frames; they are not given correctly by the Galilean transformations.

When reading Einstein's paper, the reader must keep firmly in mind that Einstein was writing for a community of physicists who were deeply inculcated in the aether theoretical framework. A present-day reader may well read the magnet and conductor thought experiment and wonder why Einstein would discuss this example in the first place. Being now imbued in the relativity frame of mind, we now have the habit of viewing physics in the relativity way and forget that here is Einstein, advocating this relativity approach for the first time.

In the same way a present-day student may be puzzled by the title of Einstein's paper: *On the electrodynamics of moving bodies*. Why was he talking about "moving bodies"? Where are the moving bodies in the paper? Again, this aspect of the paper makes sense only when one remembers that Einstein was using the standard name for a set of problems in the aether theory. There are no "moving bodies" in his paper because he was precisely advocating that the whole concept of aether should be abolished, and there was no need to discuss moving bodies in the aether-frame.

9.4.2 From "no absolute time" to the complete theory in five weeks

The above discussion indicated that Einstein was looking for ways to make electrodynamics compatible with the principle of relativity. But this goal appeared to be in direct conflict with the fact that Maxwell–Lorentz theory appeared to imply the existence of a privileged inertial frame (the aether-frame). In fact in the many years of thinking about this problem, Einstein toyed with the possibility of an electromagnetic theory based on some modified version of the Maxwell–Lorentz theory. After a long struggle with this conflict, suddenly the solution came to him. This is a well-known story. Here is how Einstein recounted it in his Kyoto Address *How I created the theory of relativity* (Einstein 1922):

Why are these two things inconsistent with each other? I felt that I was facing an extremely difficult problem. I suspected that Lorentz's ideas had to be modified somehow, but spent almost a year on fruitless thoughts. And I felt that was a puzzle not to be easily solved.

But a friend[15] of mine living in Bern helped me by chance. One beautiful day, I visited him and said to him: "I presently have a problem that I have been totally unable to solve. Today I have brought this 'struggle' with me." We then had extensive discussions, and suddenly I realized the solution. The very next day, I visited him again and immediately said to him: "Thanks to you, I have completely solved my problem."

[15] Michele Besso (1873–1955).

My solution actually concerned the concept of time. Namely, time cannot be absolutely defined by itself, and there is an unbreakable connection between time and signal velocity. Using this idea, I could now resolve the great difficulty that I previously felt.

After I had this inspiration it took only five weeks to complete what is now known as the special theory of relativity.

The key realization was based on the notion that concepts in physics, even abstract ones, must ultimately be based on experience. Whatever definition one uses, there must be a way to falsify it. In this way, he realized the absolute time was an abstract notion that could not be tested by experiment. Any realistic definition would involve the notion of simultaneity. Yet, if the signal transmission cannot be instantaneous, any definition of two simultaneous events taking place at two separate locations would be coordinate-frame dependent. If they are simultaneous to one observer, they would not be so to another in motion with respect to the first observer. Consequently, the time measurement would be coordinate-frame dependent too. Here is a way that Lorentz's local time can be realized as a physical transformation. Thus the realization of the principle of relativity would involve a new kinematics. There would be consequences for physics far beyond electromagnetism.

We will reiterate: the key idea of special relativity is the new conception of time. Its realization by Einstein was what broke up the "log-jam" that had obstructed his effort to make sense of Maxwell's electromagnetism. As we shall see, just about all the novel effects of special relativity can be traced back to this fundamental idea about time.

9.4.3   Influence of prior investigators in physics and philosophy

Lorentz

Einstein admired Lorentz greatly. He had certainly read Lorentz's 1895 paper, in which a dynamical theory of matter/aether interaction was used to show that Maxwell's equation was unchanged up to $O(v/c)$. All indications are that he was not aware of Lorentz's subsequent work, in particular the 1904 paper where the full Lorentz transformation was worked out.[16]

Michelson and Morley

The question that has repeatedly been raised is how much the Michelson–Morley experiment had influenced Einstein. On this, Einstein did not provide us with any definitive answer. But his remarks made in later years indicate that while he was aware of the Michelson–Morley result, it did not have a major impact on his thinking. He viewed this experiment mainly as another validation of the relativity principle, which he already believed anyway. Namely, the constancy of light speed was to him a manifestation of relativity in electromagnetism.

Stellar aberration and Fizeau's experiment

Most intriguingly Einstein on a number of occasions mentioned the importance of stellar aberration and Fizeau's experiment for the formulation of relativity theory. Why were these works important to Einstein? A plausible explanation[17] is that these effects, being directly related to Lorentz's local time (as our discussion in Section 9.3.2 has shown), lent important experimental support to what Einstein needed in proposing a coordinate-dependent time. Keep in mind that special relativity is all about time, and here was the experimental evidence for his new kinematics.[18]

[17]I learnt of this suggestion from Norton, (2004).

[18]See the discussion in Section 10.1. A simple and elegant resolution of stellar aberration and Fresnel's formula (hence Fizeau's experiment) by special relativity will be presented in SuppMat Sections 10.6.1 and 10.6.2.

The philosophical influence on Einstein

In the creation of special relativity, Einstein acknowledged the influence of David Hume (1711–76) and Ernst Mach (1838–1916). In his *Autobiographical Notes* (Einstein 1949) he wrote:

Today everyone knows, of course, that all attempts to clarify this paradox [of the constancy of light speed and the principle of relativity] satisfactorily were condemned to failure as long as the axiom of the absolute character of time, or of simultaneity, was rooted unrecognized in the unconscious. To recognize clearly this axiom and the arbitrary character already implies the essentials for the solution of the problem. The type of critical reasoning required for the discovery of this central point was decisively furthered, in my case, especially by the reading of David Hunme's and Ernst Mach's philosophical writings.

In other words, these philosophical readings helped to liberate Einstein's thoughts from the straightjacket of the classical concept of space and time.

9.5   SuppMat: Lorentz transformation *à la* Lorentz

In this SuppMat section, the transformation properties of the free Maxwell's equations (A.35) are worked out in some detail. In the shorthand notation for the derivatives:

$$\partial_t \equiv \frac{\partial}{\partial t}, \quad \partial_t' \equiv \frac{\partial}{\partial t'}, \quad \partial_x \equiv \frac{\partial}{\partial x}, \quad \partial_x' \equiv \frac{\partial}{\partial x'}, \quad \text{etc.,} \tag{9.32}$$

we first display these equations in terms of their components:

Faraday's law:

$$\partial_y E_z - \partial_z E_y = -\frac{1}{c}\partial_t B_x \tag{9.33}$$

$$\partial_z E_x - \partial_x E_z = -\frac{1}{c}\partial_t B_y \tag{9.34}$$

$$\partial_x E_y - \partial_y E_x = -\frac{1}{c}\partial_t B_z \tag{9.35}$$

Ampere's law with displacement current:

$$\partial_y B_z - \partial_z B_y = \frac{1}{c}\partial_t E_x \tag{9.36}$$

$$\partial_z B_x - \partial_x B_z = \frac{1}{c}\partial_t E_y \tag{9.37}$$

$$\partial_x B_y - \partial_y B_x = \frac{1}{c}\partial_t E_z \tag{9.38}$$

Gauss' law for the electric field

$$\partial_x E_x + \partial_y E_y + \partial_z E_z = 0 \tag{9.39}$$

Gauss' law for the magnetic field

$$\partial_x B_x + \partial_y B_y + \partial_z B_z = 0. \tag{9.40}$$

The source-free Maxwell's equations have the symmetry (between electricity and magnetism) under the "duality transformations":

$$E_i \to B_i \quad \text{and} \quad B_i \to -E_i. \tag{9.41}$$

Think of them as counterclockwise 90° rotations in planes spanned by the (E_i, B_i) axes.

9.5.1   Maxwell's equations are not Galilean covariant

Under the Galilean transformation (9.1) the two coordinate systems S and S' in relative motion (with constant velocity v along the x axis) are related by

$$x' = x - vt, \qquad y' = y, \qquad z' = z, \qquad \text{and} \quad t' = t. \tag{9.42}$$

We can find the relation among the derivatives by the chain rule of partial differentiation. As the time coordinate is in principle a function of the transformed coordinates (t', x', y', z'), we have

$$\frac{\partial}{\partial t} = \frac{\partial}{\partial t'}\frac{\partial t'}{\partial t} + \frac{\partial}{\partial x'}\frac{\partial x'}{\partial t} + \frac{\partial}{\partial y'}\frac{\partial y'}{\partial t} + \frac{\partial}{\partial z'}\frac{\partial z'}{\partial t} = \frac{\partial}{\partial t'} - v\frac{\partial}{\partial x'} \qquad (9.43)$$

while the spatial coordinate derivatives are unchanged

$$\frac{\partial}{\partial x} = \frac{\partial}{\partial x'}, \qquad \frac{\partial}{\partial y} = \frac{\partial}{\partial y'}, \qquad \frac{\partial}{\partial z} = \frac{\partial}{\partial z'}.$$

It is then clear that the only change of Maxwell's equations will be terms such as the RHS of (9.33)

$$-\frac{1}{c}\partial_t B_x = -\frac{1}{c}\partial'_t B_x + \frac{v}{c}\partial'_x B_x.$$

Namely, Maxwell's equations have noncovariant terms of $O(v/c)$.

9.5.2   Lorentz's local time and noncovariance at $O(v^2/c^2)$

As discussed in Section 9.3.2, Lorentz was able to find in the 1880s that the corresponding states (i.e. transformed fields) resulting from objects moving with respect to the aether-frame do satisfy Maxwell's equations—if one drops the $O(v^2/c^2)$ noncompliant terms. It was a very complicated dynamical theory in order to obtain these new fields (called corresponding states). Namely in this theory not only the space coordinates change, so do the electromagnetic fields. Nevertheless, the involved mathematics in checking their compatibility with Maxwell's equations is identical to that when checking the covariance by the transformed fields as seen by a moving observer. Since most of us are more familiar with the latter language, this is how we will phrase the following discussion.

The key to Lorentz' success is his discovery of the local time (9.11). We shall treat it as physical time being measured in the new coordinate, even though Lorentz just thought of it as some mathematical quantity that summarized a series of dynamical effects. In this manner one considers the coordinate changes of

$$x' = x - vt, \qquad y' = y, \qquad z' = z, \qquad \text{and} \quad t' = t - \frac{v}{c^2}x \qquad (9.44)$$

leading to derivative changes in a way entirely similar to (9.43)

$$\frac{\partial}{\partial x} = \frac{\partial}{\partial t'}\frac{\partial t'}{\partial x} + \frac{\partial}{\partial x'}\frac{\partial x'}{\partial x} + \frac{\partial}{\partial y'}\frac{\partial y'}{\partial x} + \frac{\partial}{\partial z'}\frac{\partial z'}{\partial x} = \frac{\partial}{\partial x'} - \frac{v}{c^2}\frac{\partial}{\partial t'}$$

or we have in general the derivative relations

$$\partial_x = \partial'_x - \frac{v}{c^2}\partial'_t, \qquad \partial_y = \partial'_y, \qquad \partial_z = \partial'_z, \qquad \text{and} \quad \partial_t = \partial'_t - v\partial'_x. \qquad (9.45)$$

We now start checking the covariance of Maxwell's equations by replacing all the derivatives in Eqs. (9.33)–(9.40). The two Gauss's laws (9.39) and (9.40) lead to

$$\partial'_x E_x + \partial'_y E_y + \partial'_z E_z = \frac{v}{c^2} \partial'_t E_x \tag{9.46}$$

$$\partial'_x B_x + \partial'_y B_y + \partial'_z B_z = \frac{v}{c^2} \partial'_t B_x. \tag{9.47}$$

One of the Faraday equations, (9.33), becomes

$$\partial'_y E_z - \partial'_z E_y = -\frac{1}{c} \partial'_t B_x + \frac{v}{c} \partial'_x B_x, \tag{9.48}$$

which upon applying (9.47) for the $\partial'_x B_x$ term becomes

$$\partial'_y \left(E_z + \frac{v}{c} B_y \right) - \partial'_z \left(E_y - \frac{v}{c} B_z \right) = -\frac{1}{c} \partial'_t B_x \left(1 - \frac{v^2}{c^2} \right). \tag{9.49}$$

In exactly the same manner, the other two components of the Faraday law become

$$\partial'_z E_x - \partial'_x \left(E_z + \frac{v}{c} B_y \right) = -\frac{1}{c} \partial'_t \left(B_y + \frac{v}{c} E_z \right) \tag{9.50}$$

$$\partial'_x \left(E_y - \frac{v}{c} B_z \right) - \partial'_y E_x = -\frac{1}{c} \partial'_t \left(B_z - \frac{v}{c} E_y \right). \tag{9.51}$$

These equations, through the duality transformation of (9.41), lead to the transformed Ampere's law:

$$\partial'_y \left(B_z - \frac{v}{c} E_y \right) - \partial'_z \left(B_y + \frac{v}{c} E_z \right) = \frac{1}{c} \partial'_t E_x \left(1 - \frac{v^2}{c^2} \right) \tag{9.52}$$

$$\partial'_z B_x - \partial'_x \left(B_z - \frac{v}{c} E_y \right) = \frac{1}{c} \partial'_t \left(E_y - \frac{v}{c} B_z \right) \tag{9.53}$$

$$\partial'_x \left(B_y + \frac{v}{c} E_z \right) - \partial'_y B_x = \frac{1}{c} \partial'_t \left(E_z + \frac{v}{c} B_y \right). \tag{9.54}$$

Furthermore, from (9.48) we obtain

$$\partial'_y B_z - \partial'_z B_y = \frac{1}{c} \partial'_t E_x - \frac{v}{c} \partial'_x E_x. \tag{9.55}$$

A comparison of these six equations (9.49)–(9.54) with the original Maxwell's Eqs. (9.33)–(9.38) suggests that the covariance of these equations requires the field transformation of

$$E'_x = E_x, \quad E'_y = E_y - \frac{v}{c} B_z, \quad E'_z = E_z + \frac{v}{c} B_y,$$

$$B'_x = B_x, \quad B'_y = B_y + \frac{v}{c} E_z, \quad B'_z = B_z - \frac{v}{c} E_y. \tag{9.56}$$

(This was also what Lorentz had for the corresponding states.) In this way, we have

$$\partial'_y E'_z - \partial'_z E'_y = -\frac{1}{c} \partial'_t B'_x \left(1 - \frac{v^2}{c^2} \right) \tag{9.57}$$

$$\partial'_z E'_x - \partial'_x E'_z = -\frac{1}{c} \partial'_t B'_y \tag{9.58}$$

$$\partial'_x E'_y - \partial'_y E'_x = -\frac{1}{c} \partial'_t B'_z \tag{9.59}$$

and

$$\partial'_y B'_z - \partial'_z B'_y = \frac{1}{c}\partial'_t E'_x\left(1 - \frac{v^2}{c^2}\right) \tag{9.60}$$

$$\partial'_z B'_x - \partial'_x B'_z = \frac{1}{c}\partial'_t E'_y \tag{9.61}$$

$$\partial'_x B'_y - \partial'_y B'_x = \frac{1}{c}\partial'_t E'_z. \tag{9.62}$$

Finally we need to check the two transformed Gauss' laws as given in (9.46) and (9.47). The application of (9.56) to (9.46) gives us:

$$\partial'_x E'_x + \partial'_y E'_y + \partial'_z E_z = \frac{v}{c}\left(\frac{1}{c}\partial'_t E_x - \partial'_y B_z + \partial'_z B_y\right) = \frac{v^2}{c^2}\partial'_x E'_x,$$

where to reach the last expression we have used the relation in (9.55). Thus the two Gauss' laws have the transformed form of

$$\partial'_x E'_x\left(1 - \frac{v^2}{c^2}\right) + \partial'_y E'_y + \partial'_z E_z = 0 \tag{9.63}$$

$$\partial'_x B'_x\left(1 - \frac{v^2}{c^2}\right) + \partial'_y B'_y + \partial'_z B'_z = 0. \tag{9.64}$$

We see that Maxwell's equations are satisfied in coordinate frame S', with the noncovariant terms in (9.57), (9.60), (9.63), and (9.64) being of order v^2/c^2.

9.5.3   Maxwell's equations are Lorentz covariant

A notable feature of the non-covariant terms obtained in the previous subsection is that they all have the form of $(1 - v^2/c^2) = \gamma^{-2}$. We have also discussed in Section 9.3.4 that the Fitzgerald–Lorentz proposal of longitudinal length contraction (9.25) involves the same factor and was implemented by Lorentz (1904) in such a way that it led to the Lorentz transformation as shown in (9.26) and (9.27).

Following the same procedure used in the previous subsection, we can easily show that Maxwell's equations retain their form, to all orders of v/c, provided that the fields transform like (9.56) but with some additional γ factors

$$E'_x = E_x, \quad E'_y = \gamma\left(E_y - \frac{v}{c}B_z\right), \quad E'_z = \gamma\left(E_z + \frac{v}{c}B_y\right),$$

$$B'_x = B_x, \quad B'_y = \gamma\left(B_y + \frac{v}{c}E_z\right), \quad B'_z = \gamma\left(B_z - \frac{v}{c}E_y\right). \tag{9.65}$$

We shall not provide the details here as a similar calculation was carried out by Einstein in his famous 1905 special relativity paper, which we report in Section 10.3 of the next chapter.

The new kinematics and $E = mc^2$

10

- Einstein created the new theory of special relativity by following the invariance principle. The principle of (special) relativity states that physics equations must be unchanged under coordinate transformations among inertial frames of reference. To this he added the principle of constancy of light speed. These two postulates formed the foundation of special relativity.

- These two postulates appeared at the outset to be contradictory: how can light speed be the same in two different reference frames which are in relativity motion? Einstein's resolution of this paradox came when he realized that the notion of simultaneity was a relative concept. Two events that are viewed to be simultaneous by one observer will appear to another observer as taking place at different times. Consequently, time is measured to run at different rates in different inertial frames.

- From these two postulates Einstein showed that, when different coordinate times are allowed, one could derive the Lorentz transformation $(t, \mathbf{r}) \rightarrow (t', \mathbf{r}')$ in a straightforward manner: If the two frames O and O' are moving with relative speed v in the x direction, one has

$$\Delta x' = \gamma(\Delta x - v\Delta t), \quad \Delta y' = \Delta y, \quad \Delta z' = \Delta z$$
$$\Delta t' = \gamma\left(\Delta t - v\Delta x/c^2\right) \quad \text{with} \quad \gamma = \left(1 - v^2/c^2\right)^{-1/2} > 1.$$

- Physical consequences follow immediately. A moving clock **appears** to run slow ("time dilation"); a moving object **appears** to contract ("length contraction"). This underscores the profound change in our conception of space and time brought about by special relativity.

- It should be noted that just about all the counter-intuitive relativistic effects spring from the new conception of time. Length contraction is such an example. To measure the length of a moving object one must specify the time when the front and back ends of the object are measured.

- From the Lorentz transformation we can immediately deduce that the familiar velocity addition rule $\mathbf{u}' = \mathbf{u} - \mathbf{v}$ will have to be modified in the relativistic regime so that the light speed remain unchanged $c' = c$. This is directly related to the invariance of a spacetime interval $s' = s$ where $s^2 \equiv x^2 + y^2 + z^2 - c^2t^2$.

- The electromagnetic fields also change when observed in different frames. Einstein derived their transformation $(\mathbf{E}, \mathbf{B}) \rightarrow (\mathbf{E}', \mathbf{B}')$ from the requirement that Maxwell's equations maintain the same form in different inertial frames.
- Einstein also showed that the Lorentz force law $\mathbf{F} = e[\mathbf{E} + \frac{1}{c}(\mathbf{v} \times \mathbf{B})]$ follows from Maxwell's equations when combined with the Lorentz transformation of the $(\mathbf{E}, \mathbf{B})$ fields.
- From a study of the relativistic work–energy theorem, Einstein discovered that one can identify a particle's energy with its inertia $E = \gamma mc^2$, where $\gamma = (1 - v^2/c^2)^{-1/2}$. In order to confirm this identification, he made a careful study showing that this energy can be converted into other forms, such as radiation energy.
- In SuppMat Section 10.6 we work out the relativistic effects for wave motion: Doppler's effect, stellar aberration, and Fresnel's formula, as well as the transformation properties of radiation energy.

10.1   The new kinematics

In the last chapter we discussed how Lorentz had introduced the notion of "local time". But he never regarded this as something physical, connected directly to the reading of a clock. Einstein's penetrating insight was that if signal transmissions were not instantaneous, simultaneity was coordinate-dependent. Namely, two events that took place at separate locations and were seen to be simultaneous in one inertial frame would be viewed in another as taking place at two different times. Since measurement of time always involves ultimately comparing simultaneous events, time measurement (just as spatial position measurement) would be coordinate-dependent. Upon the change of a coordinate frame, not only spatial variables but time as well are expected to undergo transformation. In his ground-breaking 1905 paper (Einstein 1905d), he demonstrated that the Lorentz transformation was the correct transformation among inertial frames of reference. The familiar Galilean relation was merely its low-velocity ($v \ll c$) approximation. In short, Einstein proposed a new conception of time; the resultant new kinematics deepened our understanding of electromagnetism, and altered the physics beyond.

10.1.1   Einstein's two postulates

While investigations in the nineteenth century involved the study of electromagnetic fields in the aether medium, Einstein cut right to the heart of the matter. He concluded that there was no physical evidence for the existence of aether. Physical descriptions would be much simplified and more natural if aether was dispensed with altogether.[1] From the modern perspective, what Einstein had done was to introduce a principle of symmetry, the relativity principle, as one of the fundamental laws of nature.

We have discussed Lorentz's derivation of the Lorentz transformation. He tried to use the electromagnetic theory of the electron to show that matter

[1] Recall our discussion in Section 9.4.1 of the magnet–conductor thought experiment that appeared in the introductory remarks of Einstein (1905d).

composed of electrons, when in motion, would behave in such a way to make it impossible to detect the effect of this motion on the speed of light. This was his way to explain the persistent failure of detecting any difference in the light speed in different inertial reference frames. Einstein by contrast had taken as a fundamental axiom that light speed is the same to all observers. The key feature of his new kinematics was the new conception of time that allowed the constancy of the light speed.

Einstein thus stated the two basic postulates for this new theory of relativity:

- Principle of relativity—physics equations must be the same in all inertial frames of reference.
- Constancy of light speed—the speed of light is the same in all coordinate frames, regardless of the motion of its emitter or receiver.

10.1.2   The new conception of time and the derivation of the Lorentz transformation

Time is coordinate-frame dependent

The most profound consequence of special relativity is the change it brought to our conception of time. In fact, as we shall see, all major implications of the theory can be traced back to the relativity of time. The revelation that came to Einstein was that in a world with a finite speed of light, the time interval is a frame-dependent quantity. Thus, any reference system must be specified by four coordinates (x, y, z, t)—that is, by three spatial coordinates and one time coordinate. One can picture a coordinate system as being a three-dimensional grid (to determine the position) and a set of clocks (to determine the time of an event), with a clock at every grid point in order to avoid the complication of time delay between the occurrence of an event and the registration of this event by a clock located a distance away from the event. We require all the clocks to be synchronized (say, against a master clock located at the origin). The synchronization of a clock located at a distance r from the origin can be accomplished by sending out light flashes from the master clock at $t = 0$. When the clock receives the light signal, it should be set to $t = r/c$. Equivalently, synchronization of any two clocks can be checked by sending out light flashes from these two clocks at a given time. If the two flashes arrive at their midpoint at the same time, they are synchronized. The reason that light signals are often used to set and compare time is that this ensures, in the most direct way, that the new kinematical feature of universal light velocity is properly taken into account.

Lorentz transformation derived from the two postulates

From the above stated postulates, we now find the relation between coordinates (x, y, z, t) of one inertial frame O to those (x', y', z', t') of the O' frame which moves with a constant velocity v in the positive x direction. The origins of these two systems are assumed to coincide at the initial time. Our algebraic steps are different from those taken by Einstein in his 1905 paper, but the basic assumptions and conclusion are of course the same.

[2]Equation (10.1) is just a compact way of writing $x' = a_1x + a_2t$ and $t' = b_1x + b_2t$.

1. Consider the transformation $(x, y, z, t) \rightarrow (x', y', z', t')$. Because of $y' = y$ and $z' = z$, we can simplify our equation display by concentrating on the 2D problem of $(x, t) \rightarrow (x', t')$, and can write the transformation in matrix form[2] as

$$\begin{pmatrix} x' \\ t' \end{pmatrix} = \begin{pmatrix} a_1 & a_2 \\ b_1 & b_2 \end{pmatrix} \begin{pmatrix} x \\ t \end{pmatrix}. \tag{10.1}$$

If one assumes that space is homogeneous and the progression of time is uniform, the transformation must be **linear**. Namely, the elements (a_1, a_2, b_1, b_2) of the transformation matrix $[L]$ must be independent of coordinates (x, t): we make the **same** coordinate transformation at every coordinate point (i.e. it is a global transformation). Of course these position/time independent factors can depend on the relative velocity v.

2. The origin $(x' = 0)$ of the O' frame has the trajectory of $x = vt$ in the O frame; thus reading off from Eq. (10.1) we have $0 = a_1x + a_2t$ with $x = vt$, leading to

$$a_2 = -va_1. \tag{10.2}$$

3. The origin $(x = 0)$ of the O frame has the trajectory of $x' = -vt'$ in the O' frame; reading off from Eq. (10.1) we have $x' = a_2t$ and $t' = b_2t$, or equivalently, $x'/t' = a_2/b_2 = -v$. When compared to Eq. (10.2), it implies

$$b_2 = a_1. \tag{10.3}$$

Substituting the relations (10.2) and (10.3) into the matrix equation (10.1), we have $x' = a_1(x - vt)$ and $t' = b_1x + a_1t$. Taking their ratio, we get

$$\frac{x'}{t'} = \frac{a_1\left(\dfrac{x}{t} - v\right)}{b_1\dfrac{x}{t} + a_1}. \tag{10.4}$$

4. We now impose the constancy-of-c condition: $x/t = c = x'/t'$ on Eq. (10.4): $c(b_1c + a_1) = a_1(c - v)$ leading to

$$b_1 = -\frac{v}{c^2}a_1. \tag{10.5}$$

5. Because of Eqs. (10.2), (10.3), and (10.5), the whole transformation matrix $[L]$ has only one unknown constant a_1

$$[L] = a_1 \begin{pmatrix} 1 & -v \\ -v/c^2 & 1 \end{pmatrix}, \qquad [L^{-1}] = a_1 \begin{pmatrix} 1 & v \\ v/c^2 & 1 \end{pmatrix}. \tag{10.6}$$

[2a]That the transformations $[L]$ and $[L^{-1}]$ have the same coefficient a_1 follows from the principle of relativity: the transformations of going from frame O to O' and the one from O' to O cannot be distinguished. One can in principle interchange the labels $O \leftrightarrow O'$ of the frames and keep the same coefficients. Equivalently, a_1 must necessarily be an even function of the relative velocity v.

In the above we have also written down the inverse transformation matrix $[L^{-1}]$ by simply reversing the sign of the relative velocity. The last unknown constant[2a] a_1 can then be fixed by the consistency condition of $[L][L^{-1}] = [1]$:

$$a_1^2 \begin{pmatrix} 1 & -v \\ -v/c^2 & 1 \end{pmatrix} \begin{pmatrix} 1 & v \\ v/c^2 & 1 \end{pmatrix} = \begin{pmatrix} 1 & 0 \\ 0 & 1 \end{pmatrix}$$

which implies that

$$a_1 = \sqrt{\frac{1}{1 - v^2/c^2}} \equiv \gamma. \tag{10.7}$$

This is the same Lorentz factor γ we have encountered in the previous chapter.

This concludes Einstein's derivation of the Lorentz transformation. We note that the above steps 1–3 just set up the consideration of two relative frames, and step 5 is a consistency condition. Hence the only nontrivial act is step 4 when light speed constancy is imposed.

We can rewrite the transformation in a more symmetric form by making the time coordinate have the same dimension as the other coordinate by multiplying it by a factor of c:

$$\begin{pmatrix} x' \\ ct' \end{pmatrix} = \gamma \begin{pmatrix} 1 & -\beta \\ -\beta & 1 \end{pmatrix} \begin{pmatrix} x \\ ct \end{pmatrix}, \quad \begin{pmatrix} x \\ ct \end{pmatrix} = \gamma \begin{pmatrix} 1 & \beta \\ \beta & 1 \end{pmatrix} \begin{pmatrix} x' \\ ct' \end{pmatrix} \tag{10.8}$$

where we have introduced another often-used notation

$$\beta \equiv \frac{v}{c}. \tag{10.9}$$

We note that while $0 \leq \beta \leq 1$, the Lorentz factor $\gamma = (1 - \beta^2)^{-1/2}$ is always greater than unity $\gamma \geq 1$—it approaches unity only in the low-velocity (nonrelativistic) limit, and blows up when v approaches c.

At this point we would also like to write down the transformation properties of the coordinate derivatives. For this purpose we introduce the commonly used notation of

$$x_0 \equiv ct, \qquad \partial_0 \equiv \frac{\partial}{\partial x_0} = \frac{1}{c}\frac{\partial}{\partial t} \quad \text{and} \quad \partial_0' \equiv \frac{\partial}{\partial x_0'} = \frac{1}{c}\frac{\partial}{\partial t'}.$$

Since $\partial_i x_j = \delta_{ij}$, and the Kronecker delta is unchanged under a Lorentz transformation, we can deduce that the derivatives must transform oppositely as the coordinates themselves, i.e. as their inverse, obtained by a simple sign change of β in (10.8):

$$\begin{pmatrix} \partial_x' \\ \partial_0' \end{pmatrix} = \gamma \begin{pmatrix} 1 & \beta \\ \beta & 1 \end{pmatrix} \begin{pmatrix} \partial_x \\ \partial_0 \end{pmatrix}. \tag{10.10}$$

10.1.3   Relativity of simultaneity, time dilation, and length contraction

Writing the Lorentz transformation as

$$x' = \gamma(x - vt), \quad y' = y, \quad z' = z$$

$$t' = \gamma\left(t - \frac{v}{c^2}x\right) \tag{10.11}$$

makes it clear that in the nonrelativistic limit of $\gamma \to 1$ and $v/c \to 0$, it reduces to the Galilean transformation (9.1). Let us look at some of the physical consequences that follow from the Lorentz transformation. We should keep in mind that the coordinates (x, t) are really coordinate intervals, namely, they are the interval between the coordinates (x, t) and the origin $(0, 0)$.

Relativity of simultaneity

Let us suppose the time interval is zero in one coordinate system, $t = 0$, while the position separation is nonzero, $x \neq 0$. This is the situation of two simultaneous events taking place at separate locations in the O frame. We see that, in the O' frame, we have $t' = \gamma v x/c^2 \neq 0$. Thus, simultaneity is relative.

Perhaps it will be helpful to think in terms of some concrete situation. An example is shown in Fig. 10.1. Two light pulses are sent from the midpoint towards the front and the back ends of a moving railcar. The arrival events at the two ends are viewed by an observer on the car as simultaneous. However for an observer standing on some stationary rail platform which the train passes by, the light signals would arrive at different times, resulting in the rail platform observer concluding that these two events as not simultaneous.

Time dilation and length contraction

We discuss here two other important physical implications of the special relativity (SR) postulates, **time dilation** and **length contraction**:

A moving clock **appears** to run slow;
a moving object **appears** to contract.

These physical features underscore the profound change in our conception of space and time brought about by relativity. We must give up our belief that measurements of distance and time give the same results for all observers. Special relativity makes the strange claim that observers in relative motion will have different perceptions of distance and time. This means that two identical watches worn by two observers in relative motion will tick at different rates and will not agree on the amount of time that has elapsed between two given

Fig. 10.1 Simultaneity is relative when light is not transmitted instantaneously. Two events (x_1', t_1') and (x_2', t_2') corresponding to light pulses (wavy lines) arriving at the opposite ends of a moving train, after being emitted from the midpoint (a1), are seen as simultaneous, $t_1' = t_2'$, by an observer on the train as in (a2). But to another observer standing on the rail platform, these two events (x_1, t_1) and (x_2, t_2) are not simultaneous, $t_1 \neq t_2$, because for this observer the light signals arrive at different times at the two ends of the moving railcar [(b2) and (b3)].

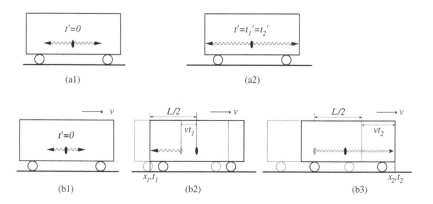

events. It's not that these two watches are defective. Rather, it's a fundamental statement about the nature of time.

Time dilation Let the O' frame be the rest frame of a clock. Because the clock has no displacement in this frame, $x' = 0$, we have from Eq. (10.11) $x = vt$ and

$$t' = \gamma\left(1 - \frac{v^2}{c^2}\right)t \quad \text{or} \quad \Delta t = \gamma\,\Delta t'. \tag{10.12}$$

With $\gamma > 1$, we have the time dilation phenomenon—a moving clock appears to run slow. Again it would be useful to illustrate this with a concrete description. In Fig. 10.2 time dilation shows up directly from the measurement by the most basic of clocks:[3] a light-pulse clock. It ticks away the time by having a light pulse bounce back and forth between two mirror-ends separated by a fixed distance d.

Length contraction

To measure the length of a moving object in the O frame, we can measure the two ends of the object simultaneously, hence $t = 0$. The measured length Δx is then related to its rest frame length $\Delta x'$ by Eq. (10.11) as

$$\Delta x = \frac{\Delta x'}{\gamma}. \tag{10.13}$$

Since γ^{-1} is always less than unity, we have the phenomenon of (longitudinal) length contraction of Lorentz and FitzGerald, discussed in Section 9.3.4.

Consider the specific example of length measurement of a moving railcar.[4] Let there be a clock attached to a fixed marker on the ground. A ground observer O, watching the train moving to the right with speed v, can measure the length L of the car by reading off the times when the front and back ends of the railcar pass this marker on the ground:

$$L = v(t_2 - t_1) \equiv v\Delta t. \tag{10.14}$$

But for an observer O' on the railcar, these two clock-reading events correspond to the passing of the two ends of the car by the (ground-) marker as the marker is seen moving to the left. O' can similarly deduce the length of the railcar in her reference frame by reading the times from the ground clock:

$$L' = v\left(t'_2 - t'_1\right) \equiv v\Delta t'. \tag{10.15}$$

These two unequal time intervals in (10.14) and (10.15) are related by the above-considered time dilation (10.12): $\Delta t' = \gamma\,\Delta t$, because Δt is the time recorded by a clock at rest, while $\Delta t'$ is the time recorded by a clock in motion (with respect to the observer O'). From this we immediately obtain

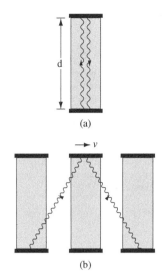

Fig. 10.2 Light-pulse clock, with mirrors at the top and bottom of a vertical vacuum chamber, (a) at rest and (b) in motion to the right in the horizontal direction. Viewed by a comoving observer O' as in (a), the time interval is $\Delta t' = d/c$. However, for an observer with respect to whom the clock is moving with a velocity v, perpendicular to the light-pulse path, the light pulse will be viewed as traversing a diagonal distance D as in (b). This involves a different time interval $\Delta t = D/c = \sqrt{d^2 + v^2\Delta t^2}/c$. Collecting Δt terms and using $\Delta t' = d/c$, we have the time dilation result of $\Delta t = \gamma\,\Delta t'$.

[3] A "basic clock" rests on some physical phenomenon that has a direct connection to the underlying laws of physics. Different clocks—mechanical clocks, biological clocks, atomic clocks, or particle decays—simply represent different physical phenomena that can be used to mark time. They are all equivalent if their time intervals are related to the basic-clock intervals by correct physical relations. We note that, while the familiar mechanical pendulum is a convenient basic clock for Newtonian physics, it is no longer so in relativity because the dynamical equation for a pendulum must be modified so that it is compatible with special relativity. In short, a phenomenon such as time dilation holds for any clock; but it is easier to see in the case of a light clock, shown in Fig. 10.2.

[4] The set-up is similar to that as shown in Fig. 10.1.

$$L = v\Delta t = \frac{v\Delta t'}{\gamma} = \frac{L'}{\gamma}, \tag{10.16}$$

which is the claimed result (10.13) of length contraction.[5]

We see that the derivation of length contraction invokes relativity of simultaneity (by way of time dilation). This follows simply from the fact that in order to deduce the length of a moving object, one must make two separate measurements of the front and back ends of that object. This means one must specify the times when they are measured, as the above-given example illustrates. Even the simplest way of making these two measurements simultaneously would still involve the change of time because of relativity of simultaneity. In fact one finds that just about all of the counter-intuitive results in special relativity are in one way or another ultimately related to the new conception of time. Thus one can conclude that the new conception of time is **the** key element of special relativity.

The new kinematics for electromagnetic waves and beyond

Relativistic wave motion   With this new kinematics, especially the notion of relative simultaneity, special relativity can provide simple and elegant explications of puzzling phenomena: stellar aberration and Fizeau's experiment, that led up to Einstein's creation of this relativity theory. The details of these SR derivations are given in SuppMat Section 10.6.

Relativity beyond electromagnetism   Einstein showed that the correct transformation among inertial frames of reference was the Lorentz transformation, which reduced to the Galilean relation in the low-velocity limit. In this way it not only solved the problem of the electromagnetism of a moving body but also pointed out that all physics must be reformulated so as to incorporate this new kinematics. For example, Newtonian mechanics had to be extended to a relativistic formulation in order that its equations be covariant under Lorentz transformation. As we shall see, the difficulty in making Newton's theory of gravity compatible with special relativity was one of the motivations behind Einstein's eventual development of general relativity, which would also be the field theory of gravitation.

10.2   The new velocity addition rule

Here we discuss how the apparent contradiction between relativity and light speed constancy is resolved by the new kinematics.

10.2.1   The invariant spacetime interval

In this new kinematics, time is no longer absolute (no longer an invariant under coordinate transformation). There is however a particular combination of space and time intervals that is invariant. This is the combination

$$s^2 \equiv x^2 + y^2 + z^2 - c^2 t^2. \tag{10.17}$$

We will demonstrate this by an explicit calculation. Plug the Lorentz transformation (10.11) into

[5] We often use the O' system for the rest-frame of whatever we are most interested in (be it a clock, or some object whose length we are measuring as in $\Delta t = \gamma \Delta t'$ or $\Delta x = \gamma^{-1} \Delta x'$). When using the results in (10.12) and (10.13), one must be certain which is the the rest-frame in the case being discussed and not blindly copy any written equation. For example, in the derivation here we have $\Delta t' = \gamma \Delta t$ rather than the usual $\Delta t = \gamma \Delta t'$ as written in Eq. (10.12).

$$s'^2 = x'^2 + y'^2 + z'^2 - c^2 t'^2 = \gamma^2 (x - vt)^2 + y^2 + z^2 - \gamma^2 c^2 \left(t - \frac{v}{c^2} x \right)^2$$

$$= \gamma^2 \left(1 - \frac{v^2}{c^2} \right) (x^2 - c^2 t^2) + y^2 + z^2 = s^2,$$

where we have used the relations in (10.11) and (10.17). What is the physical meaning of this interval? Why should one expect it to be an invariant? s is basically the time interval in the clock's rest-frame, called *proper time* (τ). In the rest-frame, there is no displacement, hence $x^2 + y^2 + z^2 = 0$, and we have

$$s^2 = -c^2 t^2 \equiv -c^2 \tau^2. \tag{10.18}$$

Since there is only one rest-frame (hence only one proper time), all observers (i.e. all coordinate frames) can agree on this value. It must be an invariant.

We of course also remember that the velocity of light is also a Lorentz invariant. So it should not be a surprise that the statements "c is absolute" and "s is absolute" are equivalent. Namely, starting from either one of them, one can prove the other, cf. Cheng (2010, Box 3.1, p. 42) and Section 10.2.2 below.

10.2.2 Adding velocities but keeping light speed constant

Writing the Lorentz transformation in terms of infinitesimal coordinate intervals

$$dx' = \gamma (dx - vdt) \tag{10.19}$$

$$dt' = \gamma \left(dt - \frac{v}{c^2} dx \right),$$

we have the ratio

$$\frac{dx'}{dt'} = \frac{dx - vdt}{dt - \frac{v}{c^2} dx} = \frac{\frac{dx}{dt} - v}{1 - \frac{v}{c^2} \frac{dx}{dt}}$$

or, in terms of particle velocities $\mathbf{u} = d\mathbf{r}/dt$,

$$u'_x = \frac{u_x - v}{1 - \frac{vu_x}{c^2}}. \tag{10.20}$$

Similarly, we can write down the expressions for u'_y and u'_z.

This new velocity addition rule replaces the Newtonian one: for a particle velocity also in the x direction ($u = u_x$), the familiar rule, $u' = u - v$, is simply the low-velocity limit of (10.20). We are particularly interested to see how this new rule can be compatible with the constancy of light velocity condition. If we have the simple case of $u_x = c$, we then have

$$u'_x = \frac{c - v}{1 - \frac{v}{c}} = c \tag{10.21}$$

expressing the result $c' = c$. For the case of light propagating in a general direction, a brute force calculation of

$$\sqrt{u'^2_x + u'^2_y + u'^2_z} = c, \tag{10.22}$$

from formulas like (10.20) would be very tedious (hence prone to algebraic mistakes). The most efficient way to prove (10.22) is through the invariance of the spacetime interval $ds' = ds$. In this case, they have vanishing invariant separation $ds' = ds = 0$. Namely, for a light signal we have

$$\frac{(ds)^2}{(dt)^2} = \frac{dx^2 + dy^2 + dz^2 - c^2 dt^2}{dt^2} = u^2 - c^2 = 0. \tag{10.23}$$

This implies $0 = \left[(ds')/(dt')\right]^2 = u'^2 - c^2$, or $u' = c$.

10.3   Maxwell's equations are Lorentz covariant

Since Maxwell's equations already have the relativistic feature of a constant speed of light, we expect they are already covariant under Lorentz transformation. We will actually use this expectation to derive the Lorentz transformation properties of electric and magnetic fields $(\mathbf{E}, \mathbf{B})$. From our basic knowledge of electromagnetism, we expect that these fields must change into each other when viewed in reference frames that are in motion with respect to each other. An electric charge at rest gives rise to an electric, but no magnetic, field. However, the same situation when seen by a moving observer is a charge in motion, which produces both electric and magnetic fields. Namely, different inertial observers will find different electric and magnetic fields, just as they would measure different position and time intervals.

10.3.1   The Lorentz transformation of electromagnetic fields

We can derive the Lorentz transformation of the $(\mathbf{E}, \mathbf{B})$ fields from the requirement that Maxwell's equations be Lorentz covariant. Namely, we ask how the electromagnetic field must change $(\mathbf{E}, \mathbf{B}) \longrightarrow (\mathbf{E}', \mathbf{B}')$ in order, for example, that the homogeneous equations maintain the same form under a Lorentz transformation: In some inertial frame O we have

$$\mathbf{\nabla} \times \mathbf{E} + \frac{1}{c}\partial_t \mathbf{B} = 0, \qquad \mathbf{\nabla} \cdot \mathbf{B} = 0, \tag{10.24}$$

while in another frame O' moving in the $+x$ direction with velocity v, we would still have

$$\mathbf{\nabla}' \times \mathbf{E}' + \frac{1}{c}\partial_t' \mathbf{B}' = 0, \qquad \mathbf{\nabla}' \cdot \mathbf{B}' = 0. \tag{10.25}$$

Let us start with the x component of Faraday's equation in the O' frame:

$$\partial_y' E_z' - \partial_z' E_y' + \partial_0' B_x' = 0. \tag{10.26}$$

Substituting in the Lorentz transformation of the derivatives, as discussed in Eq. (10.10),

$$\partial_0' = \gamma(\partial_0 + \beta\partial_x) \qquad \partial_x' = \gamma(\partial_x + \beta\partial_0), \qquad \partial_y' = \partial_y, \qquad \partial_z' = \partial_z. \tag{10.27}$$

we then have

$$\partial_y E'_z - \partial_z E'_y + \gamma \partial_0 B'_x + \gamma \beta \partial_x B'_x = 0. \qquad (10.28)$$

Or, if we start with the no-monopole equation in (10.25), we would have

$$\gamma \partial_x B'_x + \gamma \beta \partial_0 B'_x + \partial_y B'_y + \partial_z B'_z = 0. \qquad (10.29)$$

Taking a linear combination of Eqs. (10.28)–β(10.29) and noting the cancellation of the $\gamma\beta\partial_x B'_x$ terms, we get

$$\partial_y \left(E'_z - \beta B'_y \right) - \partial_z \left(E'_y + \beta B'_z \right) + \left(1 - \beta^2 \right) \gamma \partial_0 B'_x = 0.$$

Multiplying this equation by a factor of γ and noting $(1 - \beta^2)\gamma^2 = 1$, we have

$$\partial_y \left[\gamma \left(E'_z - \beta B'_y \right) \right] - \partial_z \left[\gamma \left(E'_y + \beta B'_z \right) \right] + \partial_0 B'_x = 0.$$

After comparing Eq. (10.24) in the O frame, we can then identify

$$B_x = B'_x, \quad E_y = \gamma \left(E'_y + \beta B'_z \right), \quad E_z = \gamma \left(E'_z - \beta B'_y \right). \qquad (10.30)$$

Starting with different components of Maxwell's equations we can similarly show that

$$E_x = E'_x, \quad B_y = \gamma \left(B'_y - \beta E'_z \right), \quad B_z = \gamma \left(B'_z + \beta E'_y \right). \qquad (10.31)$$

Here we have written fields in the O frame in terms of those in the O' frame; the inverse transformation is simply the same set of equations with a sign change for all the β factors. We should also note that the charge and current densities $(c\rho, \mathbf{j})$ transform in the same way as the time and position coordinates $(ct, \mathbf{r})$.

The field transformation rule displayed in (10.30) and (10.31) still seems mysterious. Why should it be of this form? Is there a simple way to understand it? In the next chapter, this rule can indeed be explained, cf. Eq. (11.37).

The different theories of Lorentz and Einstein

The Lorentz transformation was first written down by Lorentz in 1904. He arrived at his transformation in the best tradition of "constructive theory". He systematically worked out the theory based in the general framework of an aether as the medium of EM wave propagation; the theory was built up step by step to accommodate the experimental result. It is strictly an electromagnetic theory and the transformation properties of EM fields are supposedly a result of the properties of the aether. For example Lorentz's "principle of corresponding states" is applicable only to the aether and not the covariance requirement for all physics. Einstein, on the other hand, unaware of Lorentz's final accomplishment, derived the Lorentz transformation entirely differently. Namely, the transformation rule might be the same, but their respective interpretations were very different. Einstein's approach[6] was an example of "theories of principle".

[6]Further remarks about construtive theories versus theories of principle can be found at the end of Chapter 16, and in Pais (1982, p. 27).

Thermodynamics is the prototype of a theory of principle; it is based on the postulates of the impossibility of perpetual motion. The two postulates of special relativity are the over-arching principles that imply a new kinematics and dictate the development of the new physics in every aspect.

10.3.2 The Lorentz transformation of radiation energy

Recall that the EM radiation energy density is directly proportional to the field amplitude squared

$$A^2 = E_x^2 + E_y^2 + E_z^2 = B_x^2 + B_y^2 + B_z^2; \tag{10.32}$$

thus the radiation energy $U' = A'^2 V'$ in the O' frame is related to $U = A^2 V$ in the O frame by

$$U' = \frac{A'^2}{A^2} \frac{V'}{V} U \tag{10.33}$$

where V is the volume that contains the EM radiation field. The transformation relations (A', A) and (V', V) are worked out in SuppMat Section 10.6.3. They are, for a wave propagating in a direction making an angle θ with the direction of relative motion (x),

$$\frac{A'}{A} = \gamma \left(1 - \frac{v}{c} \cos \theta \right) \tag{10.34}$$

$$\frac{V'}{V} = \left[\gamma \left(1 - \frac{v}{c} \cos \theta \right) \right]^{-1}. \tag{10.35}$$

Therefore, the radiation energy transforms as

$$U' = \gamma \left(1 - \frac{v}{c} \cos \theta \right) U. \tag{10.36}$$

We shall see in Section 10.5.2 that this equation will be used in Einstein's argument for his relation $E = mc^2$.

10.4   The Lorentz force law

The field equations of electromagnetism are Maxwell's equations, and the equation of motion is given by the Lorentz force law, which describes the motion of a charge in a given electromagnetic field. At one time there was a discussion as to whether the Lorentz force law was independent of Maxwell's equations. Here Einstein showed that while the Lorentz transformation of fields followed from the field equations (as discussed above in Section 10.3.1), the Lorentz force law could be derived from the field transformation.

Einstein considered the following case of a charged particle (mass m and charge e) in an electromagnetic field $(\mathbf{E}, \mathbf{B})$. At an instant when the particle is at rest in the O frame, we have the equation of motion $md^2\mathbf{r}/dt^2 = e\mathbf{E}$. An instant later, the particle is in motion in the x direction, $v = dx/dt$. Consider another reference frame O' with respect to which the particle is then momentarily at rest; we then have

$$m\frac{d^2x'}{dt'^2} = eE'_x, \qquad m\frac{d^2y'}{dt'^2} = eE'_y, \qquad m\frac{d^2z'}{dt'^2} = eE'_z \qquad (10.37)$$

which, after using the Lorentz transformation[7] of coordinates (10.11) and (10.10) as well as fields (10.30) and (10.31), is viewed in the O frame as

[7]In the calculation we use the fact that the Lorentz transformation is linear.

$$m\gamma^3 \left(\frac{d}{dt} + v\frac{d}{dx}\right)\left(\frac{d}{dt} + v\frac{d}{dx}\right)(x - vt) = m\gamma^3\frac{d^2x}{dt^2} = eE_x$$

$$m\gamma^2 \left(\frac{d}{dt} + v\frac{d}{dx}\right)\left(\frac{d}{dt} + v\frac{d}{dx}\right)y = m\gamma^2\frac{d^2y}{dt^2} = e\gamma\left(E_y - \frac{v}{c}B_z\right)$$

$$m\gamma^2 \left(\frac{d}{dt} + v\frac{d}{dx}\right)\left(\frac{d}{dt} + v\frac{d}{dx}\right)z = m\gamma^2\frac{d^2z}{dt^2} = e\gamma\left(E_z + \frac{v}{c}B_y\right).$$

Namely,

$$m\gamma^3\frac{d^2x}{dt^2} = eE_x, \qquad m\gamma\frac{d^2y}{dt^2} = e\left(E_y - \frac{v}{c}B_z\right), \qquad m\gamma\frac{d^2z}{dt^2} = e\left(E_z + \frac{v}{c}B_y\right).$$
$$(10.38)$$

The LHS of these equations can be shown[8] (by Planck in 1906) as the three components of the relativistic force

[8]For details see SuppMat Section 10.7 below.

$$\mathbf{F} = \frac{d}{dt}\gamma m\mathbf{v} \qquad (10.39)$$

with $\gamma m\mathbf{v}$ being the relativistic momentum. The RHS of the equations in (10.38) is just the Lorentz force

$$\mathbf{F} = e\left(\mathbf{E} + \frac{1}{c}\mathbf{v} \times \mathbf{B}\right) \qquad (10.40)$$

and because $\mathbf{v} = v\hat{\mathbf{x}}$, we have $(\mathbf{v} \times \mathbf{B})_x = 0$, $(\mathbf{v} \times \mathbf{B})_y = -vB_z$, and $(\mathbf{v} \times \mathbf{B})_z = vB_y$. Namely, this matches (10.40) with the component expression given in (10.38).

10.5 The equivalence of inertia and energy

Einstein went on nest to derive, what Hendrik Lorentz termed in later years as "the most remarkable conclusions of the theory of relativity". That is, the equivalence of inertia and energy.

10.5.1 Work–energy theorem in relativity

In the last section of his 1905 relativity paper, Einstein began with a rather low key, commonplace statement: "next we determine the electron's kinetic energy". Consider a charged particle (an electron for example) being slowly accelerated from 0 to a velocity v. The change in the particle's kinetic energy K can then be attributed to the work done by the electrostatic force. The work-energy theorem, according to Eq. (10.38), states

$$
K = \int F_x dx = m \int \gamma^3 \frac{dv}{dt} dx = m \int_0^v \gamma^3 v dv
$$

$$
= \frac{m}{2} \int_0^v \frac{dv^2}{\left(1 - v^2/c^2\right)^{3/2}} = mc^2 \left[\frac{1}{\left(1 - v^2/c^2\right)^{1/2}} \right]_0^v
$$

$$
= mc^2(\gamma - 1). \tag{10.41}
$$

Some comments about this result are in order:

- Einstein noted that, since $\gamma \to \infty$ when $v \to c$, it would not be physically possible for a particle to travel faster than the speed of light.
- Einstein did not point out here that the result of (10.41) could be interpreted as indicating that a particle had the (total) energy[9] of

$$
E = \gamma mc^2. \tag{10.42}
$$

Namely, the equivalence of inertia and energy.

- The kinetic energy K has the correct nonrelativistic limit: When the velocity is nonzero but much smaller than c, the low-velocity expansion of the Lorentz factor in (10.41) leads to the familiar result

$$
K = mc^2 \left[\left(1 - v^2/c^2\right)^{-1/2} - 1 \right] \simeq \frac{1}{2} mv^2. \tag{10.43}
$$

It vanishes when the particle is at rest ($\gamma = 1$). This also suggests that the particle has an energy of mc^2 while at rest.

- Even if Einstein was immediately aware of this $E = mc^2$ interpretation, the revolutionary nature of this possibility may have given him pause. To make sure that this energy can actually be converted to other (more familiar) forms of energy he had to work out a concrete process. In any case, this realization probably came to Einstein after he had sent out his paper. Three months later he submitted another paper, which was essentially an addendum to his main relativity paper.

[9]Sometimes this relation is written by others as $E = m^* c^2$, with $m^* = \gamma m$. We work only with the rest-mass m and avoid such a velocity-dependent 'dynamical mass' altogether, as m^* is not a Lorentz invariant quantity.

10.5.2   The $E = mc^2$ paper three months later

This paper (Einstein 1905e) submitted to *Annalen* in September 1905 had the title *Does the inertia of a body depend on its energy content?* Einstein explicitly stated the result of (10.42) by considering the emission of light pulses by an atom.

Since we are interested in the relation between inertial mass and energy, let us first recall their relation in Newtonian mechanics: $K = \frac{1}{2} mv^2$. As we have this relation in the low-velocity limit of the relativistic theory, we have the general definition of mass in terms of kinetic energy:

$$
m = \lim_{v \to 0} \left(\frac{K}{v^2/2} \right). \tag{10.44}
$$

The next step is to get hold of the kinetic energy of an atom. Using this definition of the inertia, Einstein shows that the kinetic energy loss to radiation is just that given by (10.41). Einstein considered the process of an atom at rest, emitting two back-to-back radiation pulses. The energy difference of an atom before and after emission is $\Delta E = E_0 - E_1$. Let the back-to-back pulses be in directions having an angle θ and an angle $\theta + \pi$ with respect to the x axis. Each pulse carries an energy of $L/2$. Energy conservation can then be expressed as $\Delta E = L$. Since the change in kinetic energy involves the change of particle velocity, Einstein studied this process in two different reference frames: in one (the O frame), the atom is at rest; in another (the O' frame) moving with velocity v in the x direction. In the O frame, we have $\Delta E = L$; in the O' frame the corresponding energy conservation statement is $\Delta E' = L'$. From the Lorentz transformation of radiation energy derived in (10.36) we have

$$L' = \gamma \left[\left(1 - \frac{v}{c} \cos\theta \right) + \left(1 + \frac{v}{c} \cos\theta \right) \right] \frac{L}{2} = \gamma L.$$

Comparing the two energy conservation statements, we have

$$\Delta E' - \Delta E = L' - L = L(\gamma - 1). \tag{10.45}$$

The LHS of this equation may be rearranged:

$$\left(E_0' - E_1' \right) - (E_0 - E_1) = \left(E_0' - E_0 \right) - \left(E_1' - E_1 \right).$$

$(E_0' - E_0)$ is the energy difference of the atom (before light emission) as viewed in two different reference frames: the O' frame where the atom is moving, versus the O frame where the atom is at rest The difference should just be the kinetic energy, up to an additive constant C. A similar result holds for the atom after light emission:

$$E_0' - E_0 = K_0 + C$$
$$E_1' - E_1 = K_1 + C. \tag{10.46}$$

We can then conclude that the LHS of (10.45) is just the change of kinetic energy of the atom,

$$K_0 - K_1 = L(\gamma - 1). \tag{10.47}$$

Dividing both sides by $v^2/2$ and taking the low-velocity limit (we are allowed to take any relative velocity of the two reference frames), we have from (10.44) the relation

$$m_0 - m_1 = \lim_{v \to 0} L \left(\frac{\gamma - 1}{v^2/2} \right) = \frac{L}{c^2}. \tag{10.48}$$

This shows clearly that the energy L carried away by the radiation[10] is just the change of rest energy $m_0 c^2 - m_1 c^2$. In this way Einstein first stated explicitly the equivalence of energy and inertia as shown in (10.42).

[10]In later publications, such as Einstein (1907b), Einstein substantiated this claim that the equivalence between energy and inertia should hold for all forms of energy.

10.6   SuppMat: Relativistic wave motion

10.6.1   The Fresnel formula from the velocity addition rule

In Section 9.3.1 we discussed the formula (9.10) that describes light speed in a moving medium,

$$V = \frac{c}{n} + v\left(1 - \frac{1}{n^2}\right),\tag{10.49}$$

which was originally constructed by Fresnel under the hypothesis that aether was partially entrained by a moving body. It was found by Fizeau to be in agreement with his measurement of light velocity in the medium of moving water. Namely, the above relation was in agreement with the measured light speed (V) in moving ($-v$) water having index of refraction (n).

Here we show that (10.49) can be easily justified in special relativity with its new velocity addition rule of (10.20):

$$u' = \frac{u + v}{1 + \dfrac{uv}{c^2}}.\tag{10.50}$$

Since we have the light velocity in still water as $u = c/n$, the above relation becomes

$$u' = \frac{\dfrac{c}{n} + v}{1 + \dfrac{v}{cn}} \simeq \left(\frac{c}{n} + v\right)\left(1 - \frac{v}{cn}\right) \simeq \frac{c}{n} + v\left(1 - \frac{1}{n^2}\right)$$

in agreement with Fresnel's formula.

10.6.2   The Doppler effect and aberration of light

The frequency and wavevector of a propagating wave would change when viewed in reference frames in relative motion. Our calculation is entirely similar to that given in Section 9.3.2. As we shall see, relativity gives simple and elegant proofs of the previously much discussed result of stellar aberration, as well as the familiar Doppler effect. Naturally, all are generalized to the relativistic regime.

Consider a plane wave propagating in the direction $\hat{\mathbf{n}}$ (a unit vector) which lies in the *x-y* plane, making an angle θ with respect to the relative motion velocity $\mathbf{v} = v\hat{\mathbf{x}}$ of the two reference frames O and O'. The phase factor of a wave $\Phi = \omega t - \mathbf{k} \cdot \mathbf{r}$ with $k = \omega/c$ just counts the peaks and troughs of a wave; it cannot be coordinate dependent. Namely, the phase of a wave must be a Lorentz invariant.

$$\frac{\omega}{c}\left(ct - x\cos\theta - y\sin\theta\right) = \frac{\omega'}{c}\left(ct' - x'\cos\theta' - y'\sin\theta'\right).\tag{10.51}$$

Inserting the Lorentz transformation (10.11) and equating the coefficients of t, x, and y, we obtain

$$\omega = \omega'\gamma\left(1 + \beta\cos\theta'\right)\tag{10.52}$$

$$\omega\cos\theta = \omega'\gamma\left(\cos\theta' + \beta\right)\tag{10.53}$$

$$\omega\sin\theta = \omega'\sin\theta'.\tag{10.54}$$

Doppler formula

We can rewrite Eq. (10.52) by interchanging the frames of O and O', with a sign change of β:

$$\omega' = \frac{1 - \frac{v}{c} \cos \theta}{\sqrt{1 - \left(\frac{v}{c}\right)^2}} \omega. \tag{10.55}$$

This is the relativistic Doppler formula. In contrast to the classical theory where the Doppler effect is present only when the observer is moving along the direction of light propagation, we have here a new (transverse) Doppler effect $\omega' = \gamma \omega$ when light travels in the perpendicular direction $\theta = \pi/2$. This effect does not result from the compression or elongation of the wavelength, but purely from relativistic time dilation.

Stellar aberration

By taking the ratio of Eqs. (10.54) and (10.53), we obtain

$$\tan \theta = \frac{\sin \theta' \sqrt{1 - \beta^2}}{\cos \theta' + \beta} \tag{10.56}$$

which is the relativistic version of the stellar aberration result worked out in Eq. (9.21).

10.6.3   Derivation of the radiation energy transformation

The Lorentz transformation of radiation amplitude is a simple application of the transformation of EM fields derived in Section 10.3. This leads to the transformation of the radiation energy density. To obtain the energy, we also need the transformation of the volume factor. Let us assume that the EM wave originates from a point source and it spreads out with a spherical surface wavefront. This spherical wave will be seen as an ellipsoid in the moving frame. A direct application of the spacetime transformation leads to the relation between these two volumes, as will be worked out below.

The Lorentz transformation of wave amplitudes

The electromagnetic wave being a transverse wave, the fields are perpendicular to the direction of propagation $\hat{\mathbf{n}}$, and the electric field $\mathbf{E}$ and magnetic field $\mathbf{B}$ are mutually perpendicular to each other. Thus we have a right-handed system with three axes $(\mathbf{E}, \mathbf{B}, \hat{\mathbf{n}})$. Let the propagation vector $\hat{\mathbf{n}}$ be in the x-y plane making an angle θ with the x axis, and the electric field pointing in the z direction:

$$E_x = E_y = 0, \quad E_z = E, \quad B_x = B \sin \theta, \quad B_y = -B \cos \theta, \quad B_z = 0.$$

A direct application of the field transformation as given in (10.30) and (10.31) yields the transformed electric field components:

$$E'_x = E_x = 0, \quad E'_y = \gamma \left(E_y - \beta B_z \right) = 0, \quad E'_z = \gamma \left(E_z + \beta B_y \right). \tag{10.57}$$

In an EM wave, the electric and magnetic field amplitudes are equal; we shall call them A. Thus the last equation in (10.57) can be written as

$$A' = \gamma(1 - \beta \cos \theta) A \qquad (10.58)$$

because $E'_z = E' \equiv A'$, $E_z = E \equiv A$, and $B_y = -A \cos \theta$. This is the result[11] quoted in Eq. (10.34) above.

The Lorentz transformation of a spherical volume

Let us consider a spherical electromagnetic wave of radius R centered at $ct\hat{\mathbf{n}}$ (with $\hat{\mathbf{n}}$ being a unit vector) and its wavefront described in the O frame as

$$(x - ct\hat{n}_x)^2 + \left(y - ct\hat{n}_y\right)^2 + (z - ct\hat{n}_z)^2 = R^2, \qquad (10.59)$$

with a volume of $V = 4\pi R^3/3$. This sphere in (10.59) would be seen by an observer in frame O' moving with velocity $\mathbf{v} = v\hat{\mathbf{x}}$ as an ellipsoid because of length contraction in the x direction. The mathematics is a simple application of the Lorentz transformation to replace (x, y, z) by (x', y', z'). To simplify the writing, we will concentrate at the instant $t' = 0$ so that $x = \gamma x'$, $y = y'$, $z = z'$, and $ct = \gamma \beta x'$. Equation (10.59) becomes

$$\gamma^2(1 - \beta \hat{n}_x)^2 x'^2 + \left(y' - \gamma \beta \hat{n}_y x'\right)^2 + \left(z' - \gamma \beta \hat{n}_z x'\right)^2 = R^2 \qquad (10.60)$$

or, written in standard ellipsoid form,

$$\frac{x'^2}{X'^2} + \frac{\left(y' - \gamma \beta \hat{n}_y x'\right)^2}{Y'^2} + \frac{\left(z' - \gamma \beta \hat{n}_z x'\right)^2}{Z'^2} = 1. \qquad (10.61)$$

We see that this is an oblate spheroid centered at $x' = 0$, $y' = \gamma \beta \hat{n}_y x'$, and $z' = \gamma \beta \hat{n}_z x'$, with semimajor radii of

$$X' = \frac{R}{\gamma(1 - \beta \hat{n}_x)}, \qquad Y' = Z' = R. \qquad (10.62)$$

$\hat{\mathbf{n}}$ being a unit vector on the x-y plane making an angle θ with the x axis, we have $\hat{n}_x = \cos \theta$. From this we find the volume of this ellipsoid as

$$V' = \frac{4\pi}{3} X'Y'Z' = \frac{V}{\gamma(1 - \beta \cos \theta)}. \qquad (10.63)$$

This is the result quoted in Eq. (10.35) above.

10.7   SuppMat: Relativistic momentum and force

The asymmetric appearance of the force equation in (10.38) is clearly due to the fact that the x direction is the direction of relative motion $v = dx/dt$. In 1906 Max Planck argued that since one can equivalently view that, in the second law of motion, force is (instead of mass × acceleration) the rate of change of momentum, the correct relativistic momentum expression[12] had an extra factor of γ

$$\mathbf{p} = \gamma m\mathbf{v} = \frac{m\mathbf{v}}{\sqrt{1 - v^2/c^2}}. \qquad (10.64)$$

Planck arrived at this expression by a consideration of the Lagrangian formulation of mechanics of a point particle.[13] Momentum is related to the Lagrangian by the usual relation of $\mathbf{p} = \partial L / \partial \mathbf{v}$. The surprising expression for the x component of the force in (10.38) then comes about as follows:

$$
\begin{aligned}
\frac{dp_x}{dt} = \frac{d}{dt}(\gamma m v) &= \gamma m \frac{dv}{dt} + mv \frac{d}{dt}\left(1 - \frac{v^2}{c^2}\right)^{-1/2} \\
&= \gamma m \frac{dv}{dt} + \gamma^3 m \frac{v^2}{c^2}\frac{dv}{dt} = \gamma m \frac{dv}{dt}\left(1 + \gamma^2 \frac{v^2}{c^2}\right) \\
&= \gamma^3 m \frac{dv}{dt} = m\gamma^3 \frac{d^2 x}{dt^2}.
\end{aligned}
\tag{10.65}
$$

It should be noted that in the above calculation we were concerned with the variation of the coordinate x with respect to time, holding (y, z) fixed. Consider the momentum change in the other two directions. For example, in

$$
\frac{dp_x}{dt} = \frac{d}{dt}\left(m\gamma \frac{dy}{dt}\right) = m\gamma \frac{d^2 y}{dt^2}
$$

we hold (x, z) fixed. This means the relative velocity ($v = dx/dt$), hence the γ factor, are constants with respect to the differentiation.

[13] Eq. (10.64) will also be derived in Section 11.2.3 when we consider the vector transformation property of a particle's momentum.

11 Geometric formulation of relativity

- The geometric description of SR interprets the relativistic invariant interval as a length in 4D Minkowski spacetime. The Lorentz transformation is a rotation in this pseudo-Euclidean manifold, with a metric equal to $\eta_{\mu\nu} = \text{diag}(-1,1,1,1)$.

- Just about all the relativistic effects such as time dilation, length contraction, and relativity of simultaneity follow from the Lorentz transformation. Thus, the geometric formulation allows us to think of the metric $\eta_{\mu\nu}$ as embodying all of special relativity.

- The metric in general is a matrix with its diagonal elements being the length of the basis vectors and off-diagonal elements showing the deviation from orthogonality of the basis vectors. It is this geometric quantity that allows us to define the inner product of the space.

- Tensors in Minkowski space are quantities having definite transformation properties under Lorentz transformation. If a physics equation can be written as a tensor equation, it automatically respects the relativity principle.

- Examples of tensors with one index (4-vectors) include the position 4-vector, with components of time and 3D positions $x^\mu = (ct, \mathbf{x})$, and the momentum 4-vector with components of relativistic energy and 3D momentum $p^\mu = (E/c, \mathbf{p}) = \gamma m(c, \mathbf{v})$ with γ being the usual Lorentz factor. Namely, we have the relativistic momentum $\mathbf{p} = \gamma m \mathbf{v}$ and relativistic energy $E = \gamma mc^2$. From this we obtain the well-known energy–momentum relation of $E^2 = p^2 c^2 + m^2 c^4$ and the conclusion that a massless particle always travels at velocity c.

- Examples of tensors with two indices (4-tensors of rank 2) include the symmetric energy–momentum stress tensor $T_{\mu\nu} = T_{\nu\mu}$ (the source term for the relativistic gravitational field) and the antisymmetric EM field tensor $F_{\mu\nu} = -F_{\nu\mu}$ with the six components being the $(\mathbf{E}, \mathbf{B})$ fields. As a result Maxwell's equations can be written in a very compact fashion.

- The spacetime diagram is presented in Section 11.3. It is a particularly useful tool in understanding the causal structure of relativity theory.

- We also provide a summary of the various ideas and implications related to the geometric formulation of special relativity. We note in particular that such a formulation is indispensable for the eventual formulation of general relativity, which is a gravitational field theory with curved spacetime identified as the gravitational field.

While Einstein's special relativity, as presented in 1905, is an enormous simplification compared to the previous discussions of electrodynamics for a moving body, an even simpler perspective is offered by the geometric formulation of the theory. Building on the prior work of Lorentz and Poincaré, Hermann Minkowski (1864–1909), the erstwhile mathematics professor of Einstein at ETH, proposed in 1907 this version of Einstein's special relativity. Minkowski concentrated on the invariance of the theory, emphasizing that the essence of special relativity was the proposition that time should be treated on an equal footing as space. The best way to bring out this symmetry between space and time is to unite them in a single mathematical structure, now called **Minkowski spacetime**. The following are the opening words of an address he delivered at the 1908 Assembly of German National Scientists and Physicians held in Cologne, Germany (Minkowski 1908).

The views of space and time which I wish to lay before you have sprung from the soil of experimental physics, and therein lies their strength. They are radical. Henceforth space by itself, and time by itself, are doomed to fade away into mere shadows, and only a kind of union of the two will preserve an independent reality.

Minkowski emphasized that time could be regarded as the fourth coordinate of a 4D spacetime manifold. If the physics equations were written as tensor equations in this spacetime, they would be manifestly covariant under Lorentz transformation, that is, they would be automatically relativistic. Initially Einstein was not impressed by this new formulation,[1] calling it "superfluous learnedness". He started using this formulation only in 1912 for his geometric theory of gravity, in which gravity is identified as the 'structure' of spacetime. Namely, Einstein adopted this geometric language and extended it—from a flat Minkowski space to a curved one. Finally in his 1916 paper on general relativity, Einstein openly acknowledged its importance.

[1] See Pais (1982, p. 152).

11.1 Minkowski spacetime

In the last chapter we have shown that the Lorentz transformation leaves invariant the spacetime interval

$$s^2 = -c^2 t^2 + x^2 + y^2 + z^2. \tag{11.1}$$

Minkowski pointed out that this may be viewed as a (squared) length in a 4D pseudo-Euclidean space. The Lorentz transformation is the general class of length-preserving transformations (rotations) in this space.

11.1.1   Rotation in 3D space—a review

Recall the mathematical representation of rotation (say, around the z axis) in 3D space

$$
\begin{pmatrix} x' \\ y' \\ z' \end{pmatrix} = \begin{pmatrix} \cos\theta & \sin\theta & 0 \\ -\sin\theta & \cos\theta & 0 \\ 0 & 0 & 1 \end{pmatrix} \begin{pmatrix} x \\ y \\ z \end{pmatrix}. \tag{11.2}
$$

This leaves invariant the length $l^2 = x^2 + y^2 + z^2$. In the index notation ($i = 1, 2, 3$) we have

$$
x'_i = [R]_{ij} x_j \tag{11.3}
$$

with repeated indices (j) being summed over. In the geometrical perspective, rotation is a length-preserving transformation. This demands $[R]$ be an orthogonal matrix, with its inverse equal to its transpose:

$$
x'_i x'_i = [R]_{ij} [R]_{ik} x_j x_k = \delta_{jk} x_j x_k, \quad \text{or, in terms matrices } [R]^\mathsf{T}[R] = [1].
$$

This orthogonality condition can be used in turn to fix the form[2] of $[R]$, with (11.2) as an explicit example.

Rotational symmetry then requires that physics equations be covariant under any rotational transformation. This can be implemented by writing them as tensor equations. A tensor is a mathematical quantity having a definite transformation under rotation (scalars, vectors, or tensors of higher rank). For example, any two vectors such as A_i and B_i as well as a tensor of rank 2 (i.e. a tensor with two indices) must transform as

$$
A'_i = [R]_{ij} A_j \qquad B'_i = [R]_{ij} B_j \tag{11.4}
$$

and

$$
C'_{ij} = [R]_{ik} [R]_{jl} C_{kl}, \tag{11.5}
$$

all with the **same** rotation matrix $[R]$, and a factor of $[R]$ for each index. Since each term in a tensor equation must have the same transformation property, such an equation will be automatically unchanged under rotation. We illustrate this with the vector equation $\alpha A_i + \beta B_i = 0$, with ($\alpha, \beta$) being scalars. Each term transforms as a vector and will automatically keep its form in the transformed coordinate system: as $\alpha A_i + \beta B_i = 0$ in one coordinate frame implies $\alpha' A'_i + \beta' B'_i = [R]_{ij} (\alpha A_j + \beta B_j) = 0$ in the transformed frame. Thus, in order to implement rotational symmetry, all one needs to do is to write physics equations as tensor equations.

11.1.2   The Lorentz transformation as a rotation in 4D spacetime

The constancy of light speed requires[3] the invariance of the general spacetime interval s^2 of (11.1). With this identification of s as the length, the Lorentz transformation is simply "rotation" in spacetime. Concentrating on rotations in the subspace spanned by the time coordinate ($w \equiv ict$) and one of the space

[2] The steps are similar to the derivation of the Lorentz transformation given in Chapter 10.

[3] For a proof of this direct connection, see Cheng (2010, p. 42).

coordinates (x), we have the length $s^2 = w^2 + x^2$ being invariant under the rotation,

$$\begin{pmatrix} w' \\ x' \end{pmatrix} = \begin{pmatrix} \cos\theta & \sin\theta \\ -\sin\theta & \cos\theta \end{pmatrix} \begin{pmatrix} w \\ x \end{pmatrix}. \tag{11.6}$$

Putting back $w = ict$, we have

$$ct' = \cos\theta\ ct - i\sin\theta\ x$$

$$x' = -i\sin\theta\ ct + \cos\theta\ x. \tag{11.7}$$

After a re-parametrization of the rotation angle $\theta = i\psi$ [4] and using the identities $\cos(i\psi) = \cosh\psi$ and $-i\sin(i\psi) = \sinh\psi$, we recognize (11.7) as the usual Lorentz transformation as shown in Eq. (10.8):

[4] $\cos(i\psi) = \left(e^{-\psi} + e^{+\psi}\right)/2 = \cosh\psi.$

$$\begin{pmatrix} ct' \\ x' \end{pmatrix} = \begin{pmatrix} \cosh\psi & \sinh\psi \\ \sinh\psi & \cosh\psi \end{pmatrix} \begin{pmatrix} ct \\ x \end{pmatrix} = \begin{pmatrix} \gamma & -\beta\gamma \\ -\beta\gamma & \gamma \end{pmatrix} \begin{pmatrix} ct \\ x \end{pmatrix}. \tag{11.8}$$

To reach the last equality we have used

$$\cosh\psi = \gamma \quad \text{and} \quad \sinh\psi = -\beta\cosh\psi \tag{11.9}$$

with the standard notation of

$$\beta = \frac{v}{c} \quad \text{and} \quad \gamma = \frac{1}{\sqrt{1-\beta^2}}. \tag{11.10}$$

The relations (11.9) between ψ and the relative velocity v can be derived by considering, for example, the motion of the $x' = 0$ origin in the O system. Plugging $x' = 0$ into the first equation in (11.8),

$$x' = 0 = ct\sinh\psi + x\cosh\psi \quad \text{or} \quad \frac{x}{ct} = -\tanh\psi. \tag{11.11}$$

The coordinate origin $x' = 0$ moves with velocity $v = x/t$ along the x axis of the O system, thus $\beta = -\tanh\psi$. From the identity $\cosh^2\psi - \sinh^2\psi = 1$, which may be written as $\cosh\psi\sqrt{1-\tanh^2\psi} = 1$, we find the relations in (11.9).

11.2   Tensors in a flat spacetime

Once we have identified the Lorentz transformation as a rotation in spacetime, all our knowledge of rotational symmetry can be applied to this relativity coordinate symmetry. In particular 4D spacetime tensor equations are automatically relativistic. First, we need to study in some detail the geometric structure of this 4D spacetime manifold.

11.2.1   Tensor contraction and the metric

To set up a coordinate system for the 4D Minkowski space means to choose a set of four **basis vectors** $\{\mathbf{e}_\mu\}$, where $\mu = 0, 1, 2, 3$. Each $\mathbf{e}_\mu$, for a definite index value μ, is a 4D vector (Fig. 11.1). In contrast to the Cartesian coordinate system in Euclidean space, this in general is not an orthonormal

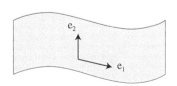

Fig. 11.1 Basis vectors for a 2D surface.

set, $\mathbf{e}_\mu \cdot \mathbf{e}_\nu \neq \delta_{\mu\nu}$, Nevertheless, we can represent such a collection of scalar products among the basis vectors as a symmetric matrix, called[5] the **metric**, or the **metric tensor**:

$$\mathbf{e}_\mu \cdot \mathbf{e}_\nu \equiv g_{\mu\nu}. \tag{11.12}$$

We can display the metric as a 4×4 matrix with elements being dot products of basis vectors:

$$[g] = \begin{pmatrix} g_{00} & g_{01} & .. \\ g_{10} & g_{11} & .. \\ \vdots & \vdots & \end{pmatrix} = \begin{pmatrix} \mathbf{e}_0 \cdot \mathbf{e}_0 & \mathbf{e}_0 \cdot \mathbf{e}_1 & .. \\ \mathbf{e}_1 \cdot \mathbf{e}_0 & \mathbf{e}_1 \cdot \mathbf{e}_1 & .. \\ \vdots & \vdots & \end{pmatrix}. \tag{11.13}$$

Thus, the diagonal elements are the (squared) lengths of the basis vectors, $|\mathbf{e}_0|^2$, $|\mathbf{e}_1|^2$, etc., while the off-diagonal elements represent their deviations from orthogonality. Any set of mutually perpendicular bases would be represented by a diagonal metric matrix.

The inverse basis vectors and the inverse metric

In Cartesian coordinate space we have "the basis vector being its own inverse" $\mathbf{e}_\mu \cdot \mathbf{e}_\nu = \delta_{\mu\nu}$, i.e. the Cartesian metric for Cartesian space is simply the identity matrix $[g] = [1]$. Minkowski spacetime being non-Cartesian, we can introduce a set of **inverse basis vectors** $\{\mathbf{e}^\mu\}$. The standard notation is to use subscript indices to label regular basis vectors and superscript indices to label inverse basis vectors. The relationship between regular basis vectors and inverse basis vectors is expressed as an orthonormality condition through their dot products:

$$\mathbf{e}_\mu \cdot \mathbf{e}^\nu = \delta_\mu{}^\nu. \tag{11.14}$$

Just like (11.12), we can define the **inverse metric tensor** $g^{\mu\nu}$:

$$\mathbf{e}^\mu \cdot \mathbf{e}^\nu \equiv g^{\mu\nu} \quad \text{and} \quad g_{\mu\nu} g^{\nu\lambda} = \delta_\mu{}^\lambda. \tag{11.15}$$

Inner products in terms of contravariant and covariant components

Because there are two sets of coordinate basis vectors, $\{\mathbf{e}_\mu\}$ and $\{\mathbf{e}^\mu\}$, there are two possible expansions for each vector $\mathbf{A}$:

Expansion of $\mathbf{A}$	Projections	Component names
$\mathbf{A} = A^\mu \mathbf{e}_\mu$	$A^\mu = \mathbf{A} \cdot \mathbf{e}^\mu$	**contravariant** components of $\mathbf{A}$
$\mathbf{A} = A_\mu \mathbf{e}^\mu$	$A_\mu = \mathbf{A} \cdot \mathbf{e}_\mu$	**covariant** components of $\mathbf{A}$

$$\tag{11.16}$$

The scalar product of any two vectors in terms of either contravariant or covariant components alone would involve the metric matrices:

$$\mathbf{A} \cdot \mathbf{B} = g_{\mu\nu} A^\mu B^\nu. \tag{11.17}$$

Similarly, $\mathbf{A} \cdot \mathbf{B} = g^{\mu\nu} A_\mu B_\nu$. One of the principal advantages of introducing these two types of tensor components is that products are simplified—they can be written without the metrics:

$$\mathbf{A} \cdot \mathbf{B} = (A_\nu \mathbf{e}^\nu) \cdot (B^\mu \mathbf{e}_\mu) = A_\nu (\mathbf{e}^\nu \cdot \mathbf{e}_\mu) B^\mu = A_\mu B^\mu. \tag{11.18}$$

The summation of a pair of a superscript and a subscript index is called a **contraction**. It reduces the index number (called the rank) of each tensor by one unit. A comparison of (11.18) with (11.17) and its subsequent equation shows that the contravariant and covariant components of a vector are related to each other through the metric:

$$A_\mu = g_{\mu\nu}A^\nu \qquad A^\mu = g^{\mu\nu}A_\nu. \tag{11.19}$$

We say that tensor indices can be lowered or raised through contractions with the metric tensor or inverse metric tensor.

11.2.2 Minkowski spacetime is pseudo-Euclidean

Consider in particular a 4D space with position coordinates being x_μ with Greek index $\mu = 0, 1, 2, 3$ so that

$$x^\mu = \left(x^0, x^1, x^2, x^3\right) = (ct, x, y, z). \tag{11.20}$$

This is a 4-vector as it satisfies the transformation property, according to (11.8):

$$x'^\mu = [L]^\mu_\nu x^\nu \qquad \text{with} \qquad [L]^\mu_\nu = \begin{pmatrix} \gamma & -\beta\gamma & 0 & 0 \\ -\beta\gamma & \gamma & 0 & 0 \\ 0 & 0 & 1 & 0 \\ 0 & 0 & 0 & 1 \end{pmatrix}. \tag{11.21}$$

For the case of the vectors being infinitesimal coordinate vectors, $\mathbf{A} = \mathbf{B} = d\mathbf{x}$, the scalar product of (11.18) is just the invariant interval $d\mathbf{x} \cdot d\mathbf{x} = ds^2$:

$$ds^2 = g_{\mu\nu}dx^\mu dx^\nu. \tag{11.22}$$

This equation relating the length to the coordinates is often taken as another definition of the metric. It is just the infinitesimal version of Eq. (11.1) from which we identify the metric for Minkowski spacetime as

$$g_{\mu\nu} = \begin{pmatrix} -1 & 0 & 0 & 0 \\ 0 & 1 & 0 & 0 \\ 0 & 0 & 1 & 0 \\ 0 & 0 & 0 & 1 \end{pmatrix} \equiv \text{diag}(-1, 1, 1, 1). \tag{11.23}$$

It can be shown that the geometry of the manifold can be determined by the metric, as geometric properties (length, angle, and shape) of the space are fixed by the metric.[6] The metric elements themselves can in turn be determined by length measurements. An inspection of (11.22) shows that once we have picked the coordinate system (x^μ) the elements of $[g]$ are related to length measurements of various ds^2 (depending on the choice the x^μ directions). The metric (11.23) tells us that Minkowski spacetime is flat because its elements are coordinate independent; it is pseudo-Euclidean as it differs from Euclidean space by a sign change of the g_{00} element. This particular pseudo-Euclidean metric is often denoted by the specific symbol $\eta_{\mu\nu} \equiv \text{diag}(-1, 1, 1, 1)$.

[6] For further discussion, see Section 12.4.1.

Position and position derivatives

We have chosen in (11.20) the position elements as components of a contravariant vector (with an upper index). Given the way one can raise and lower an index as in (11.19), we see that the covariant version of the position vector is

$x_\mu = (-ct, x, y, z)$. While the position vector is "naturally contravariant", the closely related (4-) del operator is "naturally covariant":

$$\partial_\mu \equiv \frac{\partial}{\partial x^\mu} = \left(\frac{1}{c} \frac{\partial}{\partial t} \quad \frac{\partial}{\partial x} \quad \frac{\partial}{\partial y} \quad \frac{\partial}{\partial z} \right), \tag{11.24}$$

so that $\partial_\mu x^\nu = \delta_\mu^{\ \nu}$. Also a contraction 2 version of the del operator, just like (11.22), leads to the Lorentz-invariant 4-Laplacian operator (called the **D'Alembertian**):

$$\Box \equiv \partial^\mu \partial_\mu = -\frac{1}{c^2} \frac{\partial^2}{\partial t^2} + \nabla^2, \tag{11.25}$$

with the 3-Laplacian operator being $\nabla^2 = \partial^2/\partial x^2 + \partial^2/\partial y^2 + \partial^2/\partial z^2$. Thus the relativistic wave equation has the form $\Box \psi = -c^{-2} \partial^2 \psi / \partial t^2 + \nabla^2 \psi = 0$.

Reflecting the fact that a contraction of a pair of contravariant and covariant vectors is a Lorentz scalar, the contravariant and covariant vectors must transform oppositely (i.e. under a transformation matrix and an inverse transformation matrix, respectively). In particular, we have

$$\begin{pmatrix} x_0' \\ x_1' \end{pmatrix} = \begin{pmatrix} \gamma & -\beta\gamma \\ -\beta\gamma & \gamma \end{pmatrix} \begin{pmatrix} x_0 \\ x_1 \end{pmatrix} \tag{11.26}$$

versus

$$\begin{pmatrix} \partial_0' \\ \partial_1' \end{pmatrix} = \begin{pmatrix} \gamma & \beta\gamma \\ \beta\gamma & \gamma \end{pmatrix} \begin{pmatrix} \partial_0 \\ \partial_1 \end{pmatrix}. \tag{11.27}$$

11.2.3 Relativistic velocity, momentum, and energy

Four-velocity

While the position 4-vector x^μ is given by (11.20), the 4-velocity is *not dx_μ/dt*, because t is not a Lorentz scalar. The derivative that transforms as a 4-vector is the one with respect to proper time τ, which is a scalar—directly related to the invariant spacetime separation $s^2 = -c^2 d\tau^2$ as shown in Eq. (10.18):

$$U^\mu = \frac{dx^\mu}{d\tau} = \gamma \frac{dx^\mu}{dt} = \gamma \left(c, v_x, v_y, v_z \right), \tag{11.28}$$

where we have used the time dilation relation of $t = \gamma\tau$ and $v_x = dx/dt$, etc. We can form its length by the operation of contraction (11.18)

$$U^\mu U_\mu = \gamma^2 \left(-c^2 + v^2 \right) = -c^2, \tag{11.29}$$

which is clearly an invariant.

Four-momentum

Naturally we will define the relativistic 4-momentum as

$$p^\mu \equiv m U^\mu = \gamma \left(mc, \mathbf{p}_{NR} \right) = \left(\frac{E}{c}, \mathbf{p} \right) \tag{11.30}$$

with $\mathbf{p}_{NR} = m\mathbf{v}$ being the familiar nonrelativistic momentum. We identify the relativistic 3-momentum as $\mathbf{p} = \gamma m\mathbf{v}$ (as discussed in Section 10.7) and relativistic energy $p_0 c = \gamma mc^2 \equiv E$ (as discussed in Section 10.5). Thus the ratio of relativistic momentum and energy can be expressed as that of velocity to c^2:

$$\frac{\mathbf{p}}{E} = \frac{\mathbf{v}}{c^2}. \tag{11.31}$$

The momentum and energy transform into each other under the Lorentz transformation just as space and time do. Following (11.29), we have the invariant magnitude, which leads to the well-known energy and momentum relation:

$$p_\mu p^\mu = -m^2 c^2 = -\frac{E^2}{c^2} + \mathbf{p}^2. \tag{11.32}$$

Massless particles always travel at speed c

When $m = 0$, we can no longer define the 4-momentum as $p^\mu = mU^\mu$; nevertheless, since a massless particle has energy and momentum we can still assign a 4-momentum to such a particle,[7] with components just the RHS of (11.30). When $m = 0$, the relation (11.32) with $p = |\mathbf{p}|$ becomes

$$E = pc. \tag{11.33}$$

Plugging this into the ratio of (11.31), we obtain the well-known result that massless particles such as photons and gravitons[8] always travel at the speed of $v = c$. Hence there is no rest-frame for massless particles.

11.2.4 The electromagnetic field tensor

The four spacetime coordinates form a 4-vector, and, similarly, the energy momentum components of a particle form another 4-vector. What sort of tensor can six components of the EM fields $\mathbf{E}$ and $\mathbf{B}$ form? It turns out they are the components of an antisymmetric tensor (rank 2) $F_{\mu\nu} = -F_{\nu\mu}$:

$$F_{\mu\nu} = \begin{pmatrix} 0 & -E_1 & -E_2 & -E_3 \\ E_1 & 0 & B_3 & -B_2 \\ E_2 & -B_3 & 0 & B_1 \\ E_3 & B_2 & -B_1 & 0 \end{pmatrix}. \tag{11.34}$$

Maxwell's equations can then be written in compact form: Gauss' and Ampere's laws of (9.4) are

$$\partial_\mu F^{\mu\nu} = -\frac{1}{c} j^\nu \tag{11.35}$$

where $j^\nu = (\rho c, \mathbf{j})$ is the 4-current density, while Faraday's law and the magnetic Gauss' law[9] of (9.3) are

$$\partial_\lambda F_{\mu\nu} + \partial_\nu F_{\lambda\mu} + \partial_\mu F_{\nu\lambda} = 0. \tag{11.36}$$

That electromagnetic fields form a tensor of rank 2 means that under a Lorentz transformation they must change according to the rule for higher rank tensors—a transformation matrix factor for each index, cf. Eq. (11.5). Thus we have

$$F_{\mu\nu} \longrightarrow F'_{\mu\nu} = [L]_\mu{}^\lambda [L]_\nu{}^\rho F_{\lambda\rho}. \tag{11.37}$$

[7]While we do not have $p^\mu = mU^\mu$, we still have the proportionality of the 4-momentum to its 4-velocity, $p^\mu \propto U^\mu$, with the 4-velocity defined as $U^\mu = dx^\mu/d\lambda$. Since there is no rest-frame for a massless particle, the curve parameter λ cannot be the proper time (being defined in the rest-frame). In fact one can choose λ in such a way that $p^\mu = dx^\mu/d\lambda$.

[8]Gravitons are the quanta of the gravitational field, just as photons are the quanta of the electromagnetic field.

[9]The homogeneous Maxwell's equations can also be written in an equivalent form $\partial_\mu \tilde{F}^{\mu\nu} = 0$, with the dual EM field tensor $\tilde{F}^{\mu\nu} = -\frac{1}{2} \varepsilon^{\mu\nu\lambda\rho} F_{\lambda\rho}$. In terms of their elements the replacement of $F^{\mu\nu} \to \tilde{F}^{\mu\nu}$ corresponds to the dual rotation of $\mathbf{E} \to \mathbf{B}$ and $\mathbf{B} \to -\mathbf{E}$, mentioned in Section 9.5. The task of identifying Eq. (11.36) with the more familiar Maxwell equations will be easier if we proceed through this dual field tensor $\tilde{F}^{\mu\nu}$.

[10] See Cheng (2010, Box 12.1, p. 287).

One can easily check[10] that this is precisely the transformation relations (10.30) and (10.31) derived in Section 10.4. The Maxwell equations are now written in Minkowski tensor form (11.35) and (11.36), and their covariance under Lorentz transformation is "manifest".

11.2.5 The energy–momentum–stress tensor for a field system

The structure of the charge 4-current density

The charge density ρ and 3D current density components j_i can be transformed into each other under a Lorentz transformation. This simply reflects the fact that a stationary charge will be viewed as a current by a moving observer. Thus together they form the four components of a 4-current density $j^\mu = (\rho c, j_i)$. Recall the physical meanings of charge density being charge (q) per unit volume ($\triangle x \triangle y \triangle z$) and current density, say in the x direction, being the amount of charge moved over a cross-sectional area ($\triangle y \triangle z$) in unit time $\triangle t$:

$$\rho = \frac{q}{\triangle x^1 \triangle x^2 \triangle x^3} \qquad j_x = \frac{q}{\triangle t \triangle x^2 \triangle x^3} = \frac{cq}{\triangle x^0 \triangle x^2 \triangle x^3}. \tag{11.38}$$

Thus for the interpretation of all four components j^μ we have the simple compact expression

$$j^\mu = \frac{cq}{\triangle S^\mu} \tag{11.39}$$

where the symbol $\triangle S^\mu$ stands for the Minkowski volume (a "3-surface") with one particular interval $\triangle x^\mu = 0$ (i.e. the μ component x held fixed). We also note that charge conservation of a field system is expressed in terms of the continuity equation (cf. Section 1.4)

$$\frac{d\rho}{dt} + \nabla \cdot \mathbf{j} = 0, \tag{11.40}$$

which can be compactly written in this 4-tensor formalism as a vanishing 4-divergence equation $\partial_\mu j^\mu = 0$.

The structure of the energy–momentum–stress 4-current density

The above 4-current relates the transformation and conservation of a scalar quantity, the charge q. However, if we wish to generalize the case to energy and momentum, which are themselves members of a 4-vector p_μ, the relevant currents must have two indices (hence rank 2):

$$T^{\mu\nu} = \frac{cp^\mu}{\triangle S^\nu} \tag{11.41}$$

which is a 4×4 symmetric matrix $T^{\mu\nu} = T^{\nu\mu}$. The energy–momentum conservation of a system consisting of continuum medium (e.g. a field system) can then be expressed by a vanishing divergence of $T^{\mu\nu}$:

$$\partial_\mu T^{\mu\nu} = 0. \tag{11.42}$$

The physical meaning of $T^{\mu\nu}$'s ten independent components can be worked out from (11.41):

- T_{00} is the energy density of the system.
- $T_{0i} = T_{i0}$ are the three components of the momentum density or, equivalently, the energy current density.
- T_{ij} is a 3×3 symmetric matrix with diagonal elements being the pressure (i.e. normal force per unit area) in the three directions and off-diagonal elements being the shear (parallel) force per unit area.

In short, the energy–momentum tensor, also called the stress–energy tensor, describes the distribution of matter/energy in spacetime. As we shall see, the tensor $T^{\mu\nu}$ will appear as the inhomogeneous source term for gravity in the Einstein field equation in his general theory of relativity.

Stress–energy tensor for an ideal fluid

The case of the energy–momentum tensor of an ideal fluid is a particularly important one. Here the fluid elements interact only through a perpendicular force (no shear). Thus the fluid can be characterized by two parameters: the mass density ρ and an isotropic pressure[11] p (no viscosity). In the comoving frame,[12] where each fluid element may be regarded as momentarily at rest, the stress tensor, according to (11.41), takes on a particularly simple form

[11] Do not confuse pressure with the magnitude of the 3-momentum, as both are given the symbol p.

[12] The comoving frame is defined as the coordinate system in which the fluid elements themselves carry the position coordinate labels and clocks that are synchronized.

$$T^{\mu\nu} = \begin{pmatrix} \rho c^2 & & & \\ & p & & \\ & & p & \\ & & & p \end{pmatrix} = \left(\rho + \frac{p}{c^2}\right) U^\mu U^\nu + p g^{\mu\nu}, \qquad (11.43)$$

where $g^{\mu\nu}$ is the inverse metric tensor, which for a flat spacetime is the SR invariant $\eta^{\mu\nu}$. To reach the last equality we have used the fact that in the comoving frame the fluid velocity field is simply $U^\mu = (c, \mathbf{0})$. Since ρ and p are quantities in the rest (comoving) frame (hence Lorentz scalars), the RHS is a proper tensor of rank 2 and this expression should be valid in every frame (as Lorentz transformations will not change its form).

The even simpler case, when the pressure term is absent, $p = 0$, corresponds to the case of a swamp of noninteracting particles, i.e. a "cloud of dust".

Nonrelativistic limit and the Euler equation It is instructive to consider the nonrelativistic limit ($\gamma \simeq 1$) of the energy–momentum tensor of an ideal fluid (11.43). In this limit, with $U^\mu = \gamma(c, v^i) \simeq (c, v^i)$, since the rest energy dominates over the pressure (which results from particle momenta), $\rho c^2 \gg p$, the tensor in (11.43) takes on the form[13]

[13] Keep in mind that the matrix form of $T^{\mu\nu}$ in (11.43) is valid only for the comoving frame, while that in (11.44) is for general nonrelativistic moving frames.

$$T^{\mu\nu} \overset{NR}{=} \begin{pmatrix} \rho c^2 & \rho c v^i \\ \rho c v_j & \rho v^i v^j + p \delta^{ij} \end{pmatrix}. \qquad (11.44)$$

The lower right element is actually a 3×3 matrix, which one recognizes as the ideal fluid tensor we used in Chapter 1, Eq. (1.6). From this the Euler equation emerges[14] as the nonrelativistic version of the energy–momentum conservation equation of (11.42).

[14] See Box 12.3, p. 293 in Cheng (2010) for the details of this calculation.

11.3   The spacetime diagram

Relativity brings about a profound change in the causal structure of space and time, which can be nicely visualized in terms of the **spacetime diagram** (Minkowski 1908) with time being one of the coordinates.[15] Let us first recall the corresponding causal structure of space and time in pre-relativity physics. Here the key feature of simultaneous events comes about in this way. For a given event P at a particular point in space and a particular instant of time, all the events that could in principle be reached by a particle starting from P are collectively labeled as the **future of P**, while all the events from which a particle can arrive at P form the **past of P**. Those events that are neither the future nor the past of the event P form a 3D set of events **simultaneous with P**. This notion of simultaneous events allows one to discuss in pre-relativity physics all of the space at a given instant of time, and as a corollary, allows one to study space and time separately. In relativistic physics, the events that fail to be causally connected to event P are much larger than a 3D space. As we shall see, all events outside the future and past lightcones are causally disconnected from the event P, which lies at the tip of the (the future and past) lightcones in the spacetime diagram.

11.3.1   Basic features and invariant regions

An event with coordinates (t, x, y, z) is represented by a **worldpoint** in the spacetime diagram. The history of events becomes a line of worldpoints, called a **worldline**. In Fig. 11.2, the 3D space is represented by a 1D x axis. In particular, a light signal $\Delta s^2 = \Delta x^2 + \Delta y^2 + \Delta z^2 - c^2 \Delta t^2 = 0$ passing through the origin is represented by a straight worldline at a 45° angle with respect to the axes: $\Delta x^2 - c^2 \Delta t^2 = 0$, thus $c\Delta t = \pm \Delta x$. Any line with constant velocity $v = |\Delta x / \Delta t|$ would be a straight line passing through the origin. We can clearly see that those worldlines with $v = \Delta x / \Delta t < c$, corresponding to $\Delta s^2 < 0$, would make an angle greater than 45° with respect to the spatial axis (i.e. above the worldlines for a light ray). According to relativity, no worldline can have $v > c$. If there had been such a line, it would correspond to $\Delta s^2 > 0$, and would make an angle less than 45° (i.e. below the light worldline). Since

[15]To have the same length dimension for all coordinates, the temporal axis is represented by $x^0 = ct$.

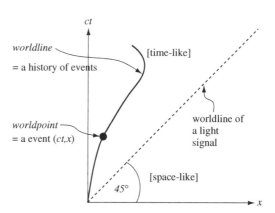

Fig. 11.2 Basic elements of a spacetime diagram, with two spatial coordinates (y, z) suppressed.

Δs^2 is invariant, it is meaningful to divide the spacetime diagram into regions, as in Fig. 11.3, corresponding to

$\Delta s^2 < 0$	time-like
$\Delta s^2 = 0$	light-like
$\Delta s^2 > 0$	space-like

where the names of the regions are listed in the right-hand column. The coordinate intervals being $c\Delta t = ct_2 - ct_1$, $\Delta x = x_2 - x_1$, etc., consider the separation of two events: one at the origin $(ct_1, \mathbf{x}_1) = (0, \mathbf{0})$, the other at a point in one of the regions $(ct_2, \mathbf{x}_2) = (ct, \mathbf{x})$:

- The light-like region has all the events which are connected to the origin with a separation of $\Delta s^2 = 0$. This corresponds to events that are connected by light signals. The 45° incline in Fig. 11.3, in which two spatial dimensions are displayed, forms a **lightcone**. It has a slope of unity, which reflects the fact that the speed of light is c. A vector that connects an event in this region to the origin, called **a light-like vector, is** a nonzero 4-vector having zero length, a **null vector**.
- The space-like region has all the events which are connected to the origin with a separation of $\Delta s^2 > 0$. (The 4-vector from the origin in this region is a **space-like vector**—having a positive squared length.) In the space-like region, it takes a signal traveling at a speed greater than c in order to connect an event to the origin. Thus, an event taking place at any point in this region cannot be influenced causally (in the sense of cause-and-effect) by an event at the origin.
- The time-like region has all the events which are connected to the origin with a separation of $\Delta s^2 < 0$. One can always find a frame O' such that such an event takes place at the same location, $x' = 0$, but at different time, $t' \neq 0$. This makes it clear that events in this region **can be causally connected** with the origin. In fact, all the worldlines passing through the origin will be confined to this region, inside the lightcone.[16]
- In Fig. 11.3, we have displayed the lightcone structure with respect to the origin of the spacetime coordinates. It should be emphasized that each point in a spacetime diagram has a lightcone. The timelike regions with respect to several worldpoints are represented by the lightcones shown in Fig. 11.4. If we consider a series of lightcones having their vertices located along a given worldline, each subsequent segment must lie within the lightcone of that point (at the beginning of that segment). It is the clear from Fig. 11.4 that any particle can only proceed in the direction of ever-increasing time. We cannot stop our biological clocks!

11.3.2 Lorentz transformation in the spacetime diagram

The nontrivial parts of the Lorentz transformation (11.26) of intervals are

$$\Delta x' = \gamma(\Delta x - \beta c\Delta t), \qquad c\Delta t' = \gamma(c\Delta t - \beta \Delta x) \qquad (11.45)$$

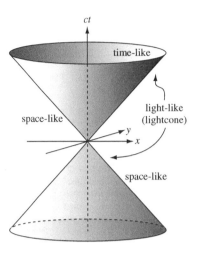

Fig. 11.3 Invariant regions in the spacetime diagram, with one of the spatial coordinates suppressed.

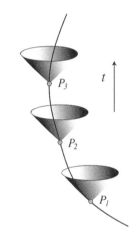

Fig. 11.4 Lightcones with respect to different worldpoints, $P_1, P_2, \ldots$, etc. along a time-like worldline, which can only proceed in the direction of ever-increasing time as each segment emanating from a given worldpoint must be contained within the lightcone with that point as its vertex.

[16]The worldline of an inertial observer (i.e. moving with constant velocity) must be a straight line inside the lightcone. This straight line is just the time axis of the coordinate system in which the inertial observer is at rest.

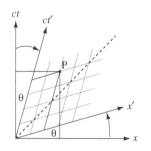

Fig. 11.5 Lorentz rotation in the spacetime diagram. The space and time axes rotate by the same amount but in opposite directions so that the lightcone (the dashed line) remains unchanged. The shaded grid represents lines of fixed x' and t'.

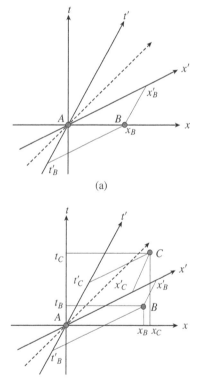

Fig. 11.6 (a) Relativity of simultaneity: $t_A = t_B$ but $t'_A > t'_B$. (b) Relativity of event order: $t_A < t_B$ but $t'_A > t'_B$. However, there is no change of event order with respect to A for all events located above the x' axis, such as the event C. This certainly includes the situation in which C is located in the forward lightcone of A (above the dashed line).

(taken, for example, with respect to the origin). We can represent these transformed axes in the spacetime diagram:

- The x' axis corresponds to the $c\Delta t' = 0$ line. This corresponds, according to the second equation above, to a line satisfying the relationship $c\Delta t = \beta\Delta x$. Hence, the x' axis is a straight line in the x-ct plane with a slope of $c\Delta t/\Delta x = \beta$.
- The ct' axis corresponds to the $\Delta x' = 0$ line. This corresponds, according to the first equation above, to a line satisfying the relationship $\Delta x = \beta c\Delta t$. Hence, the ct' axis is a straight line with a slope of $c\Delta t/\Delta x = 1/\beta$.

Depending on whether β is positive or negative, the new axes either "close in" or "open up" from the original perpendicular axes. Thus we have the **opposite-angle rule**: the two axes make opposite-signed rotations of $\pm\theta$ (Fig. 11.5). The x axis rotates by $+\theta$ relative to the x' axis; the ct axis, by $-\theta$ relative to the ct' axis. The physical basis for this rule is the need to maintain the same slope $\left(= 1; \text{i.e. } \textbf{equal angles} \text{ with respect to the two axes}\right)$ for the lightcone in every inertial frame so that light speed is the same in every frame.

Relativity of simultaneity, event-order, and causality

It is instructive to use the spacetime diagram to demonstrate some of the physical phenomena we have discussed previously. In Fig. 11.6, we have two events A and B, with A being the origin of the coordinate system O and O': $\left(x_A = t_A = 0, x'_A = t'_A = 0\right)$. In Fig. 11.6(a), the events A and B are simultaneous, $t_A = t_B$, with respect to the O system. But in the O' system, we clearly have $t'_A > t'_B$. This shows the relativity of simultaneity.

In Fig. 11.6(b), we have $t_A < t_B$ in the O frame, but we have $t'_A > t'_B$ in the O' frame. Thus, the temporal order of events can be changed by a change of reference frames. However, this change of event order can take place only if event B is located in the region below the x' axis.[17] This means that if we increase the relative speed between these two frames O and O' (with the x' axis moving ever closer to the lightcone) more and more events can have their temporal order (with respect to A at the origin) reversed as seen from the perspective of the moving observer. On the other hand, for all events above the x' axis, the temporal order is preserved. For example, with respect to event C, we have both $t_A < t_C$ and $t'_A < t'_C$. Now, of course, the region above this x' axis includes the forward lightcone of event A. This means that for two events that are causally connected (between A and any worldpoint in its forward lightcone), their temporal order cannot be changed by a Lorentz transformation. The principle of causality is safe under special relativity.

---

[17] The x' axis having a slope $1/\beta$ means that the region below it corresponds to $(\Delta x/\Delta t) > c/\beta$. This is clearly in agreement with the Lorentz transformation $\Delta t' = \gamma(\Delta t - \beta\Delta x/c)$ to have opposite sign to Δt.

11.4   The geometric formulation—a summary

Let us summarize the principal lessons we have learnt from this geometric formulation of special relativity:

- The stage on which physics takes place is Minkowski spacetime with the time coordinate being on an equal footing with spatial coordinates. "Space and time are treated symmetrically." A spacetime diagram is often useful in clarifying ideas in relativity.
- Minkowski spacetime has a pseudo-Euclidean metric $\eta_{\mu\nu} = \text{diag}(-1, 1, 1, 1)$, corresponding to an invariant length $s^2 = -c^2t^2 + x^2 + y^2 + z^2$.
- The length-preserving transformation in spacetime is a Lorentz transformation, from which all the special relativistic effects such as time dilation, length contraction, and relativity of simultaneity can be derived. Thus, in this geometric formulation, we can think of the metric as embodying all of special relativity.
- That one can understand special relativity as a theory of flat geometry in Minkowski spacetime is the crucial step in the progression towards general relativity. In general relativity, as we shall see, this geometric formulation is generalized into a warped spacetime. The corresponding metric must be position-dependent, $g_{\mu\nu}(x)$, and this metric acts as the generalized gravitational potential.
- In our historical introduction, SR seems to be all about light, but the speed c actually plays a much broader role in relativity:
  - c is the universal conversion factor between space and time coordinates that allows space and time to be treated symmetrically.
  - c is just the speed so that $\Delta s^2 = -c^2\Delta t^2 + \Delta \mathbf{x}^2$ is an invariant interval under coordinate transformations. This allows Δs to be viewed as the length in spacetime.
  - c is the universal maximal and absolute speed of signal transmission: Massless particles (e.g. photons and gravitons) travel at the speed c, while all other ($m \neq 0$) particles move at a slower speed.

GENERAL RELATIVITY

Towards a general theory of relativity

12

- Among Einstein's motivations for general relativity we will concentrate on his effort to have a deeper understanding of the empirically observed equality between gravitational and inertial masses. This equality is the manifestation in mechanics of the equivalence between inertia and gravitation. Einstein elevated this equality to a general principle, the equivalence principle (EP), in order to extract its implications beyond mechanics. A brief history of EP is presented.

- In special relativity (1905) physics laws are not changed under the (e.g. boost) coordinate transformation that connects different inertial frames of reference. It is "special" because one is still restricted to inertial frames. The equivalence between inertial and gravitation means that any reference frame, even an accelerated one, can be regarded as an inertial frame (with gravity). This loss of the privileged status of inertial frames means we cannot incorporate gravity into special relativity. Rather, Einstein taught us that we could use gravity as a means to broaden the principle of relativity from inertial frames to all coordinate systems including accelerating frames. Thus the general theory of relativity (1915) is automatically a theory of gravitation.

- The generalization of EP to electromagnetism implies gravitational redshift, gravitational time dilation, and gravitational bending of a light ray. In the process, we discuss the notion of the spacetime-dependent light velocity and the gravity-induced index of refraction.

- By thinking deeply about EP physics, Einstein came up with the idea that the gravitational effects on a body can be attributed directly to some underlying spacetime feature and gravity is nothing but the structure of warped spacetime. The relativistic gravitational field is the curved spacetime. General relativity is a field theory of gravity.

- The mathematics of a curved space is Riemannian geometry. We present some basic elements: Gaussian coordinates, metric tensor, the geodesic equation, and curvature.

Soon after completing his formulation of special relativity, Einstein started working on a relativistic theory of gravitation. In this chapter, we cover mainly the period 1907–13, when Einstein relied heavily on the **equivalence**

principle (EP) to extract some general relativity (GR) results. He also worked with his friend and colleague Marcel Grossman (1878–1936), applying Riemannian geometry to the gravity problem. Not until the end of 1915 did Einstein find the correct field equation and worked out fully his formulation of the general theory of relativity, his theory of gravitation.

12.1   Einstein's motivations for general relativity

Einstein's theory of gravitation has a unique history. It is not prompted by any failure (crisis) of Newton's theory, but resulted from the "pure thought" of one person. "Einstein just stared at his own navel and came up with GR!"

From Einstein's published papers one can infer several interconnected motivations:

1. **To have a relativistic theory of gravitation**. The Newtonian theory is not compatible with special relativity as it invokes the notion of an action-at-a-distance force, implying a signal transmission speed greater than c. Furthermore, inertia frames of reference lose their privileged status in the presence of gravity.
2. "**Space is not a thing**." This is how Einstein phrased his conviction that physics laws should not depend on reference frames, which express the relationship among physical processes in the world and do not have independent existence.[1]
3. **Why are inertial and gravitational masses equal?** One wishes for a deeper understanding of this empirical fact.

[1] In this connection one should discuss the influence of Ernst Mach (1838–1916) on Einstein. However such a study is beyond the scope of this book. One can find a brief presentation of Mach's principle, as well as a more extensive discussion of Einstein's motivation for GR in Chapter 1 of Cheng (2010). See also the comments at the beginning of Chapter 15.

12.2   The principle of equivalence between inertia and gravitation

The equality between inertial and gravitational masses is closely related to Einstein's formulation of the equivalence principle. By studying the consequences of EP, he concluded that the proper language for general relativity is Riemannian geometry and gravity can be identified with the curvature of spacetime.

12.2.1   The inertia mass vs. the gravitational mass

Let us recall the equation of motion in Newton's gravitation theory. Consider a particle with mass m in a gravitational field $\mathbf{g}$. The gravitational force then is $\mathbf{F} = m\mathbf{g}$. Newton's second law, $\mathbf{F} = m\mathbf{a}$, then becomes

$$\frac{d^2\mathbf{r}}{dt^2} = \mathbf{g}. \tag{12.1}$$

This equation of motion has the distinctive feature that it is totally independent of any of the properties of the test particle. It comes about because of the cancellation of the mass factors from both sides of the equation. However, as we shall explain, these two masses represent very different physical concepts and their observed equality is an important experimental fact for which Einstein was seeking an explanation.

- The **inertial mass** of $\mathbf{F} = m_I\mathbf{a}$ enters into the description of the response of a particle to all forces.
- The **gravitational mass** of $\mathbf{F} = m_G\mathbf{g}$ reflects the response of a particle to a particular force: gravity. The gravitational mass may be regarded as the "gravitational charge" of a particle. (Recall the relation between the force and electric charge/field: $\mathbf{F} = e\mathbf{E}$.)

Universal gravitational acceleration and $m_G = m_I$

Consider two objects A and B, composed of different materials (copper, wood, etc.). When they are dropped in a gravitational field $\mathbf{g}$ (e.g. from the top of the Leaning Tower of Pisa) we have the equations of motion

$$(\mathbf{a})_A = \left(\frac{m_G}{m_I}\right)_A \mathbf{g}, \qquad (\mathbf{a})_B = \left(\frac{m_G}{m_I}\right)_B \mathbf{g}. \qquad (12.2)$$

Part of Galileo's great legacy is the experimental observation that all bodies fall with the same acceleration $(\mathbf{a})_A = (\mathbf{a})_B$ leading to the equality,

$$\left(\frac{m_G}{m_I}\right)_A = \left(\frac{m_G}{m_I}\right)_B. \qquad (12.3)$$

The mass ratio, having been found to be universal for all substances, can then be set (by an appropriate choice of units) to unity, hence the result

$$m_G = m_I. \qquad (12.4)$$

Even at the fundamental particle physics level, matter is made up of protons, neutrons, and electrons (all having different interactions) bound together with different binding energies, and it is difficult to find an a priori reason to expect such a relation (12.3). As we shall see, this is the empirical foundation underlying the geometric formulation of the relativistic theory of gravity that is GR.

The equality of gravitational and inertial masses: A brief history

Galileo   There is no historical record of Galileo having dropped anything from the Leaning Tower of Pisa. Nevertheless, to refute the Aristotelean contention that heavier objects would fall faster than light ones, he did report performing experiments of sliding different objects on an inclined plane, Fig. 12.1(a). (The slower fall allows for more reliable measurements.) More importantly, Galileo provided a theoretical argument, "a thought experiment", in the first chapter of his *Discourse and Mathematical Demonstration of Two New Sciences*, in support of the idea that all substances should fall with the same acceleration. Consider any falling object. Without this universality of free fall, the tendency of different components of the object to fall differently would give rise to internal stress and could cause certain objects to undergo spontaneous disintegration. The nonobservation of this phenomenon could then be taken as evidence for equal accelerations.

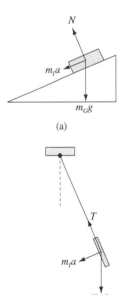

Fig. 12.1 Both the gravitational mass and inertial mass enter in phenomena: (a) sliding object on an inclined plane, and (b) oscillations of a pendulum.

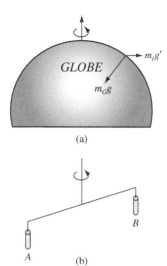

Fig. 12.2 The Eötvös experiment to detect any difference between the ratio of gravitational to inertial masses of substance A vs. B. The centrifugal acceleration $\mathbf{g}'$ can be decomposed into the vertical and horizontal components, $\mathbf{g}' = \mathbf{g}'_v + \mathbf{g}'_h$.

Newton   Newton went further by translating this universality of free fall into the universal proportionality of the inertial and gravitational masses (12.3) and built the equality $m_I = m_G$ right into the foundation of mechanics. Notably, he discussed this equality in the very beginning of his *Principia*. Furthermore, he improved upon the empirical check of the Galilean result (12.3) by experimenting with a pendulum, Fig. 12.1(b), finding:

$$\delta_{AB} \equiv \left| \frac{(m_I/m_G)_A - (m_I/m_G)_B}{(m_I/m_G)_A + (m_I/m_G)_B} \right| \leq 10^{-3}. \tag{12.5}$$

The Eötvös experiment and modern limits   At the end of the nineteenth century, the Hungarian baron Roland von Eötvös pointed out that any deviation from universality of this mass ratio (12.3) would show up as a horizontal twist τ in a torsion balance, Fig. 12.2(b). Two weights composed of different substances A and B are hung at the opposite ends of a rod, which is in turn hung from the ceiling by a fiber at a midpoint, respective distances l_A and l_B from the two ends. Because of earth's rotation, we are in a noninertial frame of reference. In order to apply Newton's laws, we must include the fictitious force, as represented by the centrifugal acceleration $\mathbf{g}'$, Fig. 12.2(a). In the vertical direction we have the gravitational acceleration $\mathbf{g}$, and the (tiny and, for our simplified calculation, negligible) vertical component $\mathbf{g}'_v$. In the horizontal direction the only nonzero torque is due to the horizontal component $\mathbf{g}'_h$. The equilibrium conditions of a vanishing total torque are:

$$\text{vertical balance}: \quad \left[l_A (m_G)_A - l_B (m_G)_B \right] g = 0 \tag{12.6}$$

$$\text{horizontal balance}: \quad \left[l_A (m_I)_A - l_B (m_I)_B \right] g'_h = \tau. \tag{12.7}$$

The equality of $l_A(m_G)_A = l_B(m_G)_B$ from (12.6) means that the twist in (12.7) is related to the sought-after nonuniversality:

$$\tau = \left[\left(\frac{m_I}{m_G} \right)_A - \left(\frac{m_I}{m_G} \right)_B \right] g'_h \, l m_G. \tag{12.8}$$

In this way Eötvös greatly improved the limit of (12.5), to $\delta_{AB} \leq 10^{-9}$. More recent experiments by others, ultimately involving the comparison of falling earth and moon in the solar gravitational field, have tightened this limit further to 1.5×10^{-13}.

12.2.2   "My happiest thought"

In the course of writing a review article on SR in 1907, Einstein came upon, what he later termed, "my happiest thought". He recalled the fundamental experimental result of Galileo that all objects fell with the same acceleration. "Since all bodies accelerate the same way, an observer in a freely falling laboratory will not be able to detect any gravitational effect (on a point particle) in this frame." Or, "gravity is transformed away in reference frames in free fall."

Principle of equivalence stated

Imagine an astronaut in a freely falling spaceship. Because all objects fall with the same acceleration, a released object in the spaceship will not be seen to fall. Thus, from the viewpoint of the astronaut, gravity is absent; everything becomes weightless. To Einstein, this vanishment of the gravitational effect is so significant that he elevated it (in order to focus on it) to a physical principle: **the equivalence principle.**

$$\left(\begin{array}{c}\text{Physics in a frame freely falling in a gravity field}\\ \text{is equivalent to}\\ \text{physics in an inertial frame without gravity}\end{array}\right).$$

Correspondingly,

$$\left(\begin{array}{c}\text{Physics in a nonaccelerating frame with gravity } \mathbf{g}\\ \text{is equivalent to}\\ \text{physics in a frame without gravity but accelerating with } \mathbf{a} = -\mathbf{g}\end{array}\right).$$

No special relativity theory of gravitation

Thus according to EP, accelerating frames of reference can be treated in exactly the same way as inertial frames. They are simply frames with gravity. Namely, in the presence of a gravitational field, inertial frames of reference lose their privileged status. This in turns means that special relativity is not applicable to any gravity theory. From this we also obtain a physical definition of an inertial frame, without reference to any external environment such as fixed stars, as the frame in which there is no gravity. Inertial forces (such as the Coriolis force) are just special kinds of gravitational forces. Einstein realized the unique position of gravitation in the theory of relativity. Namely, he understood that the question was not how to incorporate gravity into SR but rather how to use gravitation as a means to broaden the principle of relativity from inertial frames to all coordinate systems including accelerating frames.

12.3 Implications of the equivalence principle

If we confine ourselves to the physics of mechanics, EP is just a restatement of $m_I = m_G$. But once it is highlighted as a principle, it allowed Einstein to extend this equivalence between inertia and gravitation to **all physics** (not just to mechanics, but also to electromagnetism, etc.). This generalized version is sometimes called the **strong equivalence principle.** Thus the "weak EP" is just the statement $m_I = m_G$, while the "strong EP" is the principle of equivalence applied to all physics. In the following, we shall still call the strong equivalence principle EP for short.

To deduce the implications of EP, one adopts an approach similar to that used in our study of special relativity. In SR one compares the observations obtained in different frames in relative motion; here one compares the reference frame in free fall (thus no observed gravity) to the frame accelerating in a gravitational field.

12.3.1 Bending of a light ray

Let us first study the effect of gravity on a light ray traveling (horizontally) across a spaceship (Fig. 12.3) which is falling in a constant (vertical) gravitational field $\mathbf{g}$. From the viewpoint of the astronaut in the spaceship, EP

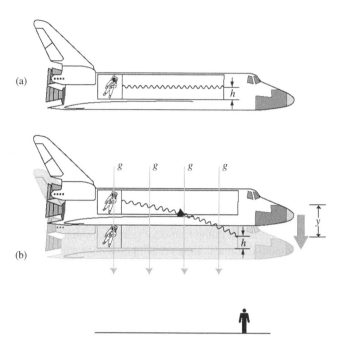

Fig. 12.3 According to the equivalence principle, a light ray will 'fall' in a gravitational field. (a) To the astronaut in the freely falling spaceship (an inertial observer), the light trajectory is straight. (b) To an observer outside the spaceship, the astronaut is accelerating (falling) in a gravitational field. The light ray will be bent so that it reaches the opposite side of the lab at a height $h - gt^2/2$ below the initial point at h.

informs us that there is no detectable effect associated either with gravity or with acceleration: the light travels straight across the spaceship from one side to the other. In this coordinate frame, the light is emitted at a height h, and received at the same height h on the opposite side of the spaceship, Fig. 12.3(a). But to an observer outside the spaceship, there is a gravitational field $\mathbf{g}$ and the spaceship is accelerating (falling) in this gravitational field. The straight trajectory of the light signal in the freely falling spaceship will appear to bend. Thus, to this outside observer, a light ray is seen to bend in the gravitational field, Fig. 12.3(b).

Ordinarily this effect is very small. However for a strong gravitational field, like that at the surface of sun, the bending of starlight grazing past the sun can be observed. In Section 12.3.4 we shall carry out the EP calculation to find this angle of light deflection to be

$$\delta\phi = \frac{2G_N M_\odot}{c^2 R_\odot} = 0.875''. \tag{12.9}$$

where G_N is Newton's constant, and $M_\odot$ and $R_\odot$ are respectively the solar mass and radius. We shall carry out this calculation after we have introduced the concept of gravitational time dilation below. As we shall explain in Section 14.1, this is exactly **half** of the correct GR prediction for the solar deflection of light from a distant star.

12.3.2   Gravitational redshift

Let us now consider the situation when the gravitational field direction is parallel (or antiparallel) to the light-ray direction. We have a receiver placed directly at a distance h above the emitter in a downward-pointing gravitational field $\mathbf{g}$.

Just as in the transverse case considered above, we first describe the situation from the viewpoint of the astronaut (in free fall). EP informs us that the spaceship in free fall is an inertial frame. Such an observer will not be able to detect any physical effects associated with gravity or acceleration. In this free-fall situation, the astronaut should not detect any frequency shift, Fig. 12.4(a): the received light frequency ω_{rec} is the same as the emitted frequency ω_{em}:

$$(\Delta\omega)_{\text{ff}} = (\omega_{\text{rec}} - \omega_{\text{em}})_{\text{ff}} = 0, \tag{12.10}$$

where the subscript "ff" reminds us that these are the values as seen by an observer in free fall.

From the viewpoint of the observer outside the spaceship, there is gravity and the spaceship is accelerating (falling) in this gravitational field, Fig. 12.4(b). Because it takes a finite amount of time $\Delta t = h/c$ for the light signal to reach the receiver on the ceiling, it will be detected by a receiver in motion towards the emitter, with a velocity $u = g\Delta t$ (g being the gravitational acceleration). The familiar Doppler formula (10.55) in the low-velocity approximation would lead us to expect a frequency shift of

$$\left(\frac{\Delta\omega}{\omega}\right)_{\text{Doppler}} = \frac{u}{c}. \tag{12.11}$$

Since the receiver has moved closer to the emitter, the light waves must have been compressed, and this shift must be toward the blue

$$(\Delta\omega)_{\text{Doppler}} = (\omega_{\text{rec}} - \omega_{\text{em}})_{\text{Doppler}} > 0. \tag{12.12}$$

We have already learned in (12.10), as deduced by the observer in free fall, that the received frequency did not deviate from the emitted frequency. Since this physical result must hold for both observers, this blueshift in (12.12) must somehow be cancelled by some other effect. To the observer outside the spaceship, gravity is also present. We can recover the null-shift result if **light is redshifted by gravity**, with just the right amount to cancel the Doppler blueshift of (12.11).

$$\left(\frac{\Delta\omega}{\omega}\right)_{\text{gravity}} = -\frac{u}{c}. \tag{12.13}$$

We now express the relative velocity on the RHS in terms of the gravitational potential difference[2] $\Delta\Phi$ at the two locations

$$u = g\Delta t = \frac{gh}{c} = \frac{\Delta\Phi}{c}. \tag{12.14}$$

When (12.13) and (12.14) are combined, we obtain the phenomenon of **gravitational frequency shift**,

$$\frac{\Delta\omega}{\omega} = -\frac{\Delta\Phi}{c^2}. \tag{12.15}$$

Namely,

$$\frac{\omega_{\text{rec}} - \omega_{\text{em}}}{\omega_{\text{em}}} = -\frac{\Phi_{\text{rec}} - \Phi_{\text{em}}}{c^2}. \tag{12.16}$$

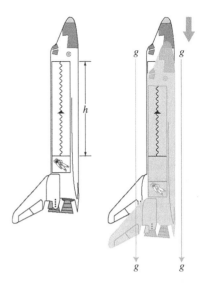

(a)   (b)

Fig. 12.4 According to the equivalence principle, the frequency of a light ray is redshifted when moving up against gravity. (a) To an inertial observer in the freely falling spaceship, there is no frequency shift. (b) To an observer outside the spaceship, this astronaut is accelerating in a gravitational field, and the null frequency shift result comes about because of the cancellation between the Doppler blueshift and the gravitational redshift.

[2] Recall the general relation $\mathbf{g} = -\nabla\Phi$ between the gravitational field $\mathbf{g}$ and the potential Φ.

A light ray emitted at a lower gravitational potential point ($\Phi_{\text{em}} < \Phi_{\text{rec}}$) with a frequency ω_{em} will be received at a higher gravitational field point as a lower frequency ($\omega_{\text{em}} > \omega_{\text{rec}}$) signal, that is, it is redshifted, even though the emitter and the receiver are not in relative motion.

In 1964 Robert Pound (1919–2010) and Glen Rebka (1931–) were able to verify this redshift effect in a terrestrial laboratory setting. Their experiment involved measurement of a minute atomic frequency shift. Normally, it is not possible to fix the frequency of an emitter or absorber to the required high accuracy because of the energy shift due to thermal recoil of the atoms. This difficulty was overcome by the discovery of the Mössbauer effect[3] just a few years earlier.

[3] **The Mössbauer effect**—When emitting light, the recoil atom can reduce the energy of the emitted photon. In reality, since the emitting atom is surrounded by other atoms in thermal motion, this brings about recoil momenta in an uncontrollable way. (We can picture the atom as being part of a vibrating lattice.) As a result, the photon energy in different emission events can vary considerably, resulting in a significant spread of their frequencies. This makes it impossible for a measurement of the atomic frequency to be carried out to high enough precision for purposes such as testing the gravitational redshift. But in 1958 Rudolf Mössbauer (1929–2011) made a breakthrough when he pointed out, and verified by observation, that crystals with high Debye–Einstein temperature (Section 5.4), that is, having a rigid crystalline structure, could pick up the recoil by the entire crystal. Namely, in such a situation, the emitting atom has an effective mass that is huge. Consequently, the atom loses no recoil energy, and the photon can pick up all the energy-change of the emitting atom, and the frequency of the emitted radiation is as precise as it can be.

12.3.3   Gravitational time dilation

At first sight, this gravitational frequency shift looks absurd. How can an observer, **stationary** with respect to the emitter, receive a different number of wave crests per unit time than the emitted rate? Here is Einstein's radical and yet simple answer: while the number of wave crests does not change, the time unit itself changes in the presence of gravity. The clocks run at different rates when situated at different gravitational field points: there is a **gravitational time dilation** effect.

Frequency being proportional to the inverse of the local proper time rate $\omega \sim 1/d\tau$, the gravitational frequency shift formula (12.16) can be converted into a time dilation formula

$$\frac{d\tau_1 - d\tau_2}{d\tau_2} = \frac{\Phi_1 - \Phi_2}{c^2}. \tag{12.17}$$

Namely, the clock at a higher gravitational potential point will run faster. This is to be contrasted with the special relativistic time dilation effect—clocks in relative motion run at different rates. Here we are saying that two clocks, even at rest with respect to each other, also run at different rates if the gravitational fields at their respective locations are different. We also note that a common choice of coordinate time t is a reading from a clock located far from the gravitational source (hence $\Phi = 0$), so the above formula can be read as the relation of this coordinate time (t) and the proper time (τ) at a location with gravitational potential Φ:

$$d\tau = \left(1 + \frac{\Phi}{c^2}\right) dt. \tag{12.18}$$

Time dilation test by atomic clocks and the GPS system

The gravitational time dilation effects have been tested directly by comparing the times kept by two cesium atomic clocks (Hafele and Keating 1972): one flown in an airplane at high altitude h (about 10 km) in a holding pattern, for a long time τ, over the ground station where the other clock sits. After the correction of the various background effects (mainly SR time dilations), the high-altitude clock was found to gain over the ground clock by a time interval of $\Delta\tau = (gh/c^2)\tau$ in agreement with the expectation given in (12.17). Furthermore it can be worked out that gravitational time dilation must be taken into account in order for the Global Position System to be accurate enough to determine distances within a few meters. Otherwise, the error would accumulate to about 10 meters in every minute or so.[4]

[4]The GPS problem is worked out as Problem 4.3 in Cheng (2010, p. 377).

12.3.4 Gravity-induced index of refraction in free space

Clocks run at different rates at locations where the gravitational field strengths are different (12.17). Since different clock rates will lead to different speed measurements, even the speed of light can be measured to have different values! Namely, one can conclude that, according to an accelerated observer (or equivalently in the presence of gravity), the speed of light deviates from c when time is measured in a reasonable way.

We are familiar with the light speed in different media being characterized by varying index of refraction. Gravitational time dilation implies that even in the vacuum there is an effective index of refraction when a gravitational field is present. Since a gravitational field is usually inhomogeneous, this index is generally a position-dependent function. At a given position r with gravitational potential $\Phi(r)$ a determination of the light speed involves the measurement of a displacement dr for a time interval $d\tau$ as recorded by a clock at rest at this position. The resultant ratio $dr/d\tau = c$ is the light speed according to the **local proper time**; this speed c is a universal constant. On the other hand, the light speed according to the **coordinate time** t will be different. A reasonable choice of time coordinate t is that given by a clock located far away from the gravitational source where the potential is set to be zero. The relation between coordinate and proper times, being given in (12.18), implies that the speed of light as measured by a remote observer is reduced by gravity as

$$c(r) \equiv \frac{dr}{dt} = \left(1 + \frac{\Phi(r)}{c^2}\right)\frac{dr}{d\tau} = \left(1 + \frac{\Phi(r)}{c^2}\right)c. \qquad (12.19)$$

Namely, the speed of light will be seen by an observer (with his coordinate clock) to vary from position to position as the gravitational potential varies from position to position. For such an observer, the effect of the gravitational field can be viewed as introducing an **index of refraction** in space:

$$n(r) \equiv \frac{c}{c(r)} = \left(1 + \frac{\Phi(r)}{c^2}\right)^{-1} \simeq \left(1 - \frac{\Phi(r)}{c^2}\right). \qquad (12.20)$$

To reiterate the key concepts behind this position-dependent speed of light, we are not suggesting that the deviation of $c(r)$ from the constant c means that the physical velocity of light has changed, or that the velocity of light is no longer a universal constant in the presence of gravitational fields. Rather, it signifies that the clocks at different gravitational points run at different rates. For an observer, with the time t measured by clocks located far from the gravitational source (taken to be the coordinate time), the velocity of light **appears to this observer** to slow down. A dramatic example is offered by the case of black holes (to be discussed in Section 14.5.2). There, as a manifestation of an infinite gravitational time dilation, it would take an infinite amount of coordinate time for a light signal to leave a black hole. Thus, to an outside observer,

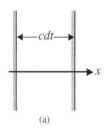

(a)

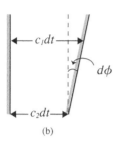

(b)

Fig. 12.5 Wavefronts of a light trajectory. (a) Wavefronts in the absence of gravity. (b) Tilting of wavefronts in a medium with an index of refraction varying in the vertical direction so that $c_1 > c_2$. The resultant light bending is signified by the small angular deflection $d\phi$.

no light can escape from a black hole, even though the corresponding proper time duration is perfectly finite.

12.3.5   Light ray deflection calculated

Here we provide the EP calculation for the light deflection angle shown in Eq. (12.9). We are familiar with the fact that a position-dependent index of refraction leads to the bending of a light ray. The amount of deflection by a transverse gravitational field can be calculated using the Huygens construction. Consider a plane light wave propagating in the $+x$ direction. At each time interval Δt, a wavefront advances a distance $c\Delta t$, see Fig. 12.5(a). The existence of a transverse gravitational field (in the y direction) means a nonvanishing derivative of the gravitational potential $d\Phi/dy \neq 0$. From (12.20) we see that a change of the gravitation potential means a change in $n(r)$, hence $c(r)$, and this leads to tilting of the wavefronts. We can then calculate the amount of the bending of the light ray by using the diagram in Fig. 12.5(b). A small angular deflection can be related to distances as

$$(d\phi) \simeq \frac{(c_1 - c_2)\,dt}{dy} \simeq \frac{d[c(r)](dx/c)}{dy}. \tag{12.21}$$

Working in the limit of weak gravity with small $\Phi(r)/c^2$ (or equivalently $n \simeq 1$), we can relate $d[c(r)]$ to a change of index of refraction as $d[c(r)] = cd[n^{-1}] = -cn^{-2}dn \simeq -cdn$. Namely, Eq. (12.21) becomes

$$(d\phi) \simeq -\frac{\partial n}{\partial y}dx. \tag{12.22}$$

But from (12.20) we have $dn(r) = -d\Phi(r)/c^2$, and thus, integrating (12.22), we obtain the total deflection angle

$$\delta\phi = \int d\phi = \frac{1}{c^2}\int_{-\infty}^{\infty}\frac{\partial\Phi}{\partial y}dx = \frac{1}{c^2}\int_{-\infty}^{\infty}(\nabla\Phi \cdot \hat{\mathbf{y}})\,dx. \tag{12.23}$$

The integrand is the gravitational acceleration perpendicular to the light path. We shall apply the above formula to the case of a spherical source $\Phi = -G_N M/r$, and $\nabla\Phi = \hat{\mathbf{r}}G_N M/r^2$. Although the gravitational field will no longer be a simple uniform field in the $\hat{\mathbf{y}}$ direction, our approximate result can still be used because the bending takes place mostly in the small region of $r \simeq r_{min}$. See Fig. 12.6.

Fig. 12.6 Angle of deflection $\delta\phi$ of light by a mass M. A point on the light trajectory (solid curve) can be labeled either as (x, y) or (r, θ). The source at S would appear to the observer at O to be located at the shifted position of S'.

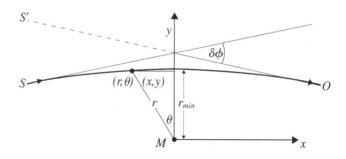

$$\delta\phi = \frac{G_N M}{c^2} \int_{-\infty}^{\infty} \frac{\hat{\mathbf{r}} \cdot \hat{\mathbf{y}}}{r^2} dx = \frac{G_N M}{c^2} \int_{-\infty}^{\infty} \frac{y}{r^3} dx, \tag{12.24}$$

where we have used $\hat{\mathbf{r}} \cdot \hat{\mathbf{y}} = \cos\theta = y/r$. An inspection of Fig. 12.6 also shows that, for small deflection, we can approximate $y \simeq r_{min}$, hence

$$r = (x^2 + y^2)^{1/2} \simeq (x^2 + r_{min}^2)^{1/2} \tag{12.25}$$

leading to

$$\delta\phi = \frac{G_N M}{c^2} \int_{-\infty}^{\infty} \frac{r_{min}}{(x^2 + r_{min}^2)^{3/2}} dx = \frac{2 G_N M}{c^2 r_{min}}. \tag{12.26}$$

With a light ray being deflected by an angle $\delta\phi$ as shown in Fig. 12.6, the light source at S would appear to the observer at O to be located at S'. Since the deflection is inversely proportional to r_{min}, one wants to maximize the amount of bending by having the smallest possible r_{min}. For light grazing the surface of the sun, $r_{min} = R_\odot$ and $M = M_\odot$, Eq. (12.26) gives an angle of deflection $\delta\phi = 0.875''$ as shown in Eq. (12.9).

12.3.6 From the equivalence principle to "gravity as the structure of spacetime"

Because the motion of a test body in a gravitational field is independent of the properties of the body, Eq. (12.1), Einstein came up with the idea that the effect on the body can be attributed directly to some spacetime feature and gravity is nothing but the structure of warped spacetime. This can be phrased as gravity not being a force; a test body in whatever gravitational field just moves freely in "spacetime with gravity". Any nontrivial motion is attributed to the structure of spacetime brought about by gravity. Gravity can cause the fabric of spacetime to warp, and the shape of spacetime responds to the matter in the environment. We must first learn the mathematical language that describes curved spacetime.

12.4 Elements of Riemannian geometry

We have mentioned that Einstein did not at first appreciate a more mathematical formulation of his relativity theory. By 1912 when he became a professor back in his Alma Mater ETH, he was fortunate to have the collaboration of his long-time friend Marcel Grossmann, who introduced him to the tensor calculus developed by E.B. Christoffel (1829–1900), G. Ricci-Curbastro (1852–1925), and T. Levi-Civita (1873–1941). This is the mathematical language Einstein needed to formulate his geometric theory of gravitation—gravity proffered as the structure of Minkowski's spacetime.

Riemannian geometry describes curved n-dimensional space. Since most of us can only visualize, and have some familiarity with, a curved surface (namely, the $n = 2$ case), we shall often use this simpler theory first pioneered by J. Carl Friedrich Gauss (1777–1855) to illustrate the more general dimensional theory studied by G.F. Bernhard Riemann (1826–66).

12.4.1 Gaussian coordinates and the metric tensor

Many of us have the habit of thinking of a curved surface as one embedded in 3D Euclidean space. Thus a spherical surface (radius R) is described by the 3D Cartesian coordinates (X, Y, Z) as

$$X^2 + Y^2 + Z^2 = R^2.$$

Namely, the embedding coordinates are subject to a constraint condition. Gauss pointed out that a much more convenient approach would be to choose a set of independent coordinates equal to the dimension of the space $(x_1, x_2, \ldots, x_n)$, for example for 2D space with coordinates such as (θ, ϕ), the polar and azimuthal angles.

The geometry (angle, length, shape of space) can then be described by length measurements through the entity called the **metric** g_{ij} (indices range from 1 to n), which is directly related to the basis vectors of the space [cf. Eqs. (11.12) and (11.22)]. It connects length measurements and the chosen Gaussian coordinates at any given point in the space by

$$ds^2 = g_{ij}dx_idx_j = \begin{pmatrix} dx_1 & dx_2 \end{pmatrix} \begin{pmatrix} g_{11} & g_{12} \\ g_{21} & g_{22} \end{pmatrix} \begin{pmatrix} dx_1 \\ dx_2 \end{pmatrix}, \qquad (12.27)$$

where we have written out explicitly, for the 2D case, the summation in the form of matrix multiplication.

Metric for a 2-sphere

To illustrate this for the case of a spherical surface (radius R), one first sets up the latitude/longitude system (i.e. a system of coordinates of polar angle θ and azimuthal angle ϕ) to label points on the globe, then measures the distances between neighboring points (Fig. 12.7). One finds that the latitudinal distances ds_ϕ (subtended by $d\phi$ between two points having the same radial distance) become ever smaller as one approaches the poles $ds_\phi = R\sin\theta d\phi$, while the longitudinal distance interval ds_θ between two points at the same longitude $(d\phi = 0)$ can be chosen to have the same value over the whole range of θ and ϕ. From such a table of distance measurements, one obtains a description of this spherical surface. Such distance measurements can be compactly expressed in terms of the metric tensor elements. Because we have chosen an orthogonal coordinate $g_{\theta\phi} = \mathbf{e}_\theta \cdot \mathbf{e}_\phi = 0$, the infinitesimal length between the origin $(0, 0)$ and a nearby point $(d\theta, d\phi)$ can be calculated[5]

[5] An infinitesimally small area on a curved surface can be thought of as a (infinitesimally small) flat plane. For such a flat surface ds can be calculated by the Pythagorean theorem.

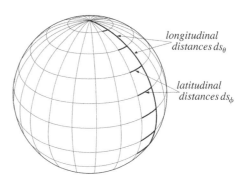

longitudinal distances ds_θ

latitudinal distances ds_ϕ

Fig. 12.7 Using distance measurements along longitudes and latitudes to specify the shape of the spherical surface.

$$[ds^2]_{(\theta,\phi)} = (ds_r)^2 + (ds_\phi)^2$$
$$= R^2 d\theta^2 + R^2 \sin^2 \theta d\phi^2. \tag{12.28}$$

A comparison of (12.28) and (12.27) leads to an expression for the metric tensor for this (θ, ϕ) coordinate system to be

$$g_{ab}^{(\theta,\phi)} = R^2 \begin{pmatrix} 1 & 0 \\ 0 & \sin^2 \theta \end{pmatrix}. \tag{12.29}$$

It is important to note that the metric is an "intrinsic" geometric quantity—that is, it can be determined within the space under discussion without ever invoking any embedding space.[6] Thus a metric description can be accomplished without referring to any embedding space.

[6]Namely, it can be completely determined by measurements by a being living within the space; cf. the discussion that follows Eq. (11.22).

12.4.2 Geodesic equation

Here we will discuss an example of using a metric to determine a geometric property of the space. Any curve in the space can be written in the form[7] of $x_i(\tau)$, where τ is some "curve parameter" having, for example, a range of $(0, 1)$. We are interested in finding, for a given space, the shortest curve, called the **geodesic**, that connects an initial $x_i(0)$ and final $x_i(1)$ positions. This can be done as follows: the length of any curve is given by Eq. (11.22)—changing now to a Greek index notation as appropriate for 4D spacetime for later discussion:

[7]For example $x_i(\tau)$ may be the trajectory of a particle with the curve parameter τ being the time variable.

$$s = \int ds = \int L(x, \dot{x}) \, d\tau \tag{12.30}$$

with $\dot{x}^\mu = dx^\mu / d\tau$ and

$$L(x, \dot{x}) = \sqrt{g_{\mu\nu} \frac{dx^\mu}{d\tau} \frac{dx^\nu}{d\tau}} \tag{12.31}$$

and the metric elements being functions of x^μ. To determine the shortest (i.e. the extremum) line in the curved space, we impose the extremization condition for variation of the path with end-points fixed:

$$\delta s = \delta \int L(x, \dot{x}) \, d\tau = 0, \tag{12.32}$$

which calculus of variation can translate (cf. Appendix A, Section A.5.2) into a partial differential equation—the **Euler–Lagrange equation**:

$$\frac{d}{d\tau} \frac{\partial L}{\partial \dot{x}^\mu} - \frac{\partial L}{\partial x^\mu} = 0. \tag{12.33}$$

The geodesic determined by the Euler–Lagrange equation

As a mathematical exercise, one can show that the **same** Euler–Lagrange equation (12.33) follows from, instead of (12.31), a Lagrangian of the form:

$$L(x, \dot{x}) = g_{\mu\nu} \dot{x}^\mu \dot{x}^\nu, \tag{12.34}$$

which without the square root is much easier to work with. With L in this form, the derivatives in Eq. (12.33) become

$$\frac{\partial L}{\partial \dot{x}^\mu} = 2g_{\mu\nu}\dot{x}^\nu, \quad \frac{\partial L}{\partial x^\mu} = \frac{\partial g_{\lambda\rho}}{\partial x^\mu}\dot{x}^\lambda\dot{x}^\rho, \tag{12.35}$$

where we have used the fact that the metric function $g_{\mu\nu}$ depends on x^μ, but not $\dot{x}^\mu$. Substituting these relations back into Eq. (12.33), we obtain the **geodesic equation**,

$$\frac{d}{d\tau}\left(g_{\mu\nu}\dot{x}^\nu\right) - \frac{1}{2}\frac{\partial g_{\lambda\rho}}{\partial x^\mu}\dot{x}^\lambda\dot{x}^\rho = 0, \tag{12.36}$$

which determines the trajectory of the "shortest curve", the **geodesic**. (One can easily use this equation to check that the geodesics in a flat space are straight lines and those on a spherical surface are great circles.)

We can cast (12.36) into a more symmetric form to facilitate a later comparison with the formal tensor differential in curved space. Carrying out the differentiation of the first term and noting that the metric's dependence on the curve parameter τ is entirely through $x^\mu(\tau)$:

$$g_{\mu\nu}\frac{d^2x^\nu}{d\tau^2} + \frac{\partial g_{\mu\nu}}{\partial x^\lambda}\frac{dx^\lambda}{d\tau}\frac{dx^\nu}{d\tau} - \frac{1}{2}\frac{\partial g_{\lambda\rho}}{\partial x^\mu}\frac{dx^\lambda}{d\tau}\frac{dx^\rho}{d\tau} = 0. \tag{12.37}$$

Since the product $(dx^\lambda/d\tau)(dx^\nu/d\tau)$ in the second term is symmetric with respect to the interchange of indices λ and ν, only the symmetric part of its coefficient:

$$\frac{1}{2}\left(\frac{\partial g_{\mu\nu}}{\partial x^\lambda} + \frac{\partial g_{\mu\lambda}}{\partial x^\nu}\right)$$

can contribute. In this way the geodesic equation (12.36), after factoring out the common $g_{\mu\nu}$ coefficient, can be cast (after relabeling some repeated indices) into the form,

$$\frac{d^2x^\nu}{d\tau^2} + \Gamma^\nu_{\lambda\rho}\frac{dx^\lambda}{d\tau}\frac{dx^\rho}{d\tau} = 0, \tag{12.38}$$

where

$$\Gamma^\mu_{\lambda\rho} = \frac{1}{2}g^{\mu\nu}\left[\frac{\partial g_{\lambda\nu}}{\partial x^\rho} + \frac{\partial g_{\rho\nu}}{\partial x^\lambda} - \frac{\partial g_{\lambda\rho}}{\partial x^\nu}\right]. \tag{12.39}$$

$\Gamma^\mu_{\lambda\rho}$, defined as this particular combination of the first derivatives of the metric tensor, are called the **Christoffel symbols** (also known as the **affine connection**). The geometric significance of this quantity will be studied in Chapter 13. They are called "symbols" because, despite its appearance with indices, $\Gamma^\mu_{\nu\lambda}$ are not tensor elements. Namely, they do not have the correct transformation property as tensor elements under a coordinate transformation, cf. (11.5). Clearly this equation is applicable for all higher dimensional spaces (just by extending the range for the indices). Of particular relevance to us, this geodesic equation turns out to be the equation of motion in the GR theory of gravitation.

12.4.3 Flatness theorem

A different choice of coordinates leads to a different metric, which is generally position-dependent. What distinguishes a flat space (from a curved one) is that for a flat space it is possible to find a coordinate system for which the metric is a constant (like the case of Euclidean space where $[\mathbf{g}] = [\mathbf{1}]$ or Minkowski space in special relativity where $[\mathbf{g}] = \text{diag}(-1, 1, 1, 1) \equiv [\boldsymbol{\eta}]$). While it is clear that flat and curved spaces are different geometric entities, they are closely related because, as is familiar from our experience with curved surfaces, locally any curved space can be approximated by a flat space. This is the content of the so-called "flatness theorem".

In a curved space with a general coordinate system x^μ and a metric value $g_{\mu\nu}$ at a given point P, we can always find a coordinate transformation $x^\mu \to \bar{x}^\mu$ and $g_{\mu\nu} \to \bar{g}_{\mu\nu}$ so that the metric is flat at this point: $\bar{g}_{\mu\nu} = \eta_{\mu\nu}$ and $\partial \bar{g}_{\mu\nu}/\partial \bar{x}^\lambda = 0$,

$$\bar{g}_{\mu\nu}(\bar{x}) = \eta_{\mu\nu} + \gamma_{\mu\nu\lambda\rho}(0)\bar{x}^\lambda \bar{x}^\rho + \cdots . \tag{12.40}$$

Namely, the metric in the neighborhood of the origin (P) will differ from $\eta_{\mu\nu}$ only by the second-order derivative. This is simply a Taylor series expansion of the metric at the origin—there is the constant $\bar{g}_{\mu\nu}(0)$ plus second-order derivative terms $\gamma_{\mu\nu\lambda\rho}(0)\bar{x}^\lambda \bar{x}^\rho$. That $\bar{g}_{\mu\nu}(0) = \eta_{\mu\nu}$ should not be a surprise; for a metric value at one point one can always find an orthogonal system so that $\bar{g}_{\mu\nu}(0) = 0$ for $\mu \neq \nu$ and the diagonal elements can be scaled to unity so that the new coordinate bases all have unit length and the metric is an identity matrix or whatever the correct flat space metric for the space with the appropriate signature. The nontrivial content of (12.40) is the absence of the first derivative.

In short, the theorem informs us that the general spacetime metric $g_{\mu\nu}(x)$ is characterized at a point (P) not so much by the value $g_{\mu\nu}|_P$ since that can always be chosen to be its flat space value, $\bar{g}_{\mu\nu}|_P = \eta_{\mu\nu}$, nor by its first derivative which can always be chosen to vanish $\partial \bar{g}_{\mu\nu}/\partial \bar{x}^\lambda|_P = 0$, but by the second derivative of the metric $\partial^2 g_{\mu\nu}/\partial x^\lambda \partial x^\rho$, which is related to the curvature to be discussed presently.

12.4.4 Curvature

In principle one can use the metric to decide whether a space is curved or not. Only for a flat space can we have a metric tensor with all its elements being constant. However, this is not a convenient tool because the values of the metric elements are coordinate-dependent. Consider a 2D flat surface; the Cartesian coordinate metric elements are constant, but they are not so had we used polar coordinates.

2D curved surface

In this connection Gauss made the discovery that it is possible to define a unique invariant second derivative of the metric tensor ($\partial^2 g$) called the **curvature** K, such that, independent of the coordinate choice, $K = 0$ for a flat surface and $K \neq 0$ for a curved surface. Since this curvature K is expressed entirely in terms of the metric and its derivatives, it is also an intrinsic geometric object. We will not present Gaussian curvature in full, but only indicate that for spaces of constant curvature (or, for any infinitesimal surface) we have

$$K = \frac{k}{R^2}$$

where R is the radius of curvature (it is simply the radius for the case of a spherical surface), and k is the curvature signature: $k = 0$ for a flat surface; $k = +1$ for a spherical surface (called "2-sphere" in geometry); and $k = -1$ for a hyperbolic surface, a "2-pseudosphere".

Unlike the cases of the plane or the sphere, there is no simple way to visualize this whole pseudosphere because its natural embedding is not into a flat 3D space with a Euclidean metric of $g_{ij} = \text{diag}(1, 1, 1)$, but into a flat 3D space[8] with a pseudo-Euclidean metric of $g_{ij} = \text{diag}(-1, 1, 1)$. Compared to the embedding of a sphere in a 3D Euclidean space (X, Y, Z) as $X^2 + Y^2 + Z^2 = R^2$, it can be worked out that the embedding of the 2D $k = -1$ surface in such a 3D pseudo-Euclidean space with coordinates (W, X, Y) corresponds to the condition $-W^2 + X^2 + Y^2 = -R^2$. While we cannot draw the whole pseudosphere in an ordinary 3D Euclidean space, the central portion of a saddle surface does represent a negative curvature surface, see Fig. 12.8(b).

In cosmology we shall encounter 4D spacetime with 3D spatial space with constant curvature: besides the flat 3D space, it is also possible to have 3D space with positive and negative curvature: a 3-sphere and 3-pseudosphere. When they are embedded in 4D space with metric $g_{ij} = \text{diag}(\pm 1, 1, 1, 1)$ with coordinates (W, X, Y, Z), these 3D spaces obey the constraint condition

$$\pm W^2 + X^2 + Y^2 + Z^2 = \pm R^2. \tag{12.41}$$

[8]One can think of it as 3D Minkowski spacetime with one time and two spatial coordinates.

Curvature measures the deviation from Euclidean relations   On a flat surface, the familiar Euclidean geometrical relations hold. For example, the circumference of a circle with radius r is $S = 2\pi r$. The curvature measures how curved a surface is because it is directly proportional to the violation of Euclidean relations. In Fig. 12.8 we show two pictures of circles with radius r drawn on surfaces with nonvanishing curvature. The circular circumference S differs from the flat surface value of $2\pi r$ by an amount controlled by the Gaussian curvature, K:

$$\lim_{r \to 0} \frac{2\pi r - S}{r^3} = \frac{\pi}{3} K. \tag{12.42}$$

For a positively curved surface the circumference is smaller than, and for a negatively curved surface larger than, that on a flat space.

Another simple example showing that curvature measures the deviation from Euclidean relations is the connection of the angular excess ϵ of a polygon (the difference of the total interior angles on a curved surface and the corresponding Euclidean value) to the curvature K times the area σ of the polygon. For a 2D surface it reads

$$\epsilon = K\sigma. \tag{12.43}$$

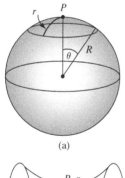

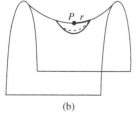

Fig. 12.8 A circle with radius r centered on point P, (a) on a spherical surface with curvature $K = 1/R^2$, (b) on the middle portion of a saddle-shaped surface, which has negative curvature $K = -1/R^2$.

In Fig. 12.9 this relation[9] is illustrated by the case of a spherical triangle. We also note that this angular excess can be measured by the directional change of a vector being parallel-transported around this polygon.

Riemann–Christoffel curvature tensor

Starting with Gauss's discovery of the curvature (a single component) in 2D space, his pupil Riemann (plus further work by Christoffel) established the existence of a rank-4 tensor, the curvature tensor, in an n-dimensional space:

$$R^{\mu}_{\lambda\alpha\beta} = \partial_{\alpha}\Gamma^{\mu}_{\lambda\beta} - \partial_{\beta}\Gamma^{\mu}_{\lambda\alpha} + \Gamma^{\mu}_{\nu\alpha}\Gamma^{\nu}_{\lambda\beta} - \Gamma^{\mu}_{\nu\beta}\Gamma^{\nu}_{\lambda\alpha}. \tag{12.44}$$

The Christoffel symbols Γ being first derivatives as shown in (12.39), the Riemann–Christoffel curvature of (12.44), $R = d\Gamma + \Gamma\Gamma$, is then a nonlinear second-derivative function of the metric, $[\partial^2 g + (\partial g)^2]$. It is independent of coordinate choice and measures the deviation from flat space relations. Thus for a flat space $R^{\mu}_{\lambda\alpha\beta} = 0$. As we shall see, a contracted version of this curvature tensor enters directly into the GR field equation, the Einstein equation.

There are many (mutually consistent) ways to derive the expression of the curvature tensor, as displayed in (12.44). One simple method involves the generalization of the 2D case to higher dimensions by calculating the parallel transport of a vector A^{μ} around an infinitesimal parallelogram spanned by two infinitesimal vectors a^{μ} and b^{ν}. The directional change $(dA)/A$ yielding the angular excess, the higher dimensional generalization of (12.43) should then be

$$dA^{\mu} = R^{\mu}_{\nu\lambda\rho}A^{\nu}a^{\lambda}b^{\rho}. \tag{12.45}$$

Namely, the vectorial change is proportional to the vector A^{ν} itself and to the two vectors $(a^{\lambda}b^{\rho})$ spanning the parallelogram. The coefficient of proportionality $R^{\mu}_{\nu\lambda\rho}$ is a quantity with four indices (antisymmetric in λ and ρ as the area $a^{\lambda}b^{\rho}$ should be antisymmetric) and we shall take this to be the definition of the curvature tensor of this n-dimensional space. For an explicit calculation, see Box 13.2, p. 311 in Cheng (2010).

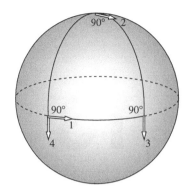

Fig. 12.9 A triangle with all interior angles being 90° on a spherical surface (radius R) has an angular excess of $\epsilon = \pi/2$. This satisfies the relation of $\epsilon = K\sigma$ with the curvature $K = R^{-2}$ and the triangular area $\sigma = \pi R^2/2$. The angular excess can be measured by the parallel-transport of a vector around this triangle (vector 1, clockwise to vector 2, to 3, and finally back to the starting point as vector 4) leading to a directional change of the vector by 90° (the angular difference between vector 4 and vector 1).

[9]For a proof of a general spherical triangle with arbitrary interial angles, see Section 5.3.2, p. 92 of Cheng (2010).

13

Curved spacetime as a gravitational field

- General relativity (GR) is a geometric theory of gravitation: a gravitational field is simply curved spacetime. The gravitational time dilation implied by the equivalence principle of the last chapter can be interpreted as showing the warpage of spacetime in the time direction.

- The effect of the gravitational interaction between two particles is described in GR as the source mass giving rise to a curved spacetime which in turn determines the motion of the test mass in such a spacetime. The 10-component spacetime metric tensor is the relativistic gravitational potential.

- This suggests that the GR equation of motion should be the equation that describes the shortest curve (a geodesic) in a curved space. The correctness of this identification is checked by its reduction to Newton's equation of motion in the limit of particles moving with nonrelativistic velocities in a weak and static gravitational field. This limit calculation clarifies the sense of how Newton's theory is extended by GR to new physical realms.

- We learn how to write a tensor equation that is covariant under general coordinate transformations (principle of general relativity). While tensors in curved spacetime are basically the same as those in flat Minkowski space, ordinary derivatives must be replaced by covariant derivatives in order to have proper transformation properties under local (i.e. spacetime dependent) transformations.

- A covariant derivative differs from an ordinary derivative with the addition of Christoffel symbols which are first derivatives of the metric tensor. The metric being the relativistic gravitational potential, the Christoffel symbols can then be identified with the gravitational field strength. Thus the replacement of the ordinary derivative in a SR equation by covariant derivatives automatically introduces gravity into physics equations. In this manner, going from the SR equation of motion for a free particle leads us to the geodesic equation, the GR equation of motion for a 'free particle'.

Einstein's general theory of relativity is a geometric theory of gravity—gravitational phenomena are attributed as reflecting the underlying curved spacetime. In this theory gravity is simply warped spacetime. In this chapter we shall mainly study the general relativity (GR) equation of motion, the geodesic equation, which describes the motion of a test particle in curved spacetime. In the next chapter we take up the GR field equation, the Einstein equation, which describes how the mass/energy source gives rise to curved spacetime.

After accepting curved spacetime and Riemannian geometry as the appropriate mathematics for its description, we then discuss the tensor calculus in such a curved space and learn how to write down the physics equations satisfying the principle of general relativity. In Section 13.4 we present the principle of general covariance, which guides us to GR equations in curved spacetime. A proper derivation of the geodesic equation as the GR equation of motion will then be presented.

13.1 The equivalence principle requires a metric description of gravity

How did Einstein get the idea for a geometric theory of gravitation? What does one mean by a geometric theory?

13.1.1 What is a geometric theory?

By a geometric theory, or a geometric description, of any physical phenomenon, we mean that the results of physical measurements can be attributed directly to the underlying geometry of space and time. This can be illustrated by the example describing a spherical surface (Fig. 12.7) that we discussed in Section 12.4.1. The length measurements on the surface of a globe are different in different directions: the east and west distances between any pairs of points separated by the same azimuthal angle $\Delta\phi$ become smaller as the pair move away from the equator, while the lengths in the north and south directions for a fixed ϕ remain the same. We could, in principle, interpret such results in two equivalent ways:

1. Without considering that the 2D space is curved, we can say that the physics (i.e. dynamics) is such that the measuring ruler changes its scale when pointing in different directions—in much the same manner as the Lorentz–FitzGerald length contraction of SR was originally interpreted.
2. The alternative description (the "geometric theory") is that we use a standard ruler with a fixed scale (defining the coordinate distance) and the varying length measurements are attributed to the underlying geometry of the curved spherical surface. This is expressed mathematically in the form of a position-dependent metric tensor $g_{ab}(x) \neq \delta_{ab}$.

In Chapter 12 we deduced several physical consequences from the empirical principle showing the equivalence of gravity and inertia. In a geometric theory gravitational phenomena are attributed as reflecting the underlying curved spacetime which has a metric as defined in (12.27)

$$ds^2 = g_{\mu\nu}dx^\mu dx^\nu. \tag{13.1}$$

For special relativity (SR) we have the geometry of a flat spacetime with a position-independent metric $g_{\mu\nu} = \eta_{\mu\nu} = \mathrm{diag}(-1, 1, 1, 1)$. The study of EP physics led Einstein to propose that gravity represents the structure of a curved spacetime $g_{\mu\nu} \neq \eta_{\mu\nu}$, and gravitational phenomena are just the effects of that curved spacetime on a test object. His theory is a geometric theory of gravity.

13.1.2 Time dilation as a geometric effect

We will first recall the EP physics of gravitational time dilation—clocks run at different rates at positions having different gravitational potential values—as summarized in Eq. (12.18)

$$d\tau(x) = \left(1 + \frac{\Phi(x)}{c^2}\right) dt \tag{13.2}$$

between the coordinate time t and the proper time $\tau(x)$ at a location x having gravitational potential $\Phi(x)$. As we shall see, this can be interpreted as a geometric effect of curved spacetime.

For the gravitational time dilation of (13.2), instead of working with a complicated scheme of clocks running at different rates, this physical phenomenon can be given a geometric interpretation as showing a nontrivial metric. Namely, a simpler way of describing the same physical situation is by using a stationary clock at $\Phi = 0$ (i.e. a location far from the source of gravity) as the "standard clock". Its fixed rate is taken to be the time coordinate t. One can then compare the time intervals $d\tau(x)$ measured by clocks located at other locations (the proper time interval at x) to this coordinate interval dt. In this instance Eq. (13.1) reduces down to $ds^2 = g_{00}dx^0 dx^0$ because $dx^i = 0$, as appropriate for a proper time interval (the time interval measured in the rest-frame, hence no displacement). The two sides of this equation can be written in terms of the proper and coordinate times. According to Eq. (10.18) the line element ds^2 is just the proper time interval $-c^2 d\tau^2$ and given that $x^0 = ct$ we then have

$$(d\tau)^2 = -g_{00}(dt)^2 \tag{13.3}$$

which by (13.2) implies

$$g_{00} = -\left(1 + \frac{\Phi(x)}{c^2}\right)^2 \simeq -\left(1 + \frac{2\Phi(x)}{c^2}\right). \tag{13.4}$$

This states that the metric element g_{00} deviates from the flat spacetime value of $\eta_{00} = -1$ because of the presence of gravity. Thus the geometric interpretation of the EP physics of gravitational time dilation is to say that gravity changes the spacetime metric element g_{00} from -1 to an x-dependent function. Gravity warps spacetime—in this case the warpage is in the time direction.

13.1.3 Further arguments for warped spacetime as the gravitational field

Adopting a geometric interpretation of EP physics, we find that the resultant geometry has all the characteristic features of a **warped** manifold of space and time: a position-dependent metric, deviations from Euclidean geometric relations, and at every location we can always transform gravity away to obtain a flat spacetime, just as one can always find a locally flat region in a curved space.

Position-dependent metric

The metric tensor in a curved space is necessarily position-dependent. Clearly, (13.4) has this property. In Einstein's geometric theory of gravitation, the metric function is all that we need to describe the gravitational field completely. The metric $g_{\mu\nu}(x)$ plays the role of relativistic gravitational potentials, just as $\Phi(x)$ is the Newtonian gravitational potential.

Non-Euclidean relations

In a curved space Euclidean relations no longer hold: for example, the ratio of the circular circumference to the radius is different from the value of 2π (cf. Section 12.4.4). As it turns out, EP does imply non-Euclidean relations among geometric measurements. We illustrate this with a simple example. Consider a cylindrical room in high-speed rotation around its axis. This acceleration case, according to EP, is equivalent to a centrifugal gravitational field. (This is one way to produce "artificial gravity".) For such a rotating frame, one finds that, because of SR (longitudinal) length contraction, the radius, which is not changed because the velocity is perpendicular to the radial direction, is no longer equal to the circular circumference of the cylinder divided by 2π (see Fig. 13.1). Thus Euclidean geometry is no longer valid in the presence of gravity. We reiterate this connection: the rotating frame, according to EP, is a frame with gravity; the rotating frame, according to SR length contraction, has a relation between its radius and circumference that is not Euclidean. Hence, we say that in the presence of gravity the measuring rods will map out a non-Euclidean geometry.

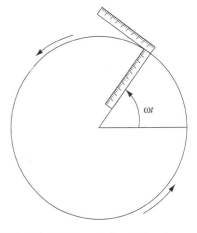

Fig. 13.1 Rotating cylinder with length contraction in the tangential direction but not in the radial direction, resulting in a non-Euclidean relation between circumference and radius.

Local flat metric and local inertial frame

In a curved space a small local region can always be described approximately as a flat space. A more precise statement is given by the flatness theorem (Section 12.4.3). Now, if we identify our spacetime as the gravitational field, self-consistency would require that the corresponding flatness theorem be satisfied. This is indeed the case because EP informs us that gravity can always be transformed away locally. In the absence of gravity, spacetime is flat. Thus Einstein put forward this elegant gravitational theory in which gravity is identified as the structure of the spacetime with EP incorporated in a fundamental way.

A 2D illustration of geometry as gravity

The possibility of using a curved space to represent a gravitational field can be illustrated with the following example involving a 2D curved surface. Two

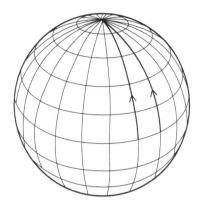

Fig. 13.2 Two particle trajectories with decreasing separation can be interpreted either as resulting from an attractive force or as reflecting the underlying geometry of a spherical surface.

masses on a spherical surface start out at the equator and move along two geodesic lines as represented by the longitudinal great circles. As they move along, the distance between them decreases (Fig. 13.2). We can attribute this to some attractive force between them, or simply to the curved space causing their trajectory to converge. That is to say, this phenomenon of two convergent particle trajectories can be thought of either as resulting from an attractive "tidal force", or from the curvature of the space. A detailed discussion will be presented in Section 14.3.2 in the next chapter.

13.2 General relativity as a field theory of gravitation

Recall that a field theoretical description of the interaction between a source and a test particle is a two-step description:

$$\boxed{\text{Source particle}} \xrightarrow[\substack{\text{Field} \\ \text{equation}}]{} \boxed{\text{Field}} \xrightarrow[\substack{\text{Equation of} \\ \text{motion}}]{} \boxed{\text{Test particle}}$$

Instead of the source particle acting directly on the test particle through some instantaneous action-at-a-distance force, the source creates a field everywhere, and the field then acts on the test particle locally. The first step is governed by the field equation which, given the source distribution, determines the field everywhere. In the case of electromagnetism this is the Maxwell equations. The second step is provided by the equation of motion, which allows us to find the motion of the test particle, once the field distribution is known. The electromagnetic equation of motion follows directly from the Lorentz force law.

Based on the study of EP phenomenology, Einstein made the conceptual leap (a logical deduction, but a startling leap nevertheless) to the idea that curved spacetime **is** the gravitational field. The effect of the gravitational interaction between two particles can be described as the source mass giving rise to a curved spacetime which in turn influences the motion of the test mass.

$$\boxed{\text{Source}} \xrightarrow[\substack{\text{Einstein field} \\ \text{equation}}]{} \boxed{\text{Curved spacetime}} \xrightarrow[\substack{\text{Geodesic} \\ \text{equation}}]{} \boxed{\text{Test particle}}$$

While spacetime in SR, like all pre-relativity physics, is fixed, it is dynamic in GR as determined by the matter/energy distribution. GR fulfills Einstein's conviction that "space is not a thing": the ever changing relation of matter and energy is reflected by an ever changing geometry. Spacetime does not have an independent existence; it is nothing but an expression of the relations among physical processes in the world.

All the equations must satisfy the principle of general relativity. It is important to note that the gravitational field is not a scalar, or a four-component vector, but a 10-component[1] symmetric tensor $g_{\mu\nu}(x)$. Here we shall first study the GR equation of motion, the geodesic equation, which describes the motion of a test

[1] A symmetric tensor $g_{\mu\nu} = g_{\nu\mu}$ has 10 independent components, in contrast to the electromagnetic field tensor which is antisymmetric, $F_{\mu\nu} = -F_{\nu\mu}$, having six components.

particle in a curved spacetime. The more difficult topic of the GR field equation, the Einstein equation, is deferred to the next chapter, after we have given a more detailed discussion of the Riemann–Christoffel curvature tensor.

13.2.1 The geodesic equation as the general relativity equation of motion

The metric function $g_{\mu\nu}(x)$ in (13.1) describes the geometry of curved spacetime. In GR, the mass/energy source determines the metric function through the field equation. Namely, the metric $g_{\mu\nu}(x)$ is the solution of the GR field equation. Knowing $g_{\mu\nu}(x)$, one can write down the equation of motion, which fixes the trajectory of the test particle. In this approach, gravity is the structure of spacetime and is not regarded as a force (that brings about acceleration). Thus a test body will move in a force-free way in such a curved spacetime; it is natural to expect[2] it to follow in this spacetime the shortest and straightest possible trajectory, the **geodesic curve**, that we have discussed in Section 12.4.2:

$$\frac{d^2 x^\nu}{d\tau^2} + \Gamma^\nu_{\lambda\rho} \frac{dx^\lambda}{d\tau} \frac{dx^\rho}{d\tau} = 0, \tag{13.5}$$

with

$$\Gamma^\mu_{\lambda\rho} = \frac{1}{2} g^{\mu\nu} \left[\frac{\partial g_{\lambda\nu}}{\partial x^\rho} + \frac{\partial g_{\rho\nu}}{\partial x^\lambda} - \frac{\partial g_{\lambda\rho}}{\partial x^\nu} \right] \tag{13.6}$$

being the Christoffel symbols. The claim is that this geodesic equation is the relativistic generalization of the Newtonian equation of motion (12.1)

$$\frac{d^2 \mathbf{r}}{dt^2} = -\nabla \Phi. \tag{13.7}$$

In the following we shall demonstrate this connection. In the process we also arrive at a more precise understanding of the sense that Einstein's theory is an extension of Newtonian gravity. As we shall see, it is much more than an extension to higher speed particle motions.

13.2.2 The Newtonian limit

Supporting our claim that the geodesic equation is the GR equation of motion, we shall now show that the geodesic equation (13.5) does reduce to the Newtonian equation of motion (13.7) in the **Newtonian limit** of

a test particle moving with nonrelativistic velocity $v \ll c$, in a weak and static gravitational field.

We now take such a limit of the GR equation of motion (13.5):

- **Nonrelativistic speed** $(dx^i/dt) \ll c$: This inequality $dx^i \ll cdt$ implies that

$$\frac{dx^i}{d\tau} \ll c \frac{dt}{d\tau} \left(= \frac{dx^0}{d\tau} \right). \tag{13.8}$$

[2]The correctness of this heuristic choice will be justified by a formal derivation of the geodesic equation in Section 13.4.2 below.

Keeping only the dominant term $\left(dx^0/d\tau\right)\left(dx^0/d\tau\right)$ in the double sum over indices λ and ρ in the geodesic equation (13.5), we have

$$\frac{d^2x^\mu}{d\tau^2} + \Gamma^\mu_{00}\frac{dx^0}{d\tau}\frac{dx^0}{d\tau} = 0. \tag{13.9}$$

- **Static field** $\left(\partial g_{\mu\nu}/\partial x^0\right) = 0$: Because all time derivatives vanish, the Christoffel symbols of (13.6) takes the simpler form

$$\Gamma^\mu_{00} = -\frac{1}{2}g^{\mu\nu}\frac{\partial g_{00}}{\partial x^\nu}. \tag{13.10}$$

- **Weak field** $h_{\mu\nu} \ll 1$: We assume that the metric is not too different from the flat spacetime metric $\eta_{\mu\nu} = \text{diag}(-1, 1, 1, 1)$

$$g_{\mu\nu} = \eta_{\mu\nu} + h_{\mu\nu} \tag{13.11}$$

where $h_{\mu\nu}(x)$ is a small correction field. $\eta_{\mu\nu}$ being a constant, we have $\partial g_{\mu\nu}/\partial x^\lambda = \partial h_{\mu\nu}/\partial x^\lambda$ and the Christoffel symbols being derivatives of the metric are of the order $h_{\mu\nu}$. To leading order, (13.10) is

$$\Gamma^\mu_{00} = -\frac{1}{2}\eta^{\mu\nu}\frac{\partial h_{00}}{\partial x^\nu}$$

which, because $\eta_{\nu\mu}$ is diagonal, has (for a static h_{00}) the following components

$$-\Gamma^0_{00} = -\frac{1}{2}\frac{\partial h_{00}}{\partial x^0} = 0 \quad\text{and}\quad \Gamma^i_{00} = -\frac{1}{2}\frac{\partial h_{00}}{\partial x^i}. \tag{13.12}$$

We can now evaluate (13.9) by using (13.12): the $\mu = 0$ equation leads to

$$\frac{dx^0}{d\tau} = \text{constant}, \tag{13.13}$$

and the three $\mu = i$ equations are

$$\frac{d^2x^i}{d\tau^2} + \Gamma^i_{00}\frac{dx^0}{d\tau}\frac{dx^0}{d\tau} = \left(\frac{d^2x^i}{c^2dt^2} + \Gamma^i_{00}\right)\left(\frac{dx^0}{d\tau}\right)^2 = 0, \tag{13.14}$$

where we have used $\left(dx^i/d\tau\right) = \left(dx^i/dx^0\right)\left(dx^0/d\tau\right)$ and the condition of (13.13) to conclude $\left(d^2x^i/d\tau^2\right) = \left(d^2x^i/dx^{0\,2}\right)\left(dx^0/d\tau\right)^2$. The above equation, together with (13.12), implies

$$\frac{d^2x^i}{c^2dt^2} - \frac{1}{2}\frac{\partial h_{00}}{\partial x^i} = 0, \tag{13.15}$$

which is to be compared with the Newtonian equation of motion (13.7). Thus $h_{00} = -2\Phi/c^2$ and using the definition of (13.11) we recover (13.4), first obtained heuristically in Section 13.1:

$$g_{00} = -\left(1 + \frac{2\Phi(x)}{c^2}\right). \tag{13.16}$$

We can indeed regard the metric tensor as the relativistic generalization of the gravitational potential. This expression also provides us with a criterion to characterize a field being weak as in (13.11):

$$\left[\, |h_{00}| \ll |\eta_{00}| \,\right] \quad \Rightarrow \quad \left[\, \frac{|\Phi|}{c^2} \ll 1 \,\right]. \tag{13.17}$$

Consider the gravitational potential at earth's surface. It is equal to the gravitational acceleration times earth's radius, $\Phi_\oplus = g \times R_\oplus = O(10^7\,\mathrm{m^2/s^2})$, or $\Phi_\oplus/c^2 = O(10^{-10})$. Thus a weak field is any gravitational field that is less than ten billion g's.

13.3   Tensors in a curved spacetime

The central point of EP, that gravity can be transformed away locally, fits snugly in the Riemannian geometrical description of a curved space as being locally flat. Einstein and Grossmann showed that tensor analysis in a curved space is just the mathematical device that can be used to implement the general principle of relativity in physics. GR requires that physics equations be covariant under any general coordinate transformation that leaves invariant the infinitesimal length of (13.1). Just as special relativity requires a physics equation to be a tensor equation with respect to Lorentz transformations,[3] GR equations must be tensor equations with respect to general coordinate transformations. In this way, the principle of general relativity can be fulfilled automatically.

Physics equations usually involve differentiation. While tensors in GR are basically the same as SR tensors as discussed in Chapter 11, the derivative operators in a curved space require considerable care. In contrast to the case of flat space, basis vectors in a curved space must change from position to position. This implies that general coordinate transformations must necessarily be position-dependent; they are local transformations. As a consequence, ordinary derivatives of tensors, except for the trivial scalars, are no longer tensors. Nevertheless it can be shown that one can construct "covariant differentiation operations" so that they result in tensor derivatives.

13.3.1   General coordinate transformations

We will first discuss the properties of general coordinate transformations and introduce the notation of expressing such transformations as partial derivative matrices.

Metric and coordinate transformations are position-dependent

The coordinate transformations in special relativity (the Lorentz transformations) are position-independent "global transformations". The rotation angles and boost velocity are the same for every spacetime point; we rotate the same amount of angle and boost with the same velocity everywhere. In GR we must deal with position-dependent "local transformations", the general coordinate transformation. This position dependence is related to the fact that in a curved

[3]Recall from Chapter 11 that the Lorentz transformation can be regarded as a rotation in 4D spacetime that leaves the length s^2 invariant.

space the basis vectors $\{\mathbf{e}_\mu\}$ must necessarily change from point to point, leading to position-dependent metric functions [see Eq. (11.12)]:

$$g_{\mu\nu} \equiv \left[\mathbf{e}_\mu(x) \cdot \mathbf{e}_\nu(x)\right] = g_{\mu\nu}(x). \tag{13.18}$$

That the metric in a curved space is always position-dependent immediately leads to the conclusion that a general coordinate transformation $[\Lambda]$ must also be position-dependent:[4]

$$\partial[\Lambda] \neq 0. \tag{13.20}$$

Having a position-dependent transformation means that an independent transformation is performed at each location. Namely, the transformation is nonlinear. The demand that physics equations be covariant under such nonlinear transformations is much more severe. As we shall see below, it not only imposes restrictions on the form of the physics equation but also requires the presence of a "force field" (in our case, the gravitational field), which is introduced via the covariant derivatives as will be presented in Section 13.4.1.

Coordinate transformation as a matrix of partial derivatives

The coordinate transformations in special relativity (the Lorentz transformations) leave invariant any finite separation $s^2 = g_{\mu\nu}x^\mu x^\nu$. In a curved space the bases and metric necessarily vary from point to point. General transformations in such a space are not expected to have such a finite invariant separation. However, since a curved space is locally flat, it is possible to demand the coordinate transformation

$$dx'^\mu = [\Lambda]^\mu_\nu dx^\nu \tag{13.21}$$

that leaves invariant the infinitesimal length. Equation (13.1) defines the metric for a given coordinate system. Let us now recall the (chain-rule) differentiation relation:

$$dx'^\mu = \frac{\partial x'^\mu}{\partial x^\nu} dx^\nu. \tag{13.22}$$

A comparison of (13.21) and (13.22) suggests that the coordinate transformation can be written as a matrix of partial derivatives:

$$[\Lambda]^\mu_\nu = \frac{\partial x'^\mu}{\partial x^\nu}. \tag{13.23}$$

Therefore, the transformation for a contravariant vector is

$$A^\mu \longrightarrow A'^\mu = \frac{\partial x'^\mu}{\partial x^\nu} A^\nu. \tag{13.24}$$

Writing out this equation explicitly, we have

$$\begin{pmatrix} A'^0 \\ A'^1 \\ A'^2 \\ A'^3 \end{pmatrix} = \begin{pmatrix} \dfrac{\partial x'^0}{\partial x^0} & \dfrac{\partial x'^0}{\partial x^1} & \dfrac{\partial x'^0}{\partial x^2} & \dfrac{\partial x'^0}{\partial x^3} \\ \dfrac{\partial x'^1}{\partial x^0} & \dfrac{\partial x'^1}{\partial x^1} & \dfrac{\partial x'^1}{\partial x^2} & \dfrac{\partial x'^1}{\partial x^3} \\ \dfrac{\partial x'^2}{\partial x^0} & \dfrac{\partial x'^2}{\partial x^1} & \dfrac{\partial x'^2}{\partial x^2} & \dfrac{\partial x'^2}{\partial x^3} \\ \dfrac{\partial x'^3}{\partial x^0} & \dfrac{\partial x'^3}{\partial x^1} & \dfrac{\partial x'^3}{\partial x^2} & \dfrac{\partial x'^3}{\partial x^3} \end{pmatrix} \begin{pmatrix} A^0 \\ A^1 \\ A^2 \\ A^3 \end{pmatrix}.$$

[4]The metric $[g]$ is a rank-2 tensor and thus, according to (13.29), transforms as $[g'] = [\Lambda][\Lambda][g]$ where we have symbolically represented the coordinate transformation by $[\Lambda]$. If we differentiate both sides of this relation, we get

$$\partial[g'] = 2[\Lambda][g](\partial[\Lambda]) + [\Lambda][\Lambda](\partial[g]). \tag{13.19}$$

For a flat space, one can always work with a coordinate system having a position-independent metric, $\partial[g'] = \partial[g] = 0$. The above relation then shows that the transformation matrix must also be position-independent, $\partial[\Lambda] = 0$. In a curve space the metric must be position-dependent $\partial[g] \neq 0$, implying that the transformation also has x-dependence, $\partial[\Lambda] \neq 0$.

This way of writing a transformation also has the advantage of preventing us from mis-identifying the transformation $[\Lambda]^\mu_\nu$ as a tensor. From now on, we shall always adopt this practice.[5] Similarly, from the transformation property of the del operator discussed in Eq. (11.27),

$$\frac{\partial}{\partial x'^\mu} = \left[\Lambda^{-1}\right]^\nu_\mu \frac{\partial}{\partial x^\nu}, \tag{13.25}$$

the chain rule of differentiation leads to the identification

$$\left[\Lambda^{-1}\right]^\nu_\mu = \frac{\partial x^\nu}{\partial x'^\mu}. \tag{13.26}$$

For covariant components of a vector, we have the transformation

$$A_\mu \longrightarrow A'_\mu = \frac{\partial x^\nu}{\partial x'^\mu} A_\nu. \tag{13.27}$$

In general, contravariant and covariant components of a tensor, $T^{\mu\nu\ldots}_{\lambda\ldots}$, transform as a direct product of contravariant and covariant vectors $T^{\mu\nu\ldots}_{\lambda\ldots} \sim A^\mu B^\nu \ldots C_\lambda \ldots$ For example, the simplest mixed tensor has the transformation

$$T^\mu_\nu \longrightarrow T'^{\;\mu}_\nu = \frac{\partial x^\lambda}{\partial x'^\nu} \frac{\partial x'^\mu}{\partial x^\rho} T^{\;\rho}_\lambda. \tag{13.28}$$

In particular, the rank-2 metric tensor with two covariant indices transforms as

$$g'_{\mu\nu} = \frac{\partial x^\alpha}{\partial x'^\mu} \frac{\partial x^\beta}{\partial x'^\nu} g_{\alpha\beta}. \tag{13.29}$$

Because we have the expansion $\mathbf{A} = A^\mu \mathbf{e}_\mu = A_\mu \mathbf{e}^\mu$, with the vector $\mathbf{A}$ being coordinate independent, the transformations of the expansion coefficients A^μ and A_μ must be "cancelled out" by those of the corresponding bases:

$$\mathbf{e}'_\mu = \frac{\partial x^\nu}{\partial x'^\mu} \mathbf{e}_\nu \quad \text{and} \quad \mathbf{e}'^\nu = \frac{\partial x'^\nu}{\partial x^\rho} \mathbf{e}^\rho. \tag{13.30}$$

This is the reason that $\{A_\mu\}$ are called the *co*variant components: they transform in the same way as the basis vectors, while the *contra*variant components transform oppositely.

We remind ourselves that the description in Riemannian geometry is through distance measurements (see Chapter 12). Coordinates in Riemannian geometry have no intrinsic meaning as they are related to distance measurement only through the metric. We are allowed to make all sorts of coordinate changes so long as the metric changes correspondingly. Such a transformation will change the individual terms in the physics equations; nevertheless, these equations still retain their forms if they are tensor equations—each is a proper tensor and transforms according to the rules as given in (13.28). Namely, the equations are covariant and obey the principle of relativity.

13.3.2 Covariant differentiation

The above discussion would seem to imply that there is no fundamental difference between tensors in flat and in curved space. But as we shall demonstrate below, this is not so when differentiation is involved.

[5]This notation is also applicable to the global transformation of SR discussed in previous chapters. As an instructive exercise, one can show that the elements of the Lorentz transformation matrix (11.21) can be recovered from partial differentiation of the Lorentz boost formulas (10.10). Namely, the familiar Lorentz transformation can also be written as a matrix of partial differentiation. For example, from $t' = \gamma\left(t - vx/c^2\right)$, or $x'^0 = \gamma\left(x^0 - \beta x^1\right)$ we have $\partial x'^0/\partial x^1 = -\gamma\beta$, etc.

Ordinary derivatives of tensor components are not tensors

In a curved space, the derivative $\partial_\nu A^\mu$ is a nontensor. Namely, even though we have A^μ and ∂_ν being good vectors, as indicated by (13.24) and (11.27),

$$\partial_\mu \longrightarrow \partial'_\mu = \frac{\partial x^\lambda}{\partial x'^\mu} \partial_\lambda, \tag{13.31}$$

the combination $\partial_\nu A^\mu$ still does not transform properly,

$$\partial_\nu A^\mu \longrightarrow \partial'_\nu A'^\mu \neq \frac{\partial x^\lambda}{\partial x'^\nu} \frac{\partial x'^\mu}{\partial x^\rho} \partial_\lambda A^\rho, \tag{13.32}$$

as required by (13.28). We can easily trace this difficulty by carrying out the differentiation

$$\partial'_\nu A'^\mu = \partial'_\nu \left([\Lambda]^\mu{}_\nu A^\nu \right)$$

and comparing with the RHS of (13.32) to find that, because of (13.20), there is an extra term. Thus $\partial_\nu A^\mu$ not being a tensor is related to the position-dependent nature of the transformation, which in turn reflects (as discussed at the beginning of this subsection) the position-dependence of the metric. Thus the root problem lies in the "moving bases", $\mathbf{e}^\mu = \mathbf{e}^\mu(x)$, of the curved space. More explicitly, because the tensor components are the projections[6] of the tensor onto the basis vectors, $A^\mu = \mathbf{e}^\mu \cdot \mathbf{V}$. The moving bases $\partial_\nu \mathbf{e}^\mu \neq 0$ produce an extra (second) term in the derivative:

$$\partial_\nu A^\mu = \mathbf{e}^\mu \cdot (\partial_\nu \mathbf{A}) + \mathbf{A} \cdot (\partial_\nu \mathbf{e}^\mu). \tag{13.33}$$

Below we discuss the two terms on the RHS separately.

Covariant derivatives as expansion coefficients of $\partial_\nu \mathbf{A}$

In order for an equation to be manifestly relativistic we must have it as a tensor equation such that the equation is unchanged under coordinate transformations. Thus, we seek a **covariant derivative** D_ν to be used in covariant physics equations. Such a differentiation is constructed so that when acting on tensor components it still yields a tensor:

$$D_\nu A^\mu \longrightarrow D'_\nu A'^\mu = \frac{\partial x^\lambda}{\partial x'^\nu} \frac{\partial x'^\mu}{\partial x^\rho} D_\lambda A^\rho. \tag{13.34}$$

As will be demonstrated below, the first term on the RHS of (13.33) is just this desired covariant derivative term.

We have suggested that the difficulty with the differentiation of vector components is due to the coordinate dependence of A^μ. By this reasoning, derivatives of a scalar function Φ should not have this complication—because a scalar tensor does not depend on the bases: $\Phi' = \Phi$,

$$\partial_\mu \Phi \longrightarrow \partial'_\mu \Phi' = \frac{\partial x^\lambda}{\partial x'^\mu} \partial_\lambda \Phi. \tag{13.35}$$

Similarly, the derivatives of the vector $\mathbf{A}$ itself (not its components) transform properly because $\mathbf{A}$ is coordinate-independent,

$$\partial_\mu \mathbf{A} \longrightarrow \partial'_\mu \mathbf{A} = \frac{\partial x^\lambda}{\partial x'^\mu} \partial_\lambda \mathbf{A}. \tag{13.36}$$

[6]We take the dot product of the inverse basis vector $\mathbf{e}^\nu$ on both sides of the expansion $\mathbf{A} = A^\mu \mathbf{e}_\mu$ to obtain the result $\mathbf{e}^\nu \cdot \mathbf{A} = A^\nu$ because of the relation $\mathbf{e}^\nu \cdot \mathbf{e}_\mu = \delta^\nu_\mu$ as displayed in (11.14).

Both (13.35) and (13.36) merely reflect the transformation of the del operator (13.31). If we dot both sides of (13.36) by the inverse basis vectors, $\mathbf{e}'^{\nu} = \left(\partial x'^{\nu}/\partial x^{\rho}\right)\mathbf{e}^{\rho}$, we obtain

$$\mathbf{e}'^{\nu} \cdot \partial'_{\mu}\mathbf{A} = \frac{\partial x^{\lambda}}{\partial x'^{\mu}}\frac{\partial x'^{\nu}}{\partial x^{\rho}}\mathbf{e}^{\rho} \cdot \partial_{\lambda}\mathbf{A}. \tag{13.37}$$

This shows that $\mathbf{e}^{\nu} \cdot \partial_{\mu}\mathbf{A}$ is a proper mixed tensor as required by (13.28), and a comparison with (13.34) demonstrates that it is just the covariant derivative we have been looking for:

$$D_{\mu}A^{\nu} = \mathbf{e}^{\nu} \cdot \partial_{\mu}\mathbf{A}. \tag{13.38}$$

This relation implies that $D_{\mu}A^{\nu}$ can be viewed as the projection of the vectors[7] $\left[\partial_{\mu}\mathbf{A}\right]$ along the direction of $\mathbf{e}^{\nu}$; we can then interpret $D_{\mu}A^{\nu}$ as the coefficient of expansion of $\left[\partial_{\mu}\mathbf{A}\right]$ in terms of the basis vectors:

$$\partial_{\mu}\mathbf{A} = \left(D_{\mu}A^{\nu}\right)\mathbf{e}_{\nu} \tag{13.39}$$

with the repeated index ν summed over.

[7] We are treating $\left[\partial_{\mu}\mathbf{A}\right]$ as a set of vectors, each being labeled by an index μ. And $D_{\mu}A^{\nu}$ is a projection of $\partial_{\mu}\mathbf{A}$ in the same way that $A^{\nu} = \mathbf{e}^{\nu} \cdot \mathbf{A}$ is a projection of the vector $\mathbf{A}$.

Christoffel symbols as expansion coefficients of $\partial_{\nu}\mathbf{e}^{\mu}$

On the other hand, we do not have a similarly simple transformation relation like (13.36) when the coordinate-independent $\mathbf{A}$ is replaced by one of the coordinate basis vectors $(\mathbf{e}_{\mu})$, which by definition change under coordinate transformations. Thus, an expansion of $\partial_{\nu}\mathbf{e}^{\mu}$ in a manner similar to (13.39):

$$\partial_{\nu}\mathbf{e}^{\mu} = -\Gamma^{\mu}_{\nu\lambda}\mathbf{e}^{\lambda} \quad \text{or} \quad \mathbf{A} \cdot (\partial_{\nu}\mathbf{e}^{\mu}) = -\Gamma^{\mu}_{\nu\lambda}A^{\lambda}, \tag{13.40}$$

does not have coefficients $\left(-\Gamma^{\mu}_{\nu\lambda}\right)$ that are tensors. Anticipating the result, we have here used the same notation for these expansion coefficients as the *Christoffel symbols* introduced in Eq. (12.39).

Plugging (13.38) and (13.40) into (13.33), we find

$$D_{\nu}A^{\mu} = \partial_{\nu}A^{\mu} + \Gamma^{\mu}_{\nu\lambda}A^{\lambda}. \tag{13.41}$$

Thus, in order to produce the covariant derivative, the ordinary derivative $\partial_{\nu}A^{\mu}$ must be supplemented by another term. This second term directly reflects the position-dependence of the basis vectors, as in (13.40). Even though both $\partial_{\nu}A^{\mu}$ and $\Gamma^{\mu}_{\nu\lambda}A^{\lambda}$ do not have the correct transformation properties, the unwanted terms produced from their respective transformations cancel each other so that their sum $D_{\nu}A^{\mu}$ is a good tensor.

Compared to the contravariant vector A^{μ} of (13.41), the covariant derivative for a covariant vector A_{μ} takes on the form

$$D_{\nu}A_{\mu} = \partial_{\nu}A_{\mu} - \Gamma^{\lambda}_{\nu\mu}A_{\lambda} \tag{13.42}$$

so that the contraction $A^{\mu}A_{\mu}$ is an invariant. A mixed tensor such as T^{μ}_{ν}, transforming in the same way as the direct product $A^{\mu}B_{\nu}$, will have a covariant derivative

$$D_{\nu}T^{\rho}_{\mu} = \partial_{\nu}T^{\rho}_{\mu} - \Gamma^{\lambda}_{\nu\mu}T^{\rho}_{\lambda} + \Gamma^{\rho}_{\nu\sigma}T^{\sigma}_{\mu}. \tag{13.43}$$

There should be a set of Christoffel symbols for each index of the tensor—a set of $(+\Gamma T)$ for a contravariant index, a $(-\Gamma T)$ for a covariant index, etc. A specific example is the covariant differentiation of the (covariant) metric tensor $g_{\mu\nu}$:

$$D_\lambda g_{\mu\nu} = \partial_\lambda g_{\mu\nu} - \Gamma^\rho_{\lambda\mu} g_{\rho\nu} - \Gamma^\rho_{\lambda\nu} g_{\mu\rho}. \tag{13.44}$$

Christoffel symbols and metric tensor

We have introduced the Christoffel symbols $\Gamma^\mu_{\nu\lambda}$ as the coefficients of expansion for $\partial_\nu \mathbf{e}^\mu$ as in (13.40). In this section we shall relate such $\Gamma^\mu_{\nu\lambda}$ to the first derivative of the metric tensor. This will justify the identification with the symbols first defined in (12.39). To derive this relation, we need an important feature of the Christoffel symbols:

$$\Gamma^\mu_{\nu\lambda} = \Gamma^\mu_{\lambda\nu} \tag{13.45}$$

i.e. they are symmetric with respect to interchange of the two lower indices.

The metric tensor is covariantly constant   While the metric tensor is position-dependent, $\partial[\mathbf{g}] \neq 0$, it is a constant with respect to covariant differentiation, $D[\mathbf{g}] = 0$ (we say, $g_{\mu\nu}$ is **covariantly constant**):

$$D_\lambda g_{\mu\nu} = 0. \tag{13.46}$$

One way to prove this is to use the expression of the metric in terms of the basis vectors: $g_{\mu\nu} = \mathbf{e}_\mu \cdot \mathbf{e}_\nu$, and apply the definition of the affine connection, $\partial_\nu \mathbf{e}_\mu = +\Gamma^\rho_{\mu\nu} \mathbf{e}_\rho$, as given in (13.40):

$$\partial_\lambda (\mathbf{e}_\mu \cdot \mathbf{e}_\nu) = (\partial_\lambda \mathbf{e}_\mu) \cdot \mathbf{e}_\nu + \mathbf{e}_\mu \cdot (\partial_\lambda \mathbf{e}_\nu)$$
$$= \Gamma^\rho_{\lambda\mu} \mathbf{e}_\rho \cdot \mathbf{e}_\nu + \Gamma^\rho_{\lambda\nu} \mathbf{e}_\mu \cdot \mathbf{e}_\rho. \tag{13.47}$$

Written in terms of the metric tensors, this relation becomes

$$\partial_\lambda g_{\mu\nu} - \Gamma^\rho_{\lambda\mu} g_{\rho\nu} - \Gamma^\rho_{\lambda\nu} g_{\mu\rho} = D_\lambda g_{\mu\nu} = 0, \tag{13.48}$$

where we have applied the definition of the covariant derivative of a covariant tensor $g_{\mu\nu}$ as in (13.44). This result of (13.46) can be understood as follows: Recall our discussion of the flatness theorem in Section 12.4.3 that the derivative of the metric vanishes in the local inertial frame $\partial_\lambda g_{\mu\nu} = 0$. However in this particular coordinate frame, the Christoffel symbols are expected to vanish (as $\partial_\nu \mathbf{e}_\mu = 0$) and there is no difference between covariant derivatives and ordinary derivatives; thus $\partial_\lambda g_{\mu\nu} = D_\lambda g_{\mu\nu} = 0$. But once written in this covariant form, it should be valid in every coordinate frame. Hence the result of (13.46). As we shall discuss (see Section 15.3), this key property allowed Einstein to introduce his "cosmological constant term" in the general relativistic field equation.

Christoffel symbols as the metric tensor derivative   In the above discussion we have used the definition (13.40) of Christoffel symbols as the coefficients of expansion of the derivative $\partial_\nu \mathbf{e}^\mu$. Here we shall derive an expression for the Christoffel symbols, as the first derivatives of the metric tensor, which agrees with the definition first introduced in (12.39). We start by using several versions of (13.48) with their indices permuted cyclically:

$$D_\lambda g_{\mu\nu} = \partial_\lambda g_{\mu\nu} - \Gamma^\rho_{\lambda\mu} g_{\rho\nu} - \Gamma^\rho_{\lambda\nu} g_{\mu\rho} = 0$$
$$D_\nu g_{\lambda\mu} = \partial_\nu g_{\lambda\mu} - \Gamma^\rho_{\nu\lambda} g_{\rho\mu} - \Gamma^\rho_{\nu\mu} g_{\lambda\rho} = 0 \qquad (13.49)$$
$$-D_\mu g_{\nu\lambda} = -\partial_\mu g_{\nu\lambda} + \Gamma^\rho_{\mu\nu} g_{\rho\lambda} + \Gamma^\rho_{\mu\lambda} g_{\nu\rho} = 0.$$

Summing over these three equations and using the symmetry property of (13.45), we obtain:

$$\partial_\lambda g_{\mu\nu} + \partial_\nu g_{\lambda\mu} - \partial_\mu g_{\nu\lambda} - 2\Gamma^\rho_{\lambda\nu} g_{\mu\rho} = 0 \qquad (13.50)$$

or, in its equivalent form,

$$\Gamma^\lambda_{\mu\nu} = \frac{1}{2} g^{\lambda\rho} \left[\partial_\nu g_{\mu\rho} + \partial_\mu g_{\nu\rho} - \partial_\rho g_{\mu\nu} \right]. \qquad (13.51)$$

This relation showing $\Gamma^\mu_{\nu\lambda}$ as the first derivative of the metric tensor is called the **fundamental theorem of Riemannian geometry**. It is just the definition stated previously in (12.39). From now on we shall often use this intrinsic geometric description of the Christoffel symbols (13.51) rather than (13.40). The symmetry property of (13.45) is explicitly displayed in (13.51).

13.4 The principle of general covariance

According to the strong principle of equivalence, gravity can always be transformed away locally. Einstein suggested an elegant formulation of the new theory of gravity based on a curved spacetime. In this way EP is a fundamental built-in feature. Local flatness (a metric structure of spacetime) means that SR (a theory of flat spacetime with no gravity) is an automatic property of the new theory. Gravity is not a force but the structure of spacetime. A particle just follows geodesics in such a curved spacetime. The physical laws, or the field equation for the relativistic potential, the metric function $g_{\mu\nu}(x)$, must have the same form no matter what generalized coordinates are used to locate, i.e. label, worldpoints (events) in spacetime. One expresses this by the requirement that the physics equations must satisfy the **principle of general covariance**. This is a two-part statement:

1. Physics equations must be covariant under general (nonlinear) coordinate transformations which leave the infinitesimal spacetime line element ds^2 invariant.
2. Physics equations should reduce to the correct special relativistic form in local inertial frames. Namely, we must have the correct SR equations in the free-fall frames, in which gravity is transformed away. Additionally, gravitational equations reduce to Newtonian equations in the limit of low-velocity particles in a weak and static field.

13.4.1 The principle of minimal substitution

This provides us with a well-defined path to go from SR equations, valid in local inertial frames with no gravity, to GR equations that are valid in every coordinate system in curved spacetime—curved because of the presence of gravity. Such GR equations must be covariant under general local transformations. To go from an SR equation to the corresponding GR equation is simple: we need to replace the ordinary derivatives $[\partial]$ in SR equations by covariant derivatives $[D]$:

$$\partial \longrightarrow D\,(= \partial + \Gamma). \tag{13.52}$$

This is known as the **minimal coupling** because we are assuming the absence of the curvature tensor terms, which vanish in the flat spacetime limit. Since the Christoffel symbols Γ are the derivatives of the metric, hence the derivatives of the gravitational potential, the introduction of covariant derivatives naturally brings the gravitational field into the physics equations. In this way we can, for example, find the equations that describe electromagnetism in the presence of a gravitational field. For example, considering the inhomogeneous Maxwell equation (11.35), we have the set of GR equations in curved spacetime,

$$D_\mu F^{\mu\nu} = \partial_\mu F^{\mu\nu} + \Gamma^\mu_{\mu\lambda} F^{\lambda\nu} + \Gamma^\nu_{\mu\lambda} F^{\mu\lambda} = -\frac{1}{c} j^\nu, \tag{13.53}$$

which are interpreted as Gauss's and Ampere's laws in the presence of a gravitational field.

13.4.2   Geodesic equation from the special relativity equation of motion

Now that we have the GR equations for electromagnetism, what about the GR equations for gravitation? Recall in the new theory, gravity is not regarded as a force. Therefore, the GR equation of motion must be the generalization of the force-free SR equation $\partial U^\mu / d\tau = 0$ leading to

$$\frac{DU^\mu}{d\tau} = 0, \tag{13.54}$$

where U^μ is the 4-velocity of the test particle, and τ is the proper time.

We now demonstrate that this equation (13.54) is just the geodesic equation (13.5). Using the explicit form of the covariant differentiation (13.41), the above equation can be written as

$$\frac{DU^\mu}{d\tau} = \frac{dU^\mu}{d\tau} + \Gamma^\mu_{\nu\lambda} U^\nu \frac{dx^\lambda}{d\tau} = 0. \tag{13.55}$$

[8]For the 4-velocity, we have $U^\mu = Dx^\mu/D\tau$ $= dx^\mu/d\tau$ because $dx^\mu/d\tau$ is already a "good vector" as can been seen from the fact that $(ds/d\tau)^2 = g_{\mu\nu}(dx^\mu/d\tau)(dx^\nu/d\tau)$ is a scalar.

Plugging in the expression of the 4-velocity in terms of the position vector[8]

$$U^\mu = \frac{dx^\mu}{d\tau}, \tag{13.56}$$

we immediately obtain

$$\frac{d^2 x^\mu}{d\tau^2} + \Gamma^\mu_{\nu\lambda} \frac{dx^\nu}{d\tau} \frac{dx^\lambda}{d\tau} = 0, \tag{13.57}$$

which is recognized as the geodesic equation (13.5). This supports our heuristic argument—"particles should follow the shortest and straightest possible trajectories"—used in Section 13.2.1 to suggest that the GR equation of motion should be the geodesic equation.

We note again that special relativity should be valid in all branches of physics: electrodynamics, mechanics, etc., but not when gravity is present. It is special because we restrict coordinate frames to inertial coordinates. It cannot be applied to gravity because inertial frames no longer have privileged status in the presence of gravity (being equivalent to accelerated frames). In fact one can define an inertial frame as one without gravity. The general principle of relativity is also applicable to all branches of physics, say, a GR formulation of Maxwell's equations. It is the correct physics description in the presence of gravity, or equivalently, the correct physics equations that automatically include gravitational effects.

The Einstein field equation

14

- We seek the GR field equation, which should be a generalization of Newton's equation $\nabla^2 \Phi = 4\pi G_N \rho$, with Φ being the gravitational potential, ρ the mass density and G_N Newton's constant. The energy–momentum tensor element $T_{00} = \rho c^2$ suggests that the RHS be extended to $T_{\mu\nu}$ while the LHS can be extended to a symmetric rank-2 curvature tensor, which is a second derivative of the metric, the relativistic potential.

- We show that the curvature tensor in a geometric theory of gravity has the physical interpretation of tidal forces. We explain this identification quantitatively through the Newtonian deviation equation, which is then extended to the GR equation of geodesic deviation.

- Following Einstein we narrow down our search for the rank-2 curvature tensor by the requirement of energy–momentum conservation in the field system. In this effort, we study the symmetry and contraction of the Riemann–Christoffel curvature as well as the Bianchi identity.

- The Einstein equation $R_{\mu\nu} - \frac{1}{2}Rg_{\mu\nu} = \left(8\pi G_N/c^4\right)T_{\mu\nu}$ is proposed to be the GR field equation that determines the metric components $g_{\mu\nu}$ once the source $T_{\mu\nu}$ is given. As a result the spacetime becomes a dynamical quantity. This GR field equation is shown to have the correct Newtonian limit. In this way Newton's $1/r^2$ gravitational force law is explained.

- The Einstein equation treats space and time on an equal footing. Its nontrivial time dependence gives rise to gravitational waves. The existence of gravitational waves has been verified through the long-term observation of the Hulse–Taylor binary pulsar system.

- The Einstein equation is a set of 10 coupled partial differential equations. Yet its exact solution for the case of a spherical source was discovered soon after its proposal. The predictions embodied in the Schwarzschild solution have been verified whenever precision measurements can be performed. The gravitation redshift, the bending of starlight by the sun and the perihelion advance of the planet Mercury are the three famous classical tests of GR.

- The phenomenon of black holes demonstrates the full power and glory of GR. Here the gravitational field is so strong that the roles of space and time are reversed when crossing the event horizon. As a result every worldline, including those of a light ray, once inside the event horizon must head into the center of the spherical mass. Alas, Einstein never believed the reality of black holes.

As discussed in the last chapter, EP physics led Einstein to the idea of a geometric description of gravity. Curved spacetime is the gravitational field; the geodesic equation is the GR equation of motion. Here we shall study the GR field equation, the Einstein equation, and its Schwarzschild solution.

In showing that the geodesic equation is the equation of motion, we have followed the straightforward procedure in Section 13.4.2 of replacing the ordinary derivatives in a SR equation by covariant derivatives to obtain their corresponding GR equation. Unfortunately this procedure cannot be used in the case of the field equation, because there is no SR field equation of gravity. (In fact, in the special relativistic flat spacetime limit, there is no gravity.) We will have to start from the Newtonian theory.

14.1   The Newtonian field equation

Newton formulated his theory of gravitation through the concept of an action-at-a-distance force

$$\mathbf{F}(\mathbf{r}) = -G_{\mathrm{N}} \frac{\mu m}{r^2} \hat{\mathbf{r}} \tag{14.1}$$

where G_{N} is **Newton's constant**, the point-source mass μ is located at the origin of the coordinate system, and the test mass m is at position $\mathbf{r}$.

Just as in the case of electrostatics $\mathbf{F}(\mathbf{r}) = q\mathbf{E}(\mathbf{r})$, we can cast this in the form $\mathbf{F} = m\mathbf{g}$. This defines the gravitational field $\mathbf{g}(\mathbf{r})$ as the gravitational force per unit mass. Newton's law, in term of this gravitational field for a point mass μ, is

$$\mathbf{g}(\mathbf{r}) = -G_{\mathrm{N}} \frac{\mu}{r^2} \hat{\mathbf{r}}. \tag{14.2}$$

Just as Coulomb's law is equivalent to Gauss's law for the electric field, this field equation (14.2) can be expressed, for an arbitrary mass distribution, as Gauss's law for the gravitational field:

$$\oint_S \mathbf{g} \cdot d\mathbf{S} = -4\pi G_{\mathrm{N}} M. \tag{14.3}$$

The area integral on the LHS is the gravitational field flux through any closed surface S, and M on the RHS is the total mass enclosed inside S. This integral representation of Gauss's law (14.3) can be converted into a differential equation. We will first turn both sides into volume integrals by using the divergence theorem on the LHS [Cf. Eq. (A.25) in Appendix A, Section A.1] and by expressing the mass on the RHS in terms of the mass density function ρ

$$\int \nabla \cdot \mathbf{g} \, dV = -4\pi G_{\mathrm{N}} \int \rho \, dV.$$

Since this relation holds for any volume, the integrands on both sides must also be equal:

$$\nabla \cdot \mathbf{g} = -4\pi G_N \rho. \tag{14.4}$$

[1] We have the familiar example of the potential for a spherically symmetric source with total mass M given by $\Phi = -G_N M/r$.

This is Newton's field equation in differential form. The gravitational potential[1] $\Phi(\mathbf{r})$ being defined through the gravitational field $\mathbf{g} \equiv -\nabla \Phi$, the field equation (14.4) becomes

$$\nabla^2 \Phi = 4\pi G_N \rho. \tag{14.5}$$

Einstein's task was to seek the GR extension of this equation.

14.2 Seeking the general relativistic field equation

We have already learned in Eq. (13.4) that the metric tensor is the relativistic generalization of the gravitational potential and in Section 11.2.5 that the mass density is the (0,0) component of the relativistic energy–momentum tensor $T_{\mu\nu}$:

$$\left(1 + \frac{2\Phi(x)}{c^2}\right) \to g_{00}(x) \quad \text{and} \quad \rho(x)c^2 \to T_{00}(x). \tag{14.6}$$

The GR field equation, being the generalization of the Newtonian field equation (14.5), must satisfy the principle of general relativity. It should be a tensor equation with the following structure,

$$[\hat{O}g] = \kappa[T]. \tag{14.7}$$

Namely, on the RHS we should have the energy–momentum tensor $T_{\mu\nu}$, which describes the distribution of matter/energy in spacetime; on the LHS some differential operator $\hat{O}$ acting on the metric $[g]$. Since we expect $[\hat{O}g]$ to have the Newtonian limit of $\nabla^2 \Phi$, then $[\hat{O}]$ must be a second-derivative operator. Besides the $\partial^2 g$ terms, we also expect it to contain nonlinear terms of the type $(\partial g)^2$. The presence of the $(\partial g)^2$ terms is suggested by the fact that energy, just like mass, is a source of gravitational fields, and gravitational fields themselves hold energy—just as electromagnetic fields have an energy density that is quadratic in the fields $(\mathbf{E}^2 + \mathbf{B}^2)$. That is, the gravitational field energy density must be quadratic in the gravitational field strength, $(\partial g)^2$. In terms of the Christoffel symbols of (13.51) $\Gamma \sim \partial g$, we anticipate $[\hat{O}g]$ to contain not only $\partial \Gamma$ but also Γ^2 terms as well.

Furthermore, because we have a rank-2 tensor $T^{\mu\nu} = T^{\nu\mu}$ that is symmetric on the RHS, $[\hat{O}g]$ on the LHS must have these properties also. Thus the task of seeking the GR field equation involves finding such a $[\hat{O}g]$ tensor:

$[\hat{O}g]$ must have the property:
symmetric rank-2 tensor $(\partial^2 g), (\partial g)^2 \smile \partial \Gamma, \Gamma^2$.

(14.8)

14.3 Curvature tensor and tidal forces

We now concentrate on the curvature tensor of spacetime. Starting with Gaussian curvature (a single component) in 2D space, Riemann (plus further work by Christoffel) established, in an n-dimensional space, the existence of a rank-4 curvature tensor:

$$R^{\mu}_{\ \lambda\alpha\beta} = \partial_\alpha \Gamma^{\mu}_{\ \lambda\beta} - \partial_\beta \Gamma^{\mu}_{\ \lambda\alpha} + \Gamma^{\mu}_{\ \nu\alpha}\Gamma^{\nu}_{\ \lambda\beta} - \Gamma^{\mu}_{\ \nu\beta}\Gamma^{\nu}_{\ \lambda\alpha}. \tag{14.9}$$

The Christoffel symbols Γ being first derivatives as shown in (13.51), the Riemann–Christoffel curvature of (14.9), $R = d\Gamma + \Gamma\Gamma$, is then a nonlinear second-derivative function of the metric, $[\partial^2 g + (\partial g)^2]$. While $R^{\mu}_{\ \lambda\alpha\beta} = 0$ for a flat space, it generally measures the deviation from flat-space relations. But what is needed in Eq. (14.7) is a rank-2 symmetric tensor. Thus some form of contraction of the Riemann–Christoffel curvature tensor is required. There can be many versions of the contracted curvature; which one is the correct one? Before answering this mathematical question, we first explain the physical meaning of curvature in spacetime.

14.3.1 Tidal forces—a qualitative discussion

What is the physical significance of the curvature? Curvature involves second derivatives of a metric. Since the metric tensor can be regarded as the relativistic gravitational potential, its first derivatives are gravitational forces, and its second derivatives must then be the relative forces between neighboring particles. Namely, they are the tidal forces.

Here we give an elementary and qualitative discussion of tidal force and its possible geometric interpretation. That will then be followed by a more mathematical presentation in terms of "the Newtonian deviation equation" which relates relative position change between two neighboring particles to tidal forces. This in turn suggests the GR generalization, "the equation of geodesic deviation" which relates the spacetime separation of two particles to the underlying spacetime curvature. Our purpose is mainly to provide a more physical feel for the curvature in GR gravitational theory.

The equivalence principle states that in a freely falling reference frame the physics is the same as that in an inertial frame with no gravity: SR applies and the metric is given by the Minkowski metric $\eta_{\mu\nu}$. As shown in the flatness theorem (Section 12.4.3), this approximation of $g_{\mu\nu}$ by $\eta_{\mu\nu}$ can be done only locally, that is, in an appropriately small region. Gravitational effects can always be detected in a **finite-sized** free-fall frame as the gravitational field is never strictly uniform in reality; the second derivatives of the metric come into play.

Consider the lunar gravitational attraction exerted on the earth. While the earth is in free fall toward the moon (and vice versa), there is still a detectable lunar gravitational effect on earth. It is so because different points on earth will feel slightly different gravitational pulls by the moon, as depicted in Fig. 14.1(a). The center-of-mass (CM) force causes the earth to "fall towards the moon" so that this CM gravitational effect is "cancelled out" in this freely falling terrestrial frame. After subtracting out this CM force, the remanent

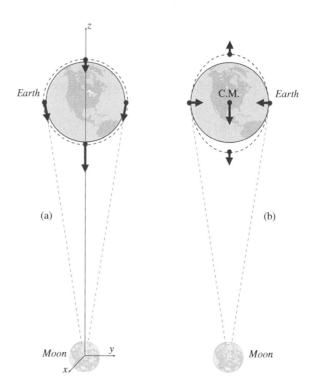

Fig. 14.1 Variations of the gravitational field as tidal forces. (a) Lunar gravitational forces on four representative points on earth. (b) After taking out the center of mass (CM) motion, the relative forces on earth are the tidal forces giving rise to longitudinal stretching and transverse compression.

[2]The ocean is pulled away in opposite directions giving rise to two tidal bulges. This explains why, as the earth rotates, there are two high tides in a day. This of course is a simplified description as there are other effects, e.g. solar tidal forces, that must be taken into account.

(relative) forces on the earth, as shown in Fig. 14.1(b), are stretching in the longitudinal direction and compression in the transverse directions. They are just the familiar tidal forces.[2] Namely, in the freely falling frame, the CM gravitational effect is transformed away, but, there are still the remnant tidal forces. They reflect the **differences** of the gravitational effects on neighboring points, and are thus proportional to the derivative of the gravitational field. Since tidal forces cannot be coordinate-transformed away, they should be regarded as the essence of gravitation.

14.3.2   Newtonian deviation equation and the equation of geodesic deviation

Here we provide first a quantitative description of the gravitational tidal forces in the Newtonian framework, which suggests an analogous GR approach showing the curvature tensor playing exactly the same role as tidal forces.

Newtonian deviation equation for tidal forces

The tidal effect concerns the relative motion of particles in a nonuniform gravitational field. Let us consider two particles: one has trajectory $\mathbf{x}(t)$ and another has $\mathbf{x}(t) + \mathbf{s}(t)$. That is, the locations of these two particles measured at the same time have a coordinate difference of $\mathbf{s}(t)$. The respective equations of motion ($i = 1, 2, 3$) obeyed by these two particles are, according to Eq. (13.7),

$$\frac{d^2x^i}{dt^2} = -\frac{\partial\Phi(x)}{\partial x^i} \quad \text{and} \quad \frac{d^2x^i}{dt^2} + \frac{d^2s^i}{dt^2} = -\frac{\partial\Phi(x+s)}{\partial x^i}. \qquad (14.10)$$

Consider the case where the separation distance $s^i(t)$ is small and we can approximate the gravitational potential $\Phi(x+s)$ by a Taylor expansion

$$\Phi(x+s) = \Phi(x) + \frac{\partial\Phi}{\partial x^j}s^j + \cdots. \qquad (14.11)$$

From the difference of the two equations in (14.10), we obtain the **Newtonian deviation equation** that describes the separation between two particle trajectories in a gravitational field

$$\frac{d^2s^i}{dt^2} = -\left(\frac{\partial^2\Phi}{\partial x^i\partial x^j}\right)s^j. \qquad (14.12)$$

Thus the relative acceleration per unit separation $(d^2s^i/dt^2)/s^j$ is given by a tensor having the second derivatives of the gravitational potential (i.e. the tidal force components) as its elements.

As an illustrative application of Eq. (14.12) we discuss the case of a spherical gravitational source, e.g. the gravity due to the moon on earth (Fig. 14.1),

$$\Phi(x) = -\frac{G_N M}{r} \qquad (14.13)$$

where the radial distance is related to the rectangular coordinates by $r = \left(x^2 + y^2 + z^2\right)^{1/2}$. Since $\partial r/\partial x^i = x^i/r$ we have

$$\frac{\partial^2\Phi}{\partial x^i\partial x^j} = \frac{G_N M}{r^3}\left(\delta_{ij} - \frac{3x^i x^j}{r^2}\right). \qquad (14.14)$$

Consider the case of the "first particle" being located along the z axis $x^i = (0,0,r)$; the Newtonian deviation equation (14.12) for the displacement of the "second particle", with the second derivative tensor given by (14.14), now takes on the form[3]

$$\frac{d^2}{dt^2}\begin{pmatrix} s_x \\ s_y \\ s_z \end{pmatrix} = \frac{-G_N M}{r^3}\begin{pmatrix} 1 & 0 & 0 \\ 0 & 1 & 0 \\ 0 & 0 & -2 \end{pmatrix}\begin{pmatrix} s_x \\ s_y \\ s_z \end{pmatrix}. \qquad (14.15)$$

[3] Since we have taken $x^i = (0,0,r)$, the term $x^i x^j$ on the RHS can be worked out with $x^1 = x^2 = 0$ and $x^3 = r$, etc.

We see that there is an attractive tidal force between the two particles in the transverse direction $T_{x,y} = -G_N M r^{-3} s_{x,y}$ that leads to compression; and a tidal repulsion $T_z = +2G_N M r^{-3} s_z$, leading to stretching, in the longitudinal (i.e. radial) direction. This of course is in agreement with the conclusion we had drawn from our qualitative discussion.

General relativistic deviation equation with curvature tensor

In GR, we shall follow a similar approach: The two equations of motion (14.10) will be replaced by the corresponding GR equations of motion, the geodesic equations; their difference, after a systematic Taylor expansion, leads to the equation of geodesic deviation[4]

[4] See Problems 14.4 and 14.5 in Cheng (2010, p. 335 and p. 407).

$$\frac{D^2s^\mu}{D\tau^2} = -\left(R^\mu{}_{\alpha\nu\beta}\dot{x}^\alpha\dot{x}^\beta\right)s^\nu, \qquad (14.16)$$

with $R^{\mu}{}_{\alpha\nu\beta}$ being the Riemann–Christoffel curvature tensor given by Eq. (14.9) and $\dot{x}^{\alpha} = \partial x^{\alpha}/\partial\tau$. Clearly Eq. (14.16) is similar to Eq. (14.12). Thus we can regard the curvature tensor as the relativistic generalization of the tidal forces. In this geometric language we see that the cause of the deviation from a flat spacetime trajectory is attributed to the curvature just as the convergent particle trajectories shown in Fig. 13.2. Such considerations led Einstein to give gravity a direct geometric interpretation by identifying these tidal forces with the curvature of spacetime.

We also record another way in which the $R^{\mu}{}_{\alpha\nu\beta}$ tensor appears in a simple equation,[5] involving the commutator of covariant derivatives acting on some arbitrary tensor (here taken to be a contravariant vector A^{μ})

[5]See Problem 13.7 in Cheng (2010, p. 316 and p. 403).

$$[D_{\alpha}, D_{\beta}]A^{\mu} = R^{\mu}{}_{\lambda\alpha\beta}A^{\lambda}. \tag{14.17}$$

This relation will be used in our derivation of the Bianchi identity which will help us to find the GR field equation.

14.3.3   Symmetries and contractions of the curvature tensor

[6]Knowing its symmetry properties, we find the number of independent components of a curvature tensor in an n-dimensional space to be $N_{(n)} = n^2(n^2 - 1)/12$. Thus $N_{(1)} = 0$: It is not possible for a one-dimensional inhabitant to see any curvature; $N_{(2)} = 1$: This is just the Gaussian curvature K for a curved surface; $N_{(4)} = 20$: There are 20 independent components in the curvature tensor for a 4D curved spacetime.

We now discuss the contraction of the Riemann curvature tensor[6] with the aim of finding the appropriate rank-2 tensor that can be used for the GR field equation. We first note that the Riemann curvature tensor with all lower indices

$$R_{\mu\nu\alpha\beta} = g_{\mu\lambda}R^{\lambda}{}_{\nu\alpha\beta} \tag{14.18}$$

has the following symmetry features:

- It is **antisymmetric** with respect to the interchange of the first and second indices, and that of the third and fourth indices, respectively:

$$R_{\mu\nu\alpha\beta} = -R_{\nu\mu\alpha\beta} \tag{14.19}$$

$$R_{\mu\nu\alpha\beta} = -R_{\mu\nu\beta\alpha}. \tag{14.20}$$

- It is **symmetric** with respect to the interchange of the pair made up of the first and second indices with the pair of third and fourth indices:

$$R_{\mu\nu\alpha\beta} = +R_{\alpha\beta\mu\nu}. \tag{14.21}$$

- It also has **cyclic** symmetry:

$$R_{\mu\nu\alpha\beta} + R_{\mu\beta\nu\alpha} + R_{\mu\alpha\beta\nu} = 0. \tag{14.22}$$

Since the symmetry properties are not changed by coordinate transformations, one can choose a particular coordinate frame to prove these symmetry relations, and once proven in one frame, we can then claim their validity in all frames. An obvious choice is the local inertial frame (with $\Gamma = 0$, $\partial\Gamma \neq 0$) where the curvature takes on simpler form, $R_{\mu\nu\alpha\beta} = g_{\mu\lambda}(\partial_{\alpha}\Gamma^{\lambda}_{\nu\beta} - \partial_{\beta}\Gamma^{\lambda}_{\nu\alpha})$. In this way, we can use the fact that $g_{\mu\nu} = g_{\nu\mu}$ and $\Gamma^{\mu}_{\nu\lambda} = \Gamma^{\mu}_{\lambda\nu}$ to check easily the validity of the symmetry properties as shown in Eqs. (14.19)–(14.22).

Contractions of the curvature tensor

We show how to reduce the rank of the Riemann tensor so it can be used in the GR field equation. Because of the symmetry properties shown above these contractions are essentially unique.

Ricci tensor $R_{\mu\nu}$ This is the Riemann curvature tensor with the first and third indices contracted,

$$R_{\mu\nu} \equiv g^{\alpha\beta} R_{\alpha\mu\beta\nu} = R^{\beta}{}_{\mu\beta\nu} \tag{14.23}$$

which is a symmetric tensor,

$$R_{\mu\nu} = R_{\nu\mu}. \tag{14.24}$$

If we had contracted different pairs of indices, the result would still be $\sigma R_{\mu\nu}$ with $\sigma = \pm 1, 0$. Thus the rank-2 curvature tensor is essentially unique.

Ricci scalar R   Contracting the Riemann curvature twice, we obtain the Ricci scalar field,

$$R \equiv g^{\alpha\beta} R_{\alpha\beta} = R^{\beta}{}_{\beta}. \tag{14.25}$$

14.3.4   The Bianchi identities and the Einstein tensor

These contractions show that there is indeed a ready-made symmetric rank-2 curvature tensor, the Ricci tensor $R_{\mu\nu}$ for the LHS of Eq. (14.7). Making this choice $[\hat{O}g]_{\mu\nu} = R_{\mu\nu}$ while working with Grossmann in 1913, Einstein found that the resultant field equation did not reduce down to Eq. (14.5) in the Newtonian limit. He then convinced himself that somehow the covariance principle had to be given up for a gravity theory.[7] This erroneous detour caused considerable delay in completing his GR program. Nevertheless, after a long struggle, finally at the end of 1915, he returned to the covariance principle and succeeded in finding the correct tensor $[\hat{O}g]_{\mu\nu} = G_{\mu\nu}$, now known as the Einstein tensor

$$G^{\mu\nu} = R^{\mu\nu} - \frac{1}{2} R g^{\mu\nu}. \tag{14.26}$$

The realization was that besides the Ricci tensor there is another rank-2 curvature tensor that satisfies the criterion for the geometry side of the field equation: it's the product of the Ricci scalar with the metric tensor itself. Thus there is actually a variety of ways to construct a rank-2 curvature tensor. Any combination of the form $R_{\mu\nu} + a R g_{\mu\nu}$ with a constant coefficient a will satisfy the search criteria. What is the correct value of a?

   To narrow down the choice, Einstein invoked the conservation law of energy–momentum in the field system. This led him to the combination (14.26). The calculation that he had to undertake is rather complicated; after he published his GR papers in 1916, others had found a way to greatly expedite the proof. It involves the mathematical relation called the Bianchi identity.

[7] In late 1915 when Einstein returned to the covariance approach (1915a) he was still working with $[\hat{O}g]_{\mu\nu} = R_{\mu\nu}$, namely the field equation $R_{\mu\nu} = \kappa T_{\mu\nu}$. But in the region where the stress tensor vanishes $T_{\mu\nu} = 0$ (i.e. outside the source) the field equation $R_{\mu\nu} = 0$ led to the correct precession value for Mercury (1915b). Afterwards Einstein found the correct field equation should be $G_{\mu\nu} = \kappa T_{\mu\nu}$ and had to withdraw the earlier proposal (1915c). The equation he was working with for the Mercury problem was correct because, with $T_{\mu\nu} = 0$ he also had the trace conditions $T = R = 0$, and the equation $R_{\mu\nu} = 0$ was the same as the correct $G_{\mu\nu} = 0$ equation.

Recall our discussion of the energy–momentum tensor in Section 11.2.5 showing that conservation is expressed in the flat spacetime as a 4-divergenceless condition $\partial_\mu T^{\mu\nu} = 0$. Thus in a warped spacetime, it must be the vanishing of the covariant divergence

$$D_\mu T^{\mu\nu} = 0.$$

Since $T^{\mu\nu}$ is covariantly constant, so must be the LHS of (14.7). Namely, a must be a value such that

$$D_\mu(R^{\mu\nu} + aRg^{\mu\nu}) = 0. \tag{14.27}$$

The Bianchi identities

Starting with the Jacobi identity (for any double commutator).

$$\left[D_\lambda, \left[D_\mu, D_\nu\right]\right] + \left[D_\nu, \left[D_\lambda, D_\mu\right]\right] + \left[D_\mu, \left[D_\nu, D_\lambda\right]\right] = 0, \tag{14.28}$$

and by applying the relation (14.17) and the cyclic symmetry of (14.22), one can prove[8] the Bianchi identities:

[8]This is worked out in Problem 13.12 of Cheng (2010, p. 317 and p. 406).

$$D_\lambda R_{\mu\nu\alpha\beta} + D_\nu R_{\lambda\mu\alpha\beta} + D_\mu R_{\nu\lambda\alpha\beta} = 0. \tag{14.29}$$

We now perform contractions on these relations. When contracting (14.29) with $g^{\mu\alpha}$, because of (13.46) showing that the metric tensor is itself covariantly constant, $D_\lambda g^{\alpha\beta} = 0$, one can push $g^{\mu\alpha}$ right through the covariant differentiation:

$$D_\lambda R_{\nu\beta} - D_\nu R_{\lambda\beta} + D_\mu g^{\mu\alpha} R_{\nu\lambda\alpha\beta} = 0. \tag{14.30}$$

Contracting another time with $g^{\nu\beta}$,

$$D_\lambda R - D_\nu g^{\nu\beta} R_{\lambda\beta} - D_\mu g^{\mu\alpha} R_{\lambda\alpha} = 0. \tag{14.31}$$

In the last two terms, the metric just raises the indices,

$$D_\lambda R - D_\nu R_\lambda^\nu - D_\mu R_\lambda^\mu = D_\lambda R - 2D_\nu R_\lambda^\nu = 0. \tag{14.32}$$

Pushing through yet another $g^{\mu\lambda}$ in order to raise the λ index in the last term,

$$D_\lambda \left(Rg^{\mu\lambda} - 2R^{\mu\lambda}\right) = 0. \tag{14.33}$$

Thus the constant in (14.27) is $a = -1/2$ and the combination (14.26) is covariantly constant

$$D_\mu G^{\mu\nu} = 0. \tag{14.34}$$

To summarize, $G^{\mu\nu}$, called the **Einstein tensor**, is a covariantly constant, rank-2, symmetric tensor involving the second derivatives of the metric $\partial^2 g$ as well as the quadratic in ∂g. This is just the sought-after mathematical quantity on the geometric side of the GR field equation.

14.4   The Einstein equation

With the identification of $G_{\mu\nu}$ with the LHS of Eq. (14.7), Einstein arrived at the GR field equation

$$G_{\mu\nu} = R_{\mu\nu} - \frac{1}{2}Rg_{\mu\nu} = \kappa T_{\mu\nu}. \tag{14.35}$$

We recall that EP informs us that gravity can always be transformed away locally (by going to a reference frame in free-fall); the essence of gravity is represented by its differentials (tidal forces). Thus the presence of the curvature of spacetime in the GR field equation can be understood. This equation can also be written in an alternative form. A contraction of the two indices in (14.35) leads to $-R = \kappa T$, where T is the trace $g^{\mu\nu}T_{\mu\nu}$. In this way we can rewrite the field equation as

$$R_{\mu\nu} = \kappa \left(T_{\mu\nu} - \frac{1}{2}Tg_{\mu\nu} \right). \tag{14.36}$$

Given that the source distribution $T_{\mu\nu}$ and the Ricci $R_{\mu\nu}$ are symmetric tensors, each of them has 10 independent elements. Thus the seemingly simple field equation is actually a set of 10 coupled nonlinear partial differential equations for the 10 components of the gravitational potential which is the spacetime metric $g_{\mu\nu}(x)$. We reiterate the central point of Einstein's theory: the spacetime geometry in GR, in contrast to SR, is not a fixed entity, but is dynamically ever-changing as determined by the mass/energy distribution.

14.4.1   The Newtonian limit for a general source

One can examine one aspect of the correctness of this proposed GR field equation by checking its Newtonian limit. A straightforward calculation similar to that carried out in Section 13.2.2 shows that it does reduce to Newton's field equation [to be displayed in (15.22)] for a particle moving nonrelativistically in a weak and static field.[9] This also gives us the identification of $\kappa = 8\pi G_{\mathrm{N}}/c^4$. The proportionality constant κ, hence Newton's constant G_{N}, is the "conversion factor" that allows us to relate the energy density on the RHS to the geometric quantity on the LHS. Recall our discussion in Section 3.4.2, Planck's discovery of Planck's constant allowed him to construct a natural unit system of mass-length-time from the constants of $h, c,$ and G_{N}. We now see that, as it turns out, each of them is a fundamental conversion factor that connects disparate types of physical phenomena: wave and particle, space and time, and, now, energy and geometry. Albert Einstein made vital contributions to each of of these surprising unifications: quantum theory, and special and general relativity.

In taking the above limit, we took the source to be nonrelativistic matter with its energy–momentum tensor completely dominated by the $T_{00} = \rho c^2$ term. Effectively we took the source to be a swarm of noninteracting particles. In certain situations, with cosmology being the notable example, we consider the source of gravity to be a plasma having mass density ρ and pressure p. Namely,

[9]For details of this limit calculation, see Section 14.2.2 in Cheng (2010, p. 323).

we need to take an ideal fluid $T_{\mu\nu}$ as in Eq. (11.43). A calculation entirely similar to the Newtonian limit calculation mentioned above leads to

$$\nabla^2 \Phi = 4\pi G_N \left(\rho + 3\frac{p}{c^2} \right).$$

(14.37)

In this way Einstein's theory makes it clear that in the relativistic theory not only mass but also pressure can be a source of a gravitational field.

14.4.2   Gravitational waves

Newton's theory of gravitation is a static theory; the field due to a source is established instantaneously. Thus, while the field has nontrivial dependence on the spatial coordinates, it does not depend on time. Einstein's theory, being relativistic, treats space and time on an equal footing. Just like Maxwell's theory, it has the feature that a field propagates outward from the source with a finite speed. Thus, just as one can shake an electric charge to generate electromagnetic waves, one can shake a mass to generate gravitational waves. By and large we are dealing with the case of a weak gravitational field. This approximation linearizes the Einstein theory. In this limit, gravitational waves may be viewed as small-curvature ripples propagating in a background of flat spacetime. It is a transverse wave having two independent polarization states, traveling at the speed of light.

These ripples of curvature can be detected as tidal forces. A major effort is underway to detect this effect using gravitational wave interferometers, which can measure the minute compression and elongation of orthogonal lengths that are caused by the passage of such a wave. In the meantime there is already indirect, but convincing, evidence for the existence of gravitational waves as predicted by general relativity (GR). This came from the observation, spanning more than 25 years, of the orbital motion of the relativistic Hulse–Taylor binary pulsar system (PSR 1913+16). Even though the binary pair is 16 light-years away from us, the basic parameters of the system can be deduced by carefully monitoring the radio pulses emitted by the pulsar, which effectively act as an accurate and stable clock. From this record we can verify a number of GR effects. According to GR the orbital period, because of the quadrupole gravitational wave radiation from the system, is predicted to decrease at a calculable rate.[10] The observed orbital rate decrease has been found to be in splendid agreement with the prediction by Einstein's theory (Fig. 14.2).

[10] See Eq. (15.71) in Cheng (2010).

14.5   The Schwarzschild solution

The Einstein equation (14.35) is a set of 10 coupled nonlinear partial differential equations. It came as a surprise that shortly after Einstein's proposal, Karl Schwarzschild (1873–1916) was able to find the exact solution[11] for the most important case of a gravitational field outside a spherical source (a star). Exterior to the source, we have $T_{\mu\nu} = 0$; the relevant equation is

[11] The solution to Eq. (14.38) is worked out in Cheng (2010, Sections 7.1 and 14.3). Our presentation only discusses the Schwarzschild exterior solution. There is also the Schwarzschild interior solution.

$$R_{\mu\nu} = 0.$$

(14.38)

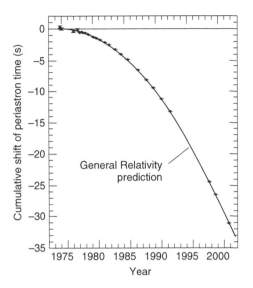

Fig. 14.2 Gravitational radiation damping causes orbital decay of the Hulse–Taylor binary pulsar. Plotted here is the accumulating shift in the epoch of periastron (Weisberg and Taylor 2003). The parabola is the GR prediction, and observations are depicted by data points. In most cases the measurement uncertainties are smaller than the linewidths. The data gap in the 1990s reflects the downtime when the Arecibo observatory was being upgraded.

Here we merely present the solution, the **Schwarzschild metric**, in spherical coordinates (ct, r, θ, ϕ):

$$g_{\mu\nu} = \begin{pmatrix} g_{00} & 0 & 0 & 0 \\ 0 & g_{rr} & 0 & 0 \\ 0 & 0 & r^2 & 0 \\ 0 & 0 & 0 & r^2 \sin^2 \theta \end{pmatrix}, \tag{14.39}$$

which differs from the flat spacetime value of $g_{00} = -g_{rr} = -1$ by

$$g_{00} = -\frac{1}{g_{rr}} = -\left(1 - \frac{r^*}{r}\right) \quad \text{with} \quad r^* = \frac{2G_N M}{c^2} \tag{14.40}$$

where M is the source mass and r^* is called the **Schwarzschild radius**. This is in general an extremely small length factor. For example, the Schwarzschild radii for the sun ($M = M_\odot$) and earth ($M = M_\oplus$) are respectively:

$$r_\odot^* \simeq 3 \text{ km} \quad \text{and} \quad r_\oplus^* \simeq 9 \text{ mm}. \tag{14.41}$$

Hence, in general, the ratio r^*/r, which signifies the modification of the flat Minkowski metric, is a very small quantity. For the Schwarzschild exterior solution to be applicable, r must have the same value at or outside of the radius of the source mass, and so the smallest value that r can take is the radius R of the spherical source: the above r^* values translate into r^*/r of

$$\frac{r_\odot^*}{R_\odot} = O\left(10^{-6}\right) \quad \text{and} \quad \frac{r_\oplus^*}{R_\oplus} = O\left(10^{-9}\right), \tag{14.42}$$

for the case of the sun and the earth, respectively.

14.5.1   Three classical tests

Experimental tests that might have been available in Einstein's day were those within the solar system which had only a weak field (namely, the deviation from flat spacetime value being a small r^*/r). A perturbative approach would be adequate for Einstein to calculate observational predictions. Here are the three famous instances where GR predictions were checked with observation.

Gravitational redshift

As we have already explained, the gravitational redshift follows directly from the equivalence principle which is embodied in the time warpage in curved spacetime as shown in Eq (13.4),

$$g_{00} = -\left(1 + \frac{2\Phi(x)}{c^2}\right), \qquad (14.43)$$

[12]For a direct derivation of the gravitational redshift (12.16) from a curved spacetime $g_{00} \neq -1$, see Box 6.2, p. 108 in Cheng (2010).

which matches (14.40) when we recall that the spherical Newtonian gravitational potential is $\Phi = -G_N M/r$. In principle, this gravitational redshift Eq. (12.16)[12] can be tested by a careful examination of the spectral emission lines from an astronomical object (hence a large gravitational potential difference). But because the effect is small, it is very difficult to separate it from the background. While there were some tentative positive confirmations of this EP prediction, conclusive data did not exist in the first few decades after Einstein's paper. Surprisingly, as we have already commented on in Section 12.3.2, this EP effect of the gravitational redshift was first verified in a series of terrestrial experiments when Robert Pound and his collaborators succeeded in the early 1960s in measuring the truly small frequency shift of radiation traveling up $h = 22.5$ m, the height of an elevator-shaft in the building housing the Harvard Physics Department:

$$\left|\frac{\Delta\omega}{\omega}\right| = \left|\frac{gh}{c^2}\right| = O\left(10^{-15}\right). \qquad (14.44)$$

Bending of starlight by the sun

We recall that the equivalence principle leads to the prediction of bending of starlight by the sun with a deflection angle of $\delta\phi_{\text{EP}}$ given in (12.26):

$$\delta\phi_{\text{EP}} = \frac{2G_N M_\odot}{c^2 r_{\text{min}}}. \qquad (14.45)$$

As explained in Section 13.1.2, this is due to the time warpage $g_{00} \neq -1$. However, now we know from (14.40) that the full GR has not only warping in time but also in space; in fact the warpages are directly related $g_{rr} = -1/g_{00}$. Here we calculate the effect of this extra warpage on the bending of the light-ray, finding a doubling of the deflection angle.

Light-ray deflection: GR vs. EP   Let us consider the light-like worldline ($ds^2 = 0$) in a fixed direction, $d\theta = d\phi = 0$:

$$ds^2 = g_{00}c^2 dt^2 + g_{rr}dr^2 = 0. \qquad (14.46)$$

To an observer far from the source, using the coordinate time and radial distance (t, r), the effective speed of light[13] according to (14.46) is

[13]Recall our discussion in Section 12.3.4.

$$c(r) \equiv \frac{dr}{dt} = c\sqrt{-\frac{g_{00}(r)}{g_{rr}(r)}}. \qquad (14.47)$$

The influence of $g_{00} \neq -1$ and $g_{rr} \neq 1$ in (14.47) are of the same size and in the same direction. Thus the deviation of the vacuum index of refraction $n(r)$ from unity is *twice* as large: compared to the result of (12.20), we now have

$$n(r) = \frac{c}{c(r)} = \sqrt{-\frac{g_{rr}(r)}{g_{00}(r)}} = \frac{1}{-g_{00}(r)} = \left(1 + 2\frac{\Phi(r)}{c^2}\right)^{-1}. \qquad (14.48)$$

Or,

$$c(r) = \left(1 + 2\frac{\Phi(r)}{c^2}\right)c. \qquad (14.49)$$

According to Equations (12.21)–(12.24), the deflection angle being directly proportional to this deviation from c, namely $d\phi \propto d[c(r)]$, is twice as large, resulting in a deflection angle $\delta\phi$ given by (14.45):

$$\delta\phi_{GR} = 2\delta\phi_{EP} = \frac{4G_N M_\odot}{c^2 r_{min}}, \qquad (14.50)$$

where r_{min} is the closest distance that the light ray comes to the massive object. We should apply $r_{min} = R_\odot$, the solar radius, for the case of a light ray grazing the edge of the sun.

The deflection of 1.74 arcseconds (about $1/4000$ of the angular width of the sun as seen from earth) that was predicted by Einstein's equations in 1915 was not easy to detect and therefore to test. One needed a solar eclipse against the background of several bright stars (so that some could be used as reference points). The angular position of a star with light grazing past the (eclipsed) sun would appear to have moved to a different position when compared to the location in the absence of the sun (see Fig. 12.6). On May 29, 1919 there was such a solar eclipse. Two British expeditions were mounted: one to Sobral in northern Brazil, and another to the island of Principe, off the coast of West Africa. The report by A.S. Eddington (1882–1944) that Einstein's prediction was successful in these tests created a worldwide sensation, partly for scientific reasons, and partly because the world was amazed that so soon after World War I the British should finance and conduct an expedition to test a theory proposed by a German citizen.

The gravitational deflection of a light ray discussed above has some similarity to the bending of light by a glass lens. The result in (14.50) becomes the basic ingredient in writing down the lens equation. The whole field of **gravitational lensing** has become an important tool in astrophysics, for detecting mass distribution in the cosmos.

Precession of Mercury's perihelion

A particle, under the Newtonian $1/r^2$ gravitational attraction, traces out a closed elliptical orbit. The presence other planets would distort such a trajectory as shown in Fig. 14.3. For the case of the planet Mercury, such perturbations could account for most[14] of the planetary perihelion advance of 574 arcseconds per century. However there was still the discrepancy of 43 arcseconds left unaccounted for. Following a similar situation involving Uranus that eventually led to the prediction and discovery of the outer planet Neptune in 1846, a new planet, named Vulcan, was predicted to lie inside Mercury's orbit. But it was never found. This is the "perihelion precession problem" that Einstein solved by applying his new theory of gravitation.

One first sets up a relativistic orbit equation, which can be solved for $r(\phi)$ by standard perturbation theory. With e being the eccentricity of the orbit, $\alpha = (1 + e)r_{\text{min}}$ and $\epsilon = 3r^*/2\alpha$, the solution (see Fig. 14.4 for the coordinate labels) is

$$r = \frac{\alpha}{1 + e\cos[(1 - \epsilon)\phi]}. \tag{14.51}$$

Thus the planet returns to its perihelion r_{min} not at $\phi = 2\pi$ (as is the case for a closed orbit) but at $\phi = 2\pi/(1 - \epsilon) \simeq 2\pi + 3\pi r^*/\alpha$. The perihelion advances (i.e. the whole orbit rotates in the same sense as the planet itself) per revolution by (Fig. 14.3)

$$\delta\phi = \frac{3\pi r^*}{\alpha} = \frac{3\pi r^*}{(1 + e)r_{\text{min}}}. \tag{14.52}$$

With the solar Schwarzschild radius $r^*_\odot = 2.95\,\text{km}$, Mercury's eccentricity $e = 0.206$, and its perihelion $r_{\text{min}} = 4.6 \times 10^7\,\text{km}$, we have the numerical value of the advance as $\delta\phi = 5 \times 10^{-7}$ radian per revolution, or $5 \times 10^{-7} \times \frac{180}{\pi} \times 60 \times 60 = 0.103''$ (arcsecond) per revolution. In terms of the advance per century,

$$0.103'' \times \frac{100\,\text{yr}}{\text{Mercury's period of } 0.241\,\text{yr}} = 43''\,\text{per century}. \tag{14.53}$$

This agrees with the observational evidence.

This calculation explaining the perihelion advance of the planet Mercury from first principles gave Einstein great of joy. This moment of elation was characterized by his biographer Pais as "by far the strongest emotional experience in Einstein's scientific life, perhaps, in all his life". The Mercury calculation and the correct prediction for the bending of starlight around the sun, as well as the correct GR field equation, were all obtained by Albert Einstein in an intense two-week period in November, 1915. Afterwards, he wrote to Arnold Sommerfeld in a, by now, famous letter.

This last month I have lived through the most exciting and the most exacting period of my life; and it would be true to say this, it has been the most fruitful. Writing letters has been out of the question. I realized that up until now my field equations of gravitation have been entirely devoid of foundation. When all my confidence in the old theory vanished, I saw clearly that a satisfactory solution could only be reached by linking it with the Riemann variations. The wonderful thing that happened then was that not only did

[14]Most of the raw observational value of 5600'' is due to the effect of rotation of our Earth-based coordinate system and it leaves the planetary perturbations to account for the remaining 574'', to which Venus contributes 277'', Jupiter 153'', Earth 90'', Mars, and the rest of the planets 10''.

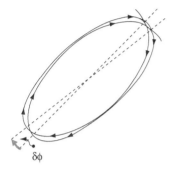

Fig. 14.3 A perturbed $1/r^2$ attraction leads to an open orbit which may be described as an elliptical orbit with a precessing axis. For planetary motion, this is usually stated as the precession of the minimal-distance point from the sun, the perihelion.

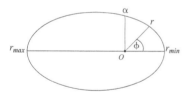

Fig. 14.4 Points on an elliptical orbit are located by the coordinates (r, ϕ), with some notable positions at $(r_{\text{min}}, 0)$, (r_{max}, π), and $(\alpha, \pi/2)$.

Newton's theory result from it, as a first approximation, but also the perihelion motion of Mercury, as a second approximation. For the deviation of light by the Sun I obtained twice the former amount.

Recall that for several years Einstein had given up on the covariance principle because the Ricci tensor $R_{\mu\nu}$ based field equation could not reproduce the Newtonian limit. The Riemann variation that he referred to above is the linear combination that we now called the Einstein tensor.

14.5.2 Black holes—the full power and glory of general relativity

In the above section, we discussed the geodesics of a particle (perihelion precession) and light (bending trajectory) in the Schwarzschild geometry. However, these GR examples by-and-large represent small corrections to the Newtonian result. Here we study the spacetime structure exterior to any object with its mass so compressed that its radius is smaller than the Schwarzschild radius $r^* = 2G_N M/c^2$. Such objects have been given the evocative name **black holes**, because it is impossible to transmit outwardly any signal, any light, from the region inside the Schwarzschild surface $r = r^*$. This necessarily involves such strong gravity and curved spacetime that the GR framework is indispensable. As we shall see, here the spacetime is so warped that the roles of space and time are reversed when one crosses the Schwarzschild surface.

Coordinate singularities

The Schwarzschild metric[15] (14.39) has singularities at $r = 0$ and $r = r^*$, and $\theta = 0$ and π. We are familiar with the notion that $\theta = 0$ and π are **coordinate singularities** associated with our choice of the spherical coordinate system. They are not physical, do not show up in physical measurements at $\theta = 0$ and π, and can be removed by a coordinate transformation. However, the $r = 0$ singularity is real. This is not surprising as the Newtonian gravitational potential for a point mass already has this feature: $-GM/r$. What about the $r = r^*$ surface? Recall that the Riemann curvature tensor $R_{\mu\nu\lambda\rho}$ is nontrivial only in a curved space. The coordinate-independent product of the curvature tensor $R_{\mu\nu\lambda\rho}R^{\mu\nu\lambda\rho} = 12r^{*2}/r^6$ is singular at $r = 0$, but not at $r = r^*$. This suggests that at $r = r^*$ there is a coordinate singularity, i.e. it is not physical and can be transformed away by a coordinate transformation.

The event horizon

While physical measurements are not singular at $r = r^*$, it does not mean that this surface is not special. It is an **event horizon**, separating events that can be viewed from afar, from those which cannot (no matter how long one waits). That is, the $r = r^*$ surface is a boundary of communication, much like earth's horizon is a boundary of our vision.

A simple way to see that no light signal can be transmitted to an outside observer is to show that any light source inside the $r = r^*$ surface would have its emitted frequency viewed by an outside observer as suffering an infinite gravitational redshift. Recall the relation (13.3) between coordinate time (t) and the proper time (τ),

[15]By "metric" we include the consideration also of the inverse metric which has a $[\sin^2 \theta]^{-1}$ term. Here we concentrate on the static Schwarzschild metric. For a rotating spherical source we have the Kerr metric which is more realistic physically but has a complicated singularity structure. For a basic introduction of the Kerr black hole, see Cheng (2010), Section 8.4.

$$d\tau = \sqrt{-g_{00}}dt. \qquad (14.54)$$

The received frequency is $\omega_{\text{rec}} \propto 1/dt$ and the emitted frequency is $\omega_{\text{em}} \propto 1/d\tau$. Hence

$$\omega_{\text{rec}} = \sqrt{-g_{00}}\omega_{\text{em}} \qquad (14.55)$$

showing that $\omega_{\text{rec}} \to 0$ as the Schwarzschild metric element $g_{00} \to 0$ as r approaches r^*. The vanishing of the received frequency means it would take an infinite amount of coordinate time to receive the next photon.

Behavior of lightcones near the event horizon

The key property of an event horizon is that any time-like worldline can pass through it only in one direction, toward the $r = 0$ singularity. Particles and light rays cannot move outward from the region $r < r^*$. Traveling inside the horizon, a particle inexorably moves towards the physical singularity at $r = 0$. To see this involves a proper study of the lightcone behavior when crossing the Schwarzschild surface.

A radial ($d\theta = d\phi = 0$) worldline for a photon in Schwarzschild spacetime in Schwarzschild coordinates has a null line-element of

$$ds^2 = -\left(1 - \frac{r^*}{r}\right) c^2 dt^2 + \left(1 - \frac{r^*}{r}\right)^{-1} dr^2 = 0. \qquad (14.56)$$

We then have

$$cdt = \pm \frac{dr}{1 - r^*/r}. \qquad (14.57)$$

This can be integrated to obtain, for some reference spacetime point of (r_0, t_0),

$$c(t - t_0) = \pm \left(r - r_0 + r^* \ln \left|\frac{r - r^*}{r_0 - r^*}\right|\right), \qquad (14.58)$$

or simply,

$$ct = \pm \left(r + r^* \ln \left|r - r^*\right| + \text{constant}\right). \qquad (14.59)$$

The $+$ sign stands for the outgoing, and the $-$ sign for the in-falling, light-like geodesics, as shown in Fig. 14.5. To aid our viewing of this spacetime diagram we have drawn in several lightcones in various spacetime regions. We note that for the region far from the source where the spacetime becomes flat, the lightcone approaches the expected form with $\pm 45°$ sides.

The most prominent feature we notice is that the lightcones "tip over" when crossing the Schwarzschild surface. Instead of opening towards the $t = \infty$ direction, they tip towards the $r = 0$ line. This can be understood by noting that the roles of space and time are interchanged when crossing over the $r = r^*$ surface in the Schwarzschild geometry.

- In the spacetime region I outside the Schwarzschild surface $r > r^*$, the time and space coordinates have the usual properties, being time-like $ds_t^2 < 0$ and space-like $ds_r^2 > 0$ (cf. Section 11.3.1). In particular, lines of fixed r (parallel to the time axis) are time-like, and lines of fixed t are space-like. Since the trajectory for any particle must necessarily be

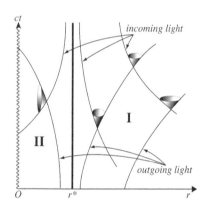

Fig. 14.5 Lightcones in Schwarzschild spacetime. Regions I and II are separated by the Schwarzschild surface. Different light rays correspond to (14.59) with different constants. Note that the outgoing light rays in region II also end at the $r = 0$ line.

time-like (i.e. every subsequent segment must be contained inside the lightcone at a given worldpoint, as shown in Fig. 11.4), the lightcones open upwards in the ever-larger t direction. In this region I of Fig. 14.5, an observer is fated to have an ever-increasing t.

- In region II, inside the Schwarzschild surface, the roles of t and r are reversed. A worldline of fixed time is now time-like. This comes about because the $(1 - r^*/r)$ factor in the metric changes sign. For a worldline to remain time-like ($ds^2 < 0$), the particle can no longer stay put at one position ($dr \neq 0$), but is forced to move inward towards the $r = 0$ singularity.

The fact that the metric becomes singular at the $r = r^*$ surface means that the Schwarzschild coordinates (t, r, θ, ϕ) are not convenient for the discussion of events near the Schwarzschild surface. Our description of the "tipping-over" of the light-cones in Fig. 14.5 obtained by using Schwarzschild coordinates is suspect as the function $t(r)$ is singular across the $r = r^*$ surface. But coordinate systems can be found so that the metric is not singular at $r = r^*$.

Black hole as a future singularity The lightcone behavior in the (retarded) Eddington–Finkelstein (EF) coordinate system is plotted in Fig. 14.6(a). We see now that lightcones tip over smoothly across the Schwarzschild surface. Inside the horizon, both sides of lightcones bend towards the $r = 0$ line; the $r = r^*$ surface is a one-way membrane allowing transmission of particles and light only inward. Particles can only move inward from the $r > r^*$ region I to the $r < r^*$ region II, towards the $r = 0$ singularity, which is in the future. This is the black hole, containing a future singularity.

White hole as a past singularity Since the Einstein equation is symmetric under time reversal $t \rightarrow -t$, there can be another set of solutions. In Fig. 14.6 (b) we have plotted the lightcones in another related system, the (advanced) Eddington–Finkelstein coordinate system. Once again, the $r = r^*$ surface only allows particles to cross in one direction—now particles can only move outward from the interior $r < r^*$ region II′ to the $r > r^*$ region I′. Thus we have here a solution to Einstein's equation that contains a past singularity at $r = 0$. Such a solution has come to be called a **white hole**.

The geometric property that makes the $r = r^*$ Schwarzschild surface an event horizon is that all points on such a surface have their lightcones entirely on one side of the surface as shown in Fig. 14.7. In the case of a black hole, they are on the side of $r < r^*$ as shown in Fig. 14.6(a); in the case of a white hole, they are on the side of $r > r^*$ as shown in Fig. 14.6(b).

The acceptance of black holes as reality

Black holes were studied extensively in the late 1930s by J.R. Oppenheimer (1904–67) and his collaborators. The physical reality of such strong-gravity objects was not at first believed by most physicists, including leading relativists such as Einstein and Eddington. General acceptance by the physics community did not occur until the 1960s. In this respect the theoretical work by teams led by J.A. Wheeler (1911–2008) in the U.S.A. and by Y.B. Zel'dovich (1914–87) in the Soviet Union, combined with modern astronomical observations, played a decisive role.

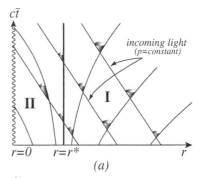

(a)

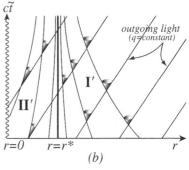

(b)

Fig. 14.6 Lightcones in Eddington–Finkelstein spacetime. (a) In retarded EF coordinates $(\tilde{t}, r)$ with a black hole, showing two regions I and II with $r = 0$ as a line of future singularities. (b) In advanced EF coordinates $(\tilde{t}, r)$ with a white hole, showing two regions I′ and II′ with $r = 0$ as past singularities.

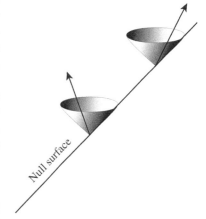

Fig. 14.7 A null surface is an event horizon. The lightcones of all points on the null surface are on one side of the surface. All time-like worldlines (samples shown as heavy arrowed lines) being contained inside lightcones can cross the null surface only in one direction. Thus, a null surface is a "one-way membrane".

15 Cosmology

- In cosmology the universe is treated as one physical system, with the galaxies being the constituent elements. On the cosmic scale the only relevant interaction among galaxies is gravitation. The general theory of relativity provides the natural framework for the study of cosmology.
- Spatial homogeneity and isotropy (the cosmological principle) leads to a spacetime described by the Robertson–Walker metric, with an overall scale factor $a(t)$ that controls the time evolution of the universe.
- In GR, space is a dynamic quantity; the metric function of spacetime is the solution to the Einstein equation. The big bang is not any sort of 'explosion of matter in space', but an expansion of space itself. The resultant scale factor $a(t)$ naturally gives rise to the Hubble relation: any two points separated from each other have a relative velocity that is proportional to the separation distance.
- In the GR description, the cosmic curvature and time evolution are directly related to the mass/energy content of the universe. In the present epoch, the familiar form of matter (composed of atoms), called 'baryonic matter', is found to be less than one-fifth of the matter content. The other portion is 'dark matter'. Its presence has been detected by its gravitational attraction. The energy density ρ is measured in units of the critical density: $\Omega \equiv \rho/\rho_c$. Adding up the baryonic and dark matter, $\Omega_M = \Omega_B + \Omega_{DM} \simeq 0.25$. According to Einstein equation, the inequality $\Omega_M < 1$ means that our universe should have negative curvature, disagreeing with observation.
- Einstein was the first one to apply GR to the study of cosmology in 1917. Like everyone then, he assumed our universe was static. This is an impossibility because in his GR theory, like the Newtonian theory, gravity is always an attraction. However Einstein discovered that GR naturally suggested a way to include a gravitational repulsion term that can counter the usual attraction. Thus motivated, he introduced on the geometric side of the Einstein field equation a term, called the cosmological constant, that corresponds to a cosmic gravitational repulsion that increases with distance.
- George Gamow claimed that Einstein said to him that the introduction of the cosmological constant was 'the biggest blunder of his life'. While

it is quite possible that Einstein had some regret of having missed the chance to predict the expanding universe, the discovery that the GR field equation can accommodate a cosmic repulsion component is certainly a great contribution to cosmology.

- A (large) cosmological constant is the key ingredient in the inflationary cosmology of the big bang. Its prediction that the spatial geometry of the universe must be, on the average, flat had been verified by observation.

- A cosmological constant, with its repulsion feature, is also needed to explain the 1998 discovery that the expansion of the universe is accelerating. Different methods of astrophysical study (supernovae redshift, CMBR anisotropy, and galaxy distribution) have arrived at a concordant picture, called ΛCDM cosmology, with the dark energy (cosmological constant Λ) and cold dark matter being the two largest energy components, with a sum $\Omega = \Omega_\Lambda + \Omega_M \simeq 0.75 + 0.25 = 1$ in agreement with the requirement of a flat universe.

Cosmology is the study of the whole universe as a physical system:

> What is its matter–energy content? How is this content organized?
> What is its history? How will it evolve in the future?

The description gives a "smeared" picture with the galaxies being the constituent elements of the system. On the cosmic scale the only relevant interaction among galaxies is gravitation; all galaxies are accelerating under their mutual gravity. Thus the study of cosmology depends crucially on our understanding of the gravitational interaction. Consequently, the proper framework for cosmology is GR. The solution of Einstein's equation describes the whole of universe because it describes the whole of spacetime. Soon after the completion of his papers on the foundation of GR, Einstein proceeded to apply his new theory to cosmology. In 1917 he published his paper, *Cosmological considerations on the general theory of relativity* showing that GR can describe an unbounded homogeneous mass distribution. Since then almost all cosmological studies have been carried out in the framework of GR.

Einstein was very much influenced by the teaching of Ernest Mach in his original motivation for general relativity.[1] Although his theory ultimately was unable to embody the strong version of Mach's principle that the total inertia of any single body was determined by the interaction with all the bodies in the universe, nevertheless he viewed his 1917 cosmology paper as the completion of his GR program because it showed that the matter in the universe determines the entire geometry of the universe.

[1] A simple discussion of Einstein's GR motivation in connection with Mach's paradox can be found in Chapter 1 of Cheng (2010).

15.1 The cosmological principle

Modern cosmology usually adopts the working hypothesis called the *cosmological principle*—at each epoch (i.e. each fixed value of cosmological time *t*) the universe is taken to be homogeneous and isotropic. It presents the **same** aspects (except for local irregularities[2]) from each point: the universe has no center and no edge. The hypothesis has been confirmed by modern astronomical observation, especially the cosmic background radiation which is homogeneous to one part in 10^5 in every direction. It gives rise to a picture of the universe as a physical system of a 'cosmic fluid'.

15.1.1   The Robertson–Walker spacetime

The cosmological principle gives rise to the geometry of the universe that can be most conveniently described in the comoving coordinate system.

Comoving coordinate system

The picture of the universe discussed above allows us to pick a privileged coordinate frame, the **comoving coordinate system**, where the coordinate time and positions are chosen to be

$t \equiv$ the proper time of each fluid element;

$x^i \equiv$ the spatial coordinates carried by each fluid element.

The comoving coordinate time can be synchronized over the whole system. A comoving observer flows with a cosmic fluid element. Because each fluid element carries its own position label the comoving coordinate is also the cosmic rest-frame—as each fluid element's position coordinates are unchanged with time. But we must remember that in GR the coordinates do not measure distance, which is a combination of the coordinates and the metric. As we shall detail below, viewed in this comoving coordinate, the expanding universe (with all galaxies rushing away from each other) is described not by changing position coordinates, but by an ever-increasing metric. Since the metric describes the geometry of space itself, this emphasizes that the physics underlying an expanding universe is not something exploding in space, but is the expansion of space itself.

The Robertson–Walker metric

Just as Schwarzschild showed that spherical symmetry restricts the metric to the form of $g_{\mu\nu} = \text{diag}(g_{00}, g_{rr}, r^2, r^2 \sin^2 \theta)$ with only two scalar functions, g_{00} and g_{rr}, here we discuss the geometry resulting from the cosmological principle for a homogeneous and isotropic universe. The resulting metric, when expressed in comoving coordinates, is the **Robertson–Walker metric**.[3] Because the coordinate time is chosen to be the proper time of fluid elements, we must have $g_{00} = -1$. The fact that the space-like slices for fixed t can be defined means that the spatial axes are orthogonal to the time axes $g_{0i} = g_{i0} = 0$. Let g_{ij} be the 3×3 spatial part of the metric; this implies the 4D metric that satisfies the cosmological principle is block-diagonal:

$$g_{\mu\nu} = \begin{pmatrix} -1 & 0 \\ 0 & g_{ij} \end{pmatrix}. \tag{15.1}$$

Namely, the invariant interval is

$$ds^2 = -c^2 dt^2 + g_{ij} dx^i dx^j \equiv -c^2 dt^2 + dl^2. \tag{15.2}$$

Because of the cosmological principle requirement (i.e. no preferred direction and position), the time-dependence in g_{ij} must be an **overall** scale factor $a(t)$, with no dependence on any of the indices:

$$dl^2 = a^2(t)\, d\tilde{l}^2 \qquad (15.3)$$

where the reduced length element $d\tilde{l}$ is t-independent, while the scale factor is normalized at the present epoch by $a(t_0) = 1$.

One has the picture of the universe as a three-dimensional map (Fig. 15.1), with cosmic fluid elements labeled by the fixed (t-independent) map coordinates $\tilde{x}_i$. Time evolution enters entirely through the time-dependence of the map scale $x_i(t) = a(t)\tilde{x}_i$—as $a(t)$ determines the size of the grids and is independent of the map coordinates $\tilde{x}_i$. As the universe expands (i.e. $a(t)$ increases), the **relative distance** relations (i.e. the shape of things) are not changed.

At a give time, 3D space being homogeneous and isotropic, one naturally expects this space to have a constant curvature (cf. Section 12.4.4). We will not work out the details here; suffice to note that there are three types of such spaces, labeled by a parameter $k = \pm 1, 0$ in g_{ij}, known as the "curvature signature", with $k = +1$ for a positively curved 3-sphere called a "**closed universe**", $k = -1$ for a negatively curved 3-pseudosphere, an "**open universe**", and $k = 0$ for a 3D flat (Euclidean) space, a "**flat universe**". As expected from GR theory, the mass/energy content of the universe will determine which geometry the universe would have.

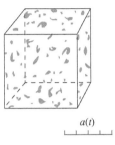

Fig. 15.1 A three-dimensional map of the cosmic fluid with elements labeled by t-independent $\tilde{x}_i$ comoving coordinates. The time dependence of any distance comes entirely from the t-dependent scale factor of the map.

Distances in the Robertson–Walker spacetime

In an expanding universe with a space that may be curved, we must be very careful in any treatment of distance. Nevertheless the time dependence of distance measurement is simple. The (proper) distance $D(\tilde{x}, t)$ to a point at the comoving distance $\tilde{x}$ and cosmic time t (i.e. the separation between emitter and receiver on the spacetime surface of a fixed time from t_0 to t), according to Eq. (15.3), is

$$D(\tilde{x}, t) = a(t)D(\tilde{x}, t_0) \qquad (15.4)$$

where $D(\tilde{x}, t_0)$ is the fixed (comoving) distance at the present epoch t_0. This implies a proper velocity of

$$V(t) = \frac{dD}{dt} = \frac{\dot{a}(t)}{a(t)}D(t). \qquad (15.5)$$

Evidently the velocity is proportional to the separation.[4] This is Hubble's relation with the proportional constant $\dot{a}(t)/a(t)$ called the Hubble constant $H(t)$; in particular its value at the present epoch is $H_0 = \dot{a}(t_0)$. Recall that the appearance of an overall scale factor in the spatial part of the Robertson–Walker metric follows from our imposition of the homogeneity and isotropy condition. The result in (15.5) confirms our expectation that in any geometrical description of a dynamical universe which satisfies the cosmological principle, Hubble's relation emerges automatically.

[4]This was first derived by G. Lemartre and independently by H. Weyl in 1925. See the reference in Peebles (1984).

15.1.2   The discovery of the expanding universe

Astronomers have devised a whole series of techniques that can be used to estimate the distances ever farther into space. Each new one, although less reliable, can be used to reach out further into the universe. During the period 1910–1930, the "cosmic distance ladder" reached out beyond 100 kpc (kiloparsec[5]). The great discovery was made that our universe was composed of a vast collection of galaxies, each resembling our own, the Milky Way. One naturally tried to study the motions of these newly discovered "island universes" by using the Doppler effect. When a galaxy is observed at visible wavelengths, its spectrum typically has absorption lines because of the relatively cool upper stellar atmospheres. For a particular absorption line measured in the laboratory to have a wavelength λ_{em}; the wavelength received by the observer may, however, be different. Such a wavelength shift

$$z \equiv \frac{\lambda_{rec} - \lambda_{em}}{\lambda_{em}} \tag{15.6}$$

is related to the emitter motion by the Doppler effect, which, for nonrelativistic motion, can be stated as

$$z = \frac{\Delta\lambda}{\lambda} \simeq \frac{V}{c}, \tag{15.7}$$

where V is the recession velocity of the emitter (away from the receiver).

A priori, for different galaxies one expects a random distribution of wavelength shifts: some positive (redshift) and some negative (blueshift). This is more or less true for the *Local Group* (galaxies near our own Milky Way). But beyond the few nearby galaxies, the measurements over a 10-year period at Arizona's Lowell Observatory by Vesto Slipher (1875–1969) of some 40 galaxies, showed that all, except a few in the Local Group, were redshifted. Edwin Hubble (1889–1953) at the Mt. Wilson Observatory, California, then attempted to correlate these redshift results to the difficult task of estimating the distances to these galaxies. He found that the redshift was proportional to the distance D to the light-emitting galaxy. In 1929, Hubble announced his result:

$$z = \frac{H_0}{c}D \tag{15.8}$$

or, substituting in the Doppler interpretation of (15.7),

$$V = H_0 D. \tag{15.9}$$

Namely, we live in an expanding universe.[6] On distance scales greater than 10 Mpc, all galaxies obey Hubble's relation: they are receding from us with speed linearly proportional to the distance (Hubble 1929). The proportionality constant H_0, the **Hubble constant**, gives the recession speed per unit separation (between the receiving and emitting galaxies). To obtain an accurate value of H_0 has been a great challenge as it requires one to ascertain great cosmic distances. Only recently has it become possible to yield consistent results among several independent methods. We have the convergent value

$$H_0 = (72 \pm 5 \text{ km/s})\text{Mpc}^{-1}, \tag{15.10}$$

[5] The astronomical distance unit, parsec, pc $= 3.1 \times 10^{16}$ m $= 3.26$ light-years.

[6] The first one to interpret Slipher's redshift result as indicating an expanding universe was the Dutch mathematician and astronomer Willem de Sitter (1872–1934) who worked with a GR model devoid of matter content.

where the subscript 0 stands for the present epoch $H_0 \equiv H(t_0)$. An inspection of Hubble's relation (15.9) shows that H_0 has the dimension of inverse time, the **Hubble time**, which has the value of

$$t_H \equiv H_0^{-1} = 13.6 \text{ Gyr}. \tag{15.11}$$

For reference, we note that the age of our earth is estimated to be around 4.6 Gyr.

Although Hubble is often claimed to be the discoverer of the expanding universe, this is not entirely an accurate attribution. Hubble deduced the Hubble relation from observational data and he never advocated the expanding universe idea. It was the Belgian priest Georges Lemaître (1894–1966) who first wrote down Hubble's relation in 1927 in the context of an expanding universe model based on general relativity[7] and compared (15.9) with observational data.

[7](Lemaître 1927, 1931). See the historical comments, for example, in Peebles (1984).

15.1.3 Big bang cosmology

If all galaxies are rushing away from each other now, presumably they must have been closer in the past. Extrapolating back in time there would be a moment, 'the big bang', when all objects were concentrated at one point of infinite density, taken to be the origin of the universe. This big bang cosmology was first advocated by Lemaître.

During the epochs immediately after the big bang, the universe was much more compact, and the energy associated with the random motions of matter and radiation was much larger. Thus, we say, the universe was much hotter. As a result, elementary particles could be in thermal equilibrium through their interactions. As the universe expanded, it also cooled. With the lowering of particle energy, particles (and antiparticles) would either disappear through annihilation, or combine into various composites of particles, or "decouple" to become free particles. As a consequence, there would be different kinds of thermal relics left behind by the hot big bang. That we can speak of the early universe with any sort of confidence rests with the idea that the universe had been in a series of equilibria. At such stages, the property of the system was determined, independent of the details of the interactions, by a few parameters such as the temperature, density, pressure, etc. Thermodynamical investigation of the cosmic history was pioneered by Richard Tolman (1881–1948). This approach to extract observable consequences from big bang cosmology was vigorously pursued by George Gamow (1904–68) and his collaborator, Ralph Alpher (1921–2007). Their theoretical work on big bang nucleosynthesis, the cosmic microwave background, received observational confirmation (Penzias and Wilson 1965), leading to the general acceptance of the big bang cosmology.

A historical aside The first estimate of the value of Hubbles constant by Hubble was too large by a factor of almost 10 (related to the calibration of distance estimate by using Cepheid variable stars as the "standard candle"). This led to an underestimation of the cosmic age that is smaller than some of the known old stars. This "cosmic age problem" was one of the motivations that led Hermann Bondi (1919–2005), Thomas Gold (1920–2004), and Fred Hoyle (1915–2001) to propose in the early 1950s a rival theory to the big bang

[8] In the steady state cosmology the expansion rate, spatial curvature, matter density, etc. are all time-independent. A constant mass density means that the universe did not have a big hot beginning; hence there cannot be a cosmic age problem. To have a constant mass density in an expanding universe requires the continuous, energy-nonconserving, creation of matter. To SSU's advocates, this spontaneous mass creation is no more peculiar than the creation of all matter at the instant of the big bang. In fact, the name 'big bang' was invented by Fred Hoyle as a somewhat dismissive description of the competing cosmology.

cosmology, the "steady state universe" (SSU).[8] The latter model was competitive until the discovery in the mid-1960s of the cosmic microwave background radiation, the afterglow of the big bang.

Age of the universe

How much time has evolved since this fiery beginning? What then is the age of our universe?

By Hubble's "constant", we mean that, at a given cosmic time, H is independent of the separation distance and the recessional velocity—the Hubble relation is a linear relation. The proportionality coefficient between distance and recessional speed is not expected to be a constant with respect to time: there is matter and energy in the universe, their mutual gravitational attraction will slow down the expansion, leading to a monotonically decreasing expansion rate $H(t)$—a **decelerating universe**. Only in an "empty universe" do we expect the expansion rate to be a constant throughout its history, $H(t) = H_0$. In that case, the age t_0 of the empty universe is given by the Hubble time

$$[t_0]_{\text{empty}} = \frac{D}{V} = \frac{1}{H_0} = t_{\text{H}}. \tag{15.12}$$

For a decelerating universe full of matter and energy, the expansion rate must be larger in the past: $[H(t < t_0)] > H_0$. Because the universe was expanding faster than the present rate, this would imply that the age of the decelerating universe must be shorter than the empty-universe age: $t_0 < t_{\text{H}}$.

What is the observational limit of the universe's age? An important approach to the phenomenological study has been the research work on systems of 10^5 or so old stars known as **globular clusters**. Astrophysical estimation of the ages of globular clusters $[t_0]_{\text{gc}}$ can then be used to set a lower bound on the cosmic age t_0:

$$12 \text{ Gyr} \lesssim [t_0]_{\text{gc}} \lesssim 15 \text{ Gyr}. \tag{15.13}$$

15.2   Time evolution of the universe

In the above, we studied the kinematics of the standard model of cosmology. The requirement of a homogeneous and isotropic space fixes the spacetime to have the Robertson–Walker metric in comoving coordinates. This geometry is specified by a curvature signature k and a t-dependent scale factor $a(t)$. Here we study the dynamics of the model universe. The unknown quantities k and $a(t)$ are to be determined by the matter/energy sources through the Einstein field equation as applied to the physical system of the cosmic fluid. The theory of the expanding universe was written down in 1922 by the Russian mathematician Alexander Friedmann (1888–1925) working with a matter-filled GR model (Friedmann 1922).

15.2.1   The FLRW cosmology

The Einstein equation (14.35) relates spacetime's geometry on one side and the mass/energy distribution on the another,

$$G_{\mu\nu} = \underbrace{\kappa}_{\text{NewtonC}} \underbrace{T_{\mu\nu}}_{\rho,p}. \qquad (15.14)$$

The spacetime must have the Robertson–Walker metric in comoving coordinates. This means $G_{\mu\nu}$ on the geometry side of Einstein's equation is expressed in terms of k and $a(t)$. The source term on the right-hand side should also be compatible with the homogeneity and isotropy requirement. The simplest plausible choice for the energy–momentum tensor $T_{\mu\nu}$ is an ideal fluid, specified by two scalar functions: mass density $\rho(t)$ and pressure $p(t)$ [as discussed in Section 11.2.5, Eq. (11.43)]. Such dynamical equations for cosmology are called the **Friedmann equations**. Because of the symmetry assumed from the cosmological principle, there are only two independent component equations. One component of the Einstein equation becomes "the first Friedmann equation",

$$\frac{\dot{a}^2(t)}{a^2(t)} + \frac{kc^2}{R_0^2 a^2(t)} = \frac{8\pi G_N}{3}\rho, \qquad (15.15)$$

where R_0 is a constant distance factor, "the radius of the universe". Another component becomes "the second Friedmann equation",

$$\frac{\ddot{a}(t)}{a(t)} = -\frac{4\pi G_N}{c^2}\left(p + \frac{1}{3}\rho c^2\right). \qquad (15.16)$$

Because the pressure and density factors are positive we have a negative second derivative $\ddot{a}(t)$: the expansion must decelerate because of mutual gravitational attraction among the cosmic fluid elements. This cosmological model is called the FLRW (Friedmann–Lemaitre–Robertson–Walker) universe.

The first Friedmann equation (15.15) can be readily rewritten as

$$-\frac{kc^2}{\dot{a}^2 R_0^2} = 1 - \Omega, \qquad (15.17)$$

where Ω is the mass density ratio, defined[9] as

$$\Omega \equiv \frac{\rho}{\rho_c} \quad \text{with the critical density} \quad \rho_c = \frac{3H^2}{8\pi G_N}. \qquad (15.18)$$

Equation (15.17) displays the connection between geometry and the matter/energy distribution. If our universe has a mass density greater than the critical density ($\Omega > 1$), the average curvature must be positive $k = +1$ (a closed universe); if the density is less than the critical density ($\Omega < 1$), then $k = -1$, the geometry of an open universe having a negative curvature; and if $\rho = \rho_c$, (i.e. $\Omega = 1$) we have the $k = 0$ flat geometry.

For a simple one-component model universe, we can solve the cosmic equation to obtain the scale factor for various curvature signature values. We merely note the qualitative behavior of $a(t)$ as depicted in Fig. 15.2. For density less than ρ_c the expansion of the open universe ($k = -1$) will continue forever; for $\rho > \rho_c$ the expansion of a closed universe ($k = +1$) will slow down to a stop then start to re-collapse—all the way to another $a = 0$ "big crunch"; for the flat universe ($k = 0$) the expansion will slow down but not enough to stop. Thus the density ratio Ω (of density to the critical density) not only determines the

[9] Plugging in Hubble's constant at the present epoch and Newton's constant, we have the critical mass density of $\rho_{c,0} = (0.97 \pm 0.08) \times 10^{-29}$ g/cm^3, and, in another convenient unit, the critical energy density is $\rho_c c^2 = (2.5 \times 10^{-3}\,\text{eV})^4/(\hbar c)^3$.

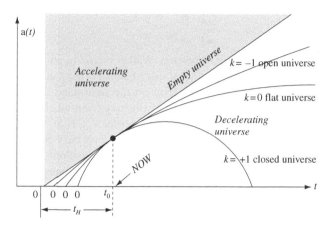

Fig. 15.2 Time dependence of the scale factor $a(t)$ for open, flat, and closed universes. The qualitative features of these curves are the same for radiation- or matter-dominated universes. All models must have the same radius a_0 and slope $\dot{a}_0$ at the present epoch t_0 in order to match the Hubble constant $H_0 = \dot{a}_0/a_0$. The origin of the cosmic time $t = 0$ is different for each curve. The straight line corresponds to an empty universe with an age $t_0 = t_H$.

geometry of the universe (from the positively curved and the negatively curved universe) it also determines its fate (from expanding forever and a contracting universe in the future).

15.2.2 Mass/energy content of the universe

What is the mass/energy density of our universe? This knowledge will determine the geometry and expansion future of our universe.

Baryonic matter

In the present epoch, the radiation energy density is negligibly small compared to the matter density $\Omega_R \ll \Omega_M$. The matter that we are familiar with is composed of atoms and hence is dominated by the mass energy of protons and neutrons. Such matter is called baryonic matter.[10] Baryonic matter has electromagnetic interaction and can emit and absorb EM radiation. The baryonic content is much larger than the visible shining stars, and its energy predominantly resides in the interstellar and intergalactic medium. Still the total baryonic density can be deduced when we compare the observed light nuclear density with the prediction of the big bang nucleosynthesis. From such phenomenology a density ratio of $\Omega_B = 0.04$ has been deduced.

[10] In particle physics, protons and neutrons belong to the class of particles called baryons, while electrons and neutrinos are examples of leptons.

Dark matter

There is compelling evidence that the vast majority of matter does not have electromagnetic interactions. Hence we cannot see it through its emission or absorption of radiation. Its presence can however be deduced from its gravitational attraction. This "dark matter" is more than five times as abundant as baryonic matter:

$$\Omega_M = \Omega_B + \Omega_{DM} \simeq 0.04 + 0.21 \simeq 0.25 \qquad (15.19)$$

That there might be a significant amount of dark matter in the universe was first pointed out in the 1930s by Fritz Zwicky (1898–1974). He noted that, given the observed radial velocities of the galaxies, the combined mass of the visible stars and gases in the Coma cluster was simply not enough to hold them together gravitationally.[11] It was reasoned that there must be a large amount of invisible mass, the dark matter, that was holding this cluster of galaxies together.

Galactic rotational curve The modern era began in 1970 when Vera Rubin and her collaborators using more sensitive techniques, were able to extend the rotation curve measurements far beyond the visible edge of gravitating systems of galaxies and clusters of galaxies. Consider the gravitational force that a spherical (or ellipsoidal) mass distribution exerts on a mass m located at a distance r from the center of a galaxy, see Fig. 15.3(a). Since the contribution outside the Gaussian sphere (radius r) cancels out, only the interior mass $M(r)$ enters into the Newtonian formula for gravitational attraction. The object is held by this gravity in a circular motion with centripetal acceleration of v^2/r. Hence

$$v(r) = \sqrt{\frac{G_N M(r)}{r}}. \tag{15.20}$$

Thus, the tangential velocity inside a galaxy is expected to rise linearly with the distance from the center ($v \sim r$) if the mass density is approximately constant. For a light source located outside the galactic mass distribution (radius R), the velocity is expected to decrease as $v \sim 1/\sqrt{r}$, see Fig. 15.3(b).

The velocity of particles located at different distances (the rotation curves) can be measured through the 21-cm lines of the hydrogen atoms. The surprising discovery was that, beyond the visible portion of the galaxies ($r > R$), instead of this fall-off, they are observed to stay at the constant peak value (as far as the measurement can be made). This indicates that the observed object is gravitationally pulled by other than the luminous matter; hence it constitutes direct evidence for the existence of dark matter. Many subsequent studies confirm this discovery. The general picture that has emerged is that of a disk of stars and gas embedded in a large halo of dark matter, see Fig. 15.4. In our simple representation, we take the halo to be spherical. In reality the dark matter halo may not be spherical and its distribution may not be smooth. In fact there are theoretical and observational ground to expect that the dark matter spreads out as an inhomogeneous web and the visible stars congregate at the potential wells of such a matter distribution. Recently a mammoth filament of dark matter stretching between two galaxy clusters has been detected through gravitational lensing (Dietrich *et al.* 2012).

Cold vs. hot dark matter Neutrinos are an example of dark matter, in fact they are called hot dark matter because (being of extremely low mass) they typically move with relativistic velocities. From the study of galactic structure formation it has been concluded that most dark matter is in the form of 'cold dark matter' (i.e. made up of heavy particles). No known particles fit this description. It is commonly held that there are yet to be discovered elementary particles, e.g. supersymmetric particles that may be hundreds or thousands of times more massive than the nucleons. Such WIMPs (weak interacting massive particles) form the cold dark matter content of universe.

With $\Omega_0 = \Omega_{M,0} \simeq 0.25$ (< 1), it would seem that we live in a negatively curved open universe (with $k = -1$). In the next section we shall discuss this

[11]In this connection, it may be useful to be reminded that a particle having a velocity greater than the "escape velocity" $v_{esc} = \sqrt{2G_N M/r}$ cannot be held gravitationally by a spherical source with mass M.

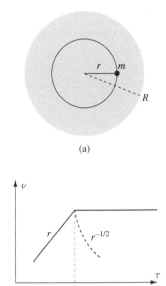

(a)

(b)

Fig. 15.3 (a) Gravitational attraction on a mass m due to a spherical mass distribution (shaded disk). The circle passing through m represents the Gaussian spherical surface. (b) The solid line is the observed rotation velocity curve $v(r)$. It does not fall as $r^{-1/2}$ beyond R, the edge of the visible portion of a galaxy.

Fig. 15.4 The dark matter halo surrounding the luminous portion of the galaxy.

topic further, and conclude that we need to modify Einstein's equation (by the addition of the cosmological constant). This theoretical input, together with new observational evidence, now suggests that we in fact live in a $k = 0$ flat universe, with the energy/mass density in the universe just equal to the critical value.

15.3   The cosmological constant

One of Einstein's great contributions to cosmology is his discovery that GR can accommodate a form of cosmic gravitational repulsion when he introduced the cosmological constant in his field equation.

15.3.1   Einstein and the static universe

Before the discovery by Lemaître and by Hubble of an expanding universe in the late 1920s, just about everyone, Einstein included, believed that we lived in a static universe. Recall that the then-observed universe consisted essentially of stars within the Milky Way galaxy. But gravity, whether Newtonian or relativistic, is a universal attraction. Hence, theoretically speaking, a static universe is an impossibility. With the goal of obtaining a static universe solution from general relativity (GR), Einstein altered his field equation to include a repulsion component. This could, in principle, balance the usual gravitational attraction to yield a static cosmic solution. Einstein discovered that the geometry side (second derivatives of the metric) of his field equation (15.14) could naturally accommodate an additional term. Mathematically it must be a symmetric rank-2 tensor with a vanishing covariant divergence constructed out of the metric, and he noted that the metric tensor $g_{\mu\nu}$ itself possessed all these properties, in particular $D^{\mu} g_{\mu\nu} = 0$. In this way the field equation (14.35) is modified to become

$$G_{\mu\nu} + \Lambda g_{\mu\nu} = \kappa T_{\mu\nu}. \tag{15.21}$$

Such a modification will, however, alter its nonrelativistic limit to differ from Newton's equation

$$\nabla^2 \Phi = 4\pi G_{\mathrm{N}} \rho. \tag{15.22}$$

In order that this alteration is still compatible with known phenomenology, it must have a coefficient Λ so small as to be unimportant in all situations except on truly large cosmic scales. Hence, this additional constant Λ has come to be called the **cosmological constant**. As we shall explain below, the new term does have the feature of increasing size with distance.

Properties of the cosmological constant term

Λ as vacuum energy and pressure   While we have introduced this term as an additional geometric term, we could just as well move it to the RHS of Eq. (15.21) and view it as an additional source term of gravity. In particular, when the regular energy–momentum tensor is absent $T_{\mu\nu} = 0$ (i.e. we are in the vacuum state), the field equation becomes

$$G_{\mu\nu} = -\Lambda g_{\mu\nu} \equiv \kappa T^{\Lambda}_{\mu\nu}. \tag{15.23}$$

Namely we can interpret the new term as the 'energy–momentum tensor of the vacuum' $T^{\Lambda}_{\mu\nu} = -\kappa^{-1}\Lambda g_{\mu\nu} = (-c^4\Lambda/8\pi G_{\mathrm{N}})g_{\mu\nu}$. Just as $T_{\mu\nu}$ for the cosmic fluid's ordinary radiation and matter depends on two functions of the energy density ρ and pressure p, this vacuum-energy–momentum tensor $T^{\Lambda}_{\mu\nu}$ can be similarly parametrized by the "**vacuum-energy density**" ρ_{Λ} and "**vacuum pressure**" p_{Λ}. From the structure of the metric $g_{\mu\nu}$ these two quantities are related to a positive cosmological constant Λ as follows: the vacuum-energy[12] per unit volume,

$$\rho_{\Lambda} = \frac{\Lambda c^2}{8\pi G_{\mathrm{N}}} > 0, \tag{15.24}$$

is a constant (in space and in time) and the corresponding vacuum-pressure,

$$p_{\Lambda} = -\rho_{\Lambda}c^2 < 0, \tag{15.25}$$

is negative. Such a negative pressure, as we shall show, gives rise to a gravitational repulsion.

Λ as gravitational repulsion that increases with distance   In order to have a physical feel for the cosmological constant, we shall take the Newtonian limit of the Einstein field equation (cf. Section 14.4.1). For nonrelativistic matter having a negligible pressure term, we recover the familiar equation (15.22) corresponding to a gravitational attraction for a point mass M at the origin

$$\mathbf{g}_M = -G_{\mathrm{N}}M\frac{\mathbf{r}}{r^3}; \tag{15.26}$$

but for a source having pressure comparable to its mass density, instead of Eq. (15.22), the Newtonian limit becomes Eq. (14.37):

$$\nabla^2\Phi = 4\pi G_{\mathrm{N}}\left(\rho + 3\frac{p}{c^2}\right). \tag{15.27}$$

In GR not only mass, but also pressure, can be a source of a gravitational field. Explicitly displaying together the contributions from ordinary matter and vacuum energy, the density and pressure each has two parts: $\rho = \rho_{\mathrm{M}} + \rho_{\Lambda}$ and $p = p_{\mathrm{M}} + p_{\Lambda}$. The Newton/Poisson equation (15.27) becomes

$$\nabla^2\Phi = 4\pi G_{\mathrm{N}}\left(\rho_{\mathrm{M}} + 3\frac{p_{\mathrm{M}}}{c^2} + \rho_{\Lambda} + 3\frac{p_{\Lambda}}{c^2}\right)$$
$$= 4\pi G_{\mathrm{N}}\rho_{\mathrm{M}} - 8\pi G_{\mathrm{N}}\rho_{\Lambda} = 4\pi G_{\mathrm{N}}\rho_{\mathrm{M}} - \Lambda c^2, \tag{15.28}$$

where we have used (15.25) and set $p_{\mathrm{M}} = 0$ because $\rho_{\mathrm{M}}c^2 \gg p_{\mathrm{M}}$. For the vacuum energy dominated case of $\Lambda c^2 \gg 4\pi G_{\mathrm{N}}\rho_{\mathrm{M}}$, the Poisson equation can be solved (after setting the potential to zero at the origin) by $\Phi_{\Lambda}(r) = -\Lambda c^2 r^2/6$. Between any two mass points, this potential corresponds to a repulsive force (per unit mass) that increases with separation $\mathbf{r}$,

$$\mathbf{g}_{\Lambda} = -\nabla\Phi_{\Lambda} = +\frac{\Lambda c^2}{3}\mathbf{r}, \tag{15.29}$$

[12] In nonrelativistic physics only the relative value of energy is meaningful—the motion of a particle with potential energy $V(x)$ is exactly the same as one with $V(x) + C$, where C is a constant. In GR, since the whole energy–momentum tensor is the source of gravity, the actual value of energy makes a difference.

in contrast to the familiar gravitational attraction of (15.26). With this pervasive repulsion that increases with distance, even a small Λ can have a significant effect in truly large dimensions.

Einstein's static universe

We now consider the Friedmann equations (15.15) and (15.16) with a nonvanishing cosmological constant,

$$\frac{\dot{a}^2 + kc^2/R_0^2}{a^2} = \frac{8\pi G_N}{3}(\rho_M + \rho_\Lambda), \tag{15.30}$$

$$\frac{\ddot{a}}{a} = -\frac{4\pi G_N}{c^2}\left[(p_M + p_\Lambda) + \frac{1}{3}(\rho_M + \rho_\Lambda)c^2\right]. \tag{15.31}$$

The RHS of (15.31) need not necessarily be negative because of the presence of the negative pressure term $p_\Lambda = -\rho_\Lambda c^2$. Consequently, a decelerating universe is no longer the inevitable outcome. For nonrelativistic matter ($p_M = 0$), we have

$$\frac{\ddot{a}}{a} = -\frac{4\pi G_N}{3}(\rho_M - 2\rho_\Lambda). \tag{15.32}$$

The static condition of $\ddot{a} = 0$ now leads to the constraint:

$$\rho_M = 2\rho_\Lambda = \frac{\Lambda c^2}{4\pi G_N}. \tag{15.33}$$

Namely, the mass density ρ_M of the universe is fixed by the cosmological constant. The other static condition of $\dot{a} = 0$ implies, through (15.30), the static solution $a = a_0 = 1$:

$$\frac{kc^2}{R_0^2} = 8\pi G_N\rho_\Lambda = \Lambda c^2. \tag{15.34}$$

Since the RHS is positive, we must have

$$k = +1. \tag{15.35}$$

Namely, the static universe has a positive curvature (a closed universe) and finite size. (Just like the case of a 2D spherical surface it is finite in size but has no boundary.) The "radius of the universe" is also determined, according to (15.34), by the cosmological constant:

$$R_0 = \frac{1}{\sqrt{\Lambda}}. \tag{15.36}$$

Thus, the basic features of such a static universe, the density and size, are determined by the arbitrary input parameter Λ. Not only is this a rather artificial arrangement, but also the solution is, in fact, unstable. That is, a small variation will cause the universe to deviate from this static point.

Einstein's biggest blunder?

Having missed the chance of predicting an expanding universe before its discovery, Einstein came up with a solution which did not really solve the perceived difficulty. (His static solution is unstable.) It had often been said that later in life Einstein considered the introduction of the cosmological constant to be "the biggest blunder of his life!" This originated from a characterization by George Gamow in his autobiography (Gamow, 1970):

Thus, Einstein's original gravity equation was correct, and changing it was a mistake. Much later, when I was discussing cosmological problems with Einstein, he remarked that the introduction of the cosmological term was the biggest blunder he ever made in his life.

Then Gamow went on to say,

But this blunder, rejected by Einstein, is still sometimes used by cosmologists even today, and the cosmological constant Λ rears its ugly head again and again and again.

What we can conclude for sure is that Gamow himself considered the cosmological constant 'ugly' (because this extra term made the field equation less simple). Generations of cosmologists kept on including it because the quantum vacuum energy[13] gives rise to such a term and there was no physical principle one could invoke to exclude this term. (If it is not forbidden, it must exist!) In fact the discovery of the cosmological constant as the source of a new cosmic repulsive force must be regarded as one of Einstein's great achievements.

[13] See the discussion at the end of this chapter of the quantum vacuum state having an energy with just the properties of the cosmological constant.

15.3.2   The Inflationary epoch

As we shall explain, the idea of a nonzero cosmological constant has been the key in solving a number of fundamental problems in cosmology. Einstein taught us the way to bring about a gravitational repulsion. Although the original motivation (the static universe) may be invalid, this repulsive force was needed to account for the explosion that was the big bang (inflationary epoch), and was needed to explain how the expansion of the universe could in the present epoch accelerate (dark energy).

The inflationary cosmology

The FLRW cosmology (with a small or vanishing Λ) has acute problems related to its initial conditions:

- **The flatness problem**: gravitational attraction always enhances any initial curvature.[14] In light of this property, it is puzzling that the present mass density Ω_0 has been found observationally to be not too different from the critical density value $(1 - \Omega_0) = O(1)$. This means that Ω must have been extremely close to unity (extremely flat) in the cosmic past. Such a fine-tuned initial condition would require an explanation.
- **The horizon problem**: Our universe is observed to be very homogeneous and isotropic. In fact, we can say that it is "too" homogeneous and isotropic. Consider two different parts of the universe that are so far apart that no light signal sent from one at the beginning of the universe could have reached the other. Namely they are outside of each other's horizons.

[14] From Eq. (15.17) we see that if $k = 0$ (a flat geometry), we must have the density ratio $\Omega = 1$ exactly; when $k \neq 0$ for a universe having curvature, then $[1 - \Omega(t)]$ must be **ever-increasing** because the denominator on the LHS is ever decreasing. Thus, the condition for a flat universe $\Omega = 1$ is an **unstable equilibrium point**—if Ω ever deviates from 1, this deviation will increase with time.

Yet they are observed to have similar properties. This suggests their being in thermal contact sometime in the past. How can this be possible?

Exponential expansion in a Λ-dominated universe   What type of cosmic expansion would result from the gravitational repulsion as represented by a dominant cosmological constant? As the energy density ρ_Λ is a constant and unchanged by an increasing volume, the more the space expands, the greater is the vacuum energy and negative pressure, causing the space to expand even faster. This self-reinforcing feature implies an exponential increase[15] of the scale factor and its derivative, with the time constant Δt related to Λ:

$$a(t) \sim e^{t/\Delta t} \quad \text{and} \quad \dot{a}(t) \sim e^{t/\Delta t}. \tag{15.37}$$

In fact, we can think of this Λ repulsive force as residing in the space itself, so as the universe expands, the push from this Λ energy increases as well. (The total energy was conserved during the rapid expansion because of the concomitant creation of the gravitational field, which has a **negative** potential energy.)

Solving the initial condition problems   The initial condition problems of the FLRW cosmology can be solved if, in the early moments, the universe had gone through an epoch of extraordinarily rapid expansion—the inflationary epoch. This can solve the flatness problem, as any initial curvature could be stretched flat by the burst of expansion.[16] This can solve the horizon problem if the associated expansion rate could reach superluminal speed. If the expansion rate could be greater than light speed,[17] then one horizon volume could have been stretched out to a large volume that corresponded to many horizon volumes after this burst of expansion. This rapid expansion could happen if there existed then a large cosmological constant Λ, which could supply a huge repulsion to the system. Some ideas from particle physics, as first noted by Alan Guth (1947–) suggests such a large vacuum energy can indeed come about during the "spontaneous symmetry breaking" phase transitions associated with a "grand unified theory" (cf. Chapter 16). Although the general predictions of inflation have checked out with observation, we should note that inflationary cosmology, rather than just a specific theory, is really a framework[18] in which to think of the big bang. The inflationary epoch almost certainly occurred and there are many competing theories of inflation.

The most natural theory for the origin   It is beyond the scope of this book to give a proper quantum field theoretical discussion of inflationary cosmology. Still, we should point out that to many investigators in this field the inflationary theory of the big bang is the most 'natural' theory for the beginning of the universe. It provides the framework to understand the origin of all matter and energy, and the associated vacuum fluctuations are inflated to be the initial inhomogeneity (the density fluctuation) needed to start the formation, through gravitational clumping, of the observed structures—the stars, the galaxies, and the clusters of galaxies, etc. Such a universe (with a flat spatial geometry) has a vanishing total energy: all forms of energy are exactly balanced by the negative gravitational energy.

---

[15]For a Λ-dominated source, the Friedmann equation (15.30) becomes a rate equation: $da/dt = a/\Delta t$, having an exponential solution with $\Delta t = \sqrt{3/(8\pi G_N \rho_\Lambda)}$.

[16]From Eq. (15.17) we see that, starting from some initial density $\Omega(t_1)$, the exponential expansion of (15.37) would make the subsequent density ratio extremely close to unity after several periods as can be seen in $\frac{1-\Omega(t_2)}{1-\Omega(t_1)} = \left[\frac{\dot{a}(t_2)}{\dot{a}(t_1)}\right]^{-2} = e^{-2(t_2-t_1)/\Delta\tau}$.

[17]As shown in Eq. (15.37), in an exponential expansion, the rate of expansion also increases exponentially. This does not contradict special relativity, which says that an object cannot pass another one faster than c in one fixed frame. Putting it in another way, while an object cannot travel faster than the speed of light through space, there is no restriction stipulating that space itself cannot expand faster than c.

[18]In many ways it is like evolution in biology as paradigms rather than just one specific theory.

15.3.3 The dark energy leading to an accelerating universe

Inflation predicts a spatially flat universe

One of the firm predictions of inflationary cosmology is that the spatial geometry of our universe must be flat, which was in fact confirmed by a whole
series of observations, starting with the high-altitude balloon observations of
CMBR anisotropy (by the Boomerang and Maxima collaborations) in the mid-
1990s. But there were problems associated with the implications of such a
geometrically flat universe:

- **The missing energy problem**: GR requires a flat universe to have a
  mass/energy density exactly equal to the critical density, $\Omega_0 = 1$. Yet
  observationally, including both the baryonic and dark matter as shown in
  Eq. (15.19), we only find less than a third of this value. Thus, it appears
  that to have a flat universe we would have to solve a "missing energy
  problem".
- **The cosmic age problem**: Our universe is now matter dominated, and its
  expansion, as we suggested in the previous discussion, should be decelerating. From the study of the time evolution of the universe, one learns
  that the age of a matter-dominated flat universe should be two-thirds of
  the Hubble time,

$$(t_0)_{\text{flat}} = \frac{2}{3}t_{\text{H}} \lesssim 9\,\text{Gyr}, \qquad (15.38)$$

which is shorter than the estimated age of old stars. Notably the globular clusters have been deduced to be older than 12 Gyr as discussed in
(15.13). Thus, it appears that to have a flat universe we would also have
to solve a "cosmic age problem".

Possible resolution of problems in a flat universe with dark energy

A possible resolution of these phenomenological difficulties of a flat universe (hence inflationary cosmology) would be to assume the presence of
dark energy: any form of energy having a negative pressure to give rise to
gravitational repulsion. The simplest example of dark energy[19] is Einstein's
cosmological constant. Such assumed presence of a cosmological constant
must be completely different from the immense sized Λ present during the
inflation epoch. Rather, the constant dark energy density ρ_Λ should now be
only a little more than two-thirds of the critical density to provide the required
missing energy.

$$\Omega = \Omega_{\text{M}} + \Omega_\Lambda \overset{?}{=} 1. \qquad (15.39)$$

A nonvanishing Λ would also provide the repulsion to accelerate the expansion
of the universe. In such an accelerating universe the expansion rate in the past
must have been smaller than the current rate H_0. This means that it would take
a longer period (as compared to a decelerating or empty universe) to reach the
present era, thus a longer age $t_0 > 2t_{\text{H}}/3$ even though the geometry is flat. This
just might as well possibly solve the cosmic age problem mentioned above.

[19]One should not confuse dark energy with
the energies of neutrinos, black holes, etc.
which are also 'dark', but are counted as
parts of the 'dark matter', as the associated
pressure is not negative and does not lead to
gravitational repulsion. From Eq. (15.27) we
find that dark energy is defined by the condition of pressure being more negative than
$-\rho c^2/3$.

[20]The luminosity distance d_L is the distance deduced from the relation between the measured flux f and the intrinsic luminosity L of the emitting object, $f \equiv L/(4\pi d_L^2)$.

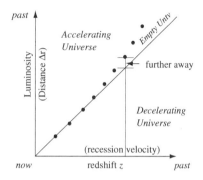

Fig. 15.5 A Hubble diagram: The Hubble curve for an accelerating universe bends upwards. A supernova on this curve at a given redshift would be further out in distance than anticipated.

The Hubble curve for an accelerating universe should turn upward

A Hubble curve (as in Fig. 15.5) is a plot of the luminosity distance[20] versus the redshift (measuring recession velocity) of the light-emitting star. A straight Hubble curve means a cosmic expansion that is coasting. This can only happen in an empty universe (cf. Fig. 15.2). If the expansion is accelerating, the expansion rate H must be smaller in the past. From Eq. (15.8): $HD = z$, we see that, for a given redshift z, the distance D to the light-emitting supernova must be larger than that for an empty or decelerating universe when $H \geq H_0$. Thus the Hubble curve for an accelerating universe would bend upwards.

Distant supernovae and the 1998 discovery

By 1998 two collaborations, "the Supernova Cosmological Project", led by Saul Perlmutter and "the High-z Supernova Search Team", led by Adam Riess and Brian Schmidt, each had accumulated some 50 Type-1a supernovae at high redshifts—z: 0.4–0.7 corresponding to supernovae occurring five to eight billion years ago. They made the astonishing discovery that the expansion of the universe was actually accelerating, as indicated by the fact that the measured luminosities were on the average 25% less than anticipated, and the Hubble curve bent upward.

Extracting Ω_M and Ω_Λ from the measured Hubble curve   Since the matter content Ω_M would lead to a decelerated expansion and a downward bending

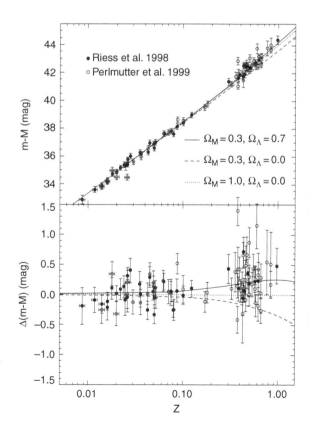

Fig. 15.6 Discovery of an accelerating universe. The Hubble plot showing the data points from Riess *et al.* (1998) and Perlmutter *et al.* (1999) lying above the empty universe (dotted) line. The dashed curve represents the prediction of a flat universe without cosmological constant, the solid curve being the best fit of the observational data. The vertical axes are distances (expressed in terms of "distance modulus"). In the lower panel $\triangle(m - M)$ is the difference after subtracting out the then expected value for a decelerating universe with $\Omega_M = 0.3$ and $\Omega_\Lambda = 0$. The graph is based on Fig. 1 of Reiss (2000).

Hubble curve and Ω_Λ would lead to an accelerated expansion and a upward bending Hubble curve, one can then extract the matter and dark energy contents in the present epoch by fitting the Hubble curve to the observational data of Fig. 15.6 with the result shown in Fig. 15.7.

The extracted values not only solve the missing energy problem,

$$\Omega = \Omega_M + \Omega_\Lambda \simeq 0.25 + 0.75 = 1,$$

but from their values one can also calculate the age of the universe to find it very close to that of an empty universe:

$$t_0 = 1.02\, t_H \simeq 14\,\text{Gyr}. \tag{15.40}$$

Remark   We have mentioned earlier that the cosmological constant Λ is expected to have such a small magnitude that it should be negligible on ordinary astronomical scales. We can demonstrate this by showing that in the solar system the gravitation repulsion due to Ω_Λ, compared to the familiar gravitational attraction, is totally unimportant. We take the ratio of the gravitational repulsion (15.29) due to the dark energy $g_\Lambda = +r\Lambda c^2/3 = +r8\pi G_N \rho_{c,0}\Omega_\Lambda$ to the usual solar gravitational attraction (15.26) $g_M = -rG_N M_\odot/r^3$ to have:

$$\left|\frac{\mathbf{g}_\Lambda}{\mathbf{g}_M}\right| = \frac{8\pi\Omega_\Lambda}{3}\frac{\rho_{c,0} r^3}{M_\odot} = O(10^{-22}), \tag{15.41}$$

where we have plugged in the critical mass density of (15.18) $\rho_{c,0} = (0.97 \pm 0.08) \times 10^{-29}$ g/cm^3 and the average separation between the sun and earth, the "astronomical unit", $r = AU = 1.5 \times 10^{13}$cm.

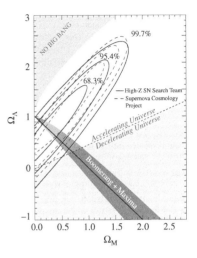

Fig. 15.7 Fitting Ω_Λ and Ω_M to the discovery data of Fig. 15.6 obtained by the High-Z SN Search Team and Supernova Cosmology Project. The favored values of Ω_Λ and Ω_M follow from the central values of CMB anisotropy $\Omega_\Lambda + \Omega_M \simeq 1$ (the straight line) and those of the Supernova data represented by confidence contours (ellipses) around $\Omega_\Lambda - \Omega_M \simeq 0.5$. The graph is based on Fig. 2 of Riess (2000).

Further evidence of dark energy and the mystery of its origin

After the supernovae discovery, the presence of dark energy $\Omega_\Lambda = 0.75$ was further confirmed by the analysis of the anisotropy power spectrum of the cosmic microwave background radiation (CMBR), as well as of the distribution of galaxies. Namely a consistent set of cosmic parameters has been derived from totally different modes of astrophysical observations: Supernova redshifts, CMBR, and the galaxy distribution.[21]

The problem of interpreting Λ as quantum vacuum energy   The introduction of the cosmological constant in the GR field equation does not explain its physical origin. In the inflation model one postulates that it is the energy of the "false vacuum" of an inflation/Higgs field that acts like an effective cosmological constant driving the inflationary expansion. What is the physical origin of the dark energy that brings about the accelerating expansion of the present epoch? A natural candidate is the quantum vacuum energy (cf. Section 6.4.2). The zero-point energy of a quantized field automatically has the property of having an energy density that is a constant thus giving rise to a negative pressure. However such a quantum vacuum energy $\Omega_{\text{q.vac}}$, whilehaving the

[21] For example, it has been shown that such a dark energy is just the agent needed to explain the observed slow-down of galaxy clusters' growth (Vikhlinin *et al.* 2009).

[22]This involves an integration of all the momentum degrees of freedom up to the natural quantum gravity scale, which is assumed to the Planck energy/momentum scale of Section 3.4.2. This yields a ratio of $\Omega_{q.vac}/\Omega_\Lambda = O(10^{120})$. Qualitatively we can understand this result on dimensional grounds:

$$\frac{\Omega_{q.vac}}{\Omega_\Lambda} = \frac{\rho_{q.vac}}{\rho_{c,0}} = \left[\frac{(energy)_{q.vac}}{(energy)_{c,0}}\right]^4$$

because $\rho c^2 = (energy)^4/(\hbar c)^3$ and the natural scale for quantum gravity is the Planck energy $(energy)_{q.vac} = m_P c^2 = 1.22 \times 10^{28}$ eV and the critical energy density is $(energy)_{c,0} = 2.5 \times 10^{-3}$ eV. For a more detailed calculation, see Section 11.7, p. 271 in Cheng (2010), where the possibility of cancellation between boson and fermion vacuum energies is also discussed.

correct property, is expected to be much too large[22] to account for the observed $\Omega_\Lambda = O(1)$.

The ΛCDM universe as the standard model of cosmology

Cosmology has seen major achievements over the past decade, to the extent that something like a standard model for the origin and development of the universe is now in place: the FLRW cosmology proceeded by an inflationary epoch. Many of the basic cosmological parameters have been deduced in several independent ways, arriving at a consistent set of results. These data are compatible with our universe being infinite and spatially flat, having matter/energy density equal to the critical density, $\Omega_0 = 1$. The largest energy component is consistent with it being Einstein's cosmological constant $\Omega_\Lambda \simeq 0.75$. In the present epoch this dark energy content is comparable in size to the matter density $\Omega_M \simeq 0.25$, which is made up mostly of cold dark matter. Thus this standard model is often call the ΛCDM cosmology model. The expansion of the universe will never stop—in fact having entered the accelerating phase, the expansion will be getting faster and faster.

WALKING IN EINSTEIN'S STEPS

Internal symmetry and gauge interactions

16

- An introduction to the gauge theory is presented in this chapter. We start with a review of gauge invariance in electromagnetism. That is followed by a discussion of gauge symmetry in quantum mechanics, showing that the gauge transformation must involve a spacetime-dependent change of the phase of a charged particle's wavefunction (i.e. field). This can be viewed as a transformation in the (internal) charge space by changing the particle field's label.

- If we reverse the above procedure, instead of going from the change in EM potential then to the wavefunction transformation, we start first with the phase transformation in quantum mechanics. This initial step may be understood as changing a spacetime-independent symmetry of the quantum mechanics equation to a local symmetry—a procedure called "gauging a symmetry". EM potentials are viewed then as compensating factors needed to implement such a local symmetry—the presence of potentials (with appropriate transformation properties as gauge fields) is required so that the physics equations are covariant under such local symmetry transformations.

- We demonstrate how Maxwell's electrodynamics can be "derived" from the requirement of a local $U(1)$ symmetry in the internal charge space. In this way we understand the essence of Maxwell's theory as special relativity and gauge invariance. Much like the elevating by Einstein of the equality of gravitational and inertial masses to the equivalence principle of gravitation and inertia, we call the approach of finding dynamics by promoting a global symmetry to a local symmetry, the gauge principle. Using the gauge principle, we can then generalize this approach to electromagnetism to the investigation of other fundamental interactions.

- In 1919 Hermann Weyl first attempted to derive electromagnetism from a local scale invariance. He was inspired by the success of general relativity, Einstein's new theory of gravity formulated as a local spacetime symmetry. Weyl was ultimately successful in this endeavor; this came after the advent of modern quantum mechanics (QM) in 1926 when Vladimir Fock discovered that QM wave equations with electromagnetic coupling are invariant under local phase transformations. It was pointed out that Weyl's scale change in spacetime should be understood as a spacetime-dependent $U(1)$ phase change in the charge space. However, Weyl's original terminology of gauge (i.e. scale) transformation has been retained in common usage.

- A $U(1)$ phase change, being commutative, is an abelian transformation. This was extended by C. N. Yang and R. L. Mills to the case of non-commutative symmetries. The resultant equations are nonlinear—the gauge fields themselves are charged (unlike the abelian case of the electromagnetic field being electrically neutral, but like gravity where the gravitational field is itself a source of gravity). This richness is one of the key ingredients that allowed nonabelian gauge theories (also called Yang–Mills theories) to be the framework for modern particle theory.

- We describe briefly the steps of going from quantum electrodynamics to the formulation of the new theory of fundamental strong interaction, quantum chromodynamics. The gauge theory of electroweak interactions has a more complicated structure because its local symmetry must be spontaneously broken (via the Higgs mechanism) to account for the short-range nature of weak interactions. In sum, the successful formulation of the Standard Model shows that fundamental particle interactions are all gauge interactions. This is a mighty generalization of Einstein's symmetry principle, from spacetime to internal charge spaces. It allowed us first to have a deeper understanding of electromagnetism, which was crucial to our finding new theories for the strong and weak interactions.

- Abelian gauge symmetry is discussed in detail (Sections 16.1–16.3)—up to the point of seeing how Maxwell's equations follow from gauge symmetry. Quantum field theories of QED, QCD, and the Standard Model are described qualitatively in the subsequent sections.

16.1 Einstein and the symmetry principle

One of Einstein's greatest legacies in physics has been his bringing about of our realization of the importance of symmetry in physics. His theory of relativity was built on the foundation of invariance principles. Before Einstein, symmetries were generally regarded as mathematical curiosities of great value to crystallographers, but hardly worthy to be included among the fundamental laws of physics. We now understand that a symmetry principle is not only an organizational device, but also a method to discover new dynamics. Einstein's relativity theories based on coordinate symmetries have given us a deeper appreciation of the structure of physics. His formulation of the symmetry among inertial frames of reference showed us the true meaning of the Lorentz transformation; this allowed us to deduce all the (special) relativistic effects in a compact way and to discover new equations for other branches of physics (relativistic mechanics, etc.) so they could be compatible with the symmetry principle of relativity. The extension of this principle from a special class of coordinates to all reference frames, his creation of general relativity, showed us the way of using spacetime-dependent (local) symmetry to generate dynamics—in the case of general relativity (GR), its gravitational interaction.

Ever since Einstein, a symmetry principle has been an essential guiding light in our effort to make new discoveries in theoretical physics. The topic of symmetry in physics is a rich one, especially in quantum physics. In this chapter we shall concentrate[1] on the gauge symmetry. It is one of the most important principles in fundamental physics. Local symmetry in some "internal" (or "charge") space has been the key to our discoveries of new basic physics, leading to the formulation of the Standard Model of particle physics. Starting from the work of Hermann Weyl (inspired by Einstein's GR discovery), we gradually learnt that electromagnetism could be understood as arising from a spacetime-dependent local symmetry (gauge symmetry) in the charge space. Namely, we discovered another profound lesson contained in Maxwell's equations: Besides teaching us the proper relation among inertial frames of reference (as given by the Lorentz transformation), these equations have such a structure as showing that electromagnetism is a gauge interaction. This simple $U(1)$ local symmetry associated with electromagnetism was later generalized to the noncommuting (Yang–Mills) gauge symmetry, which is a key element in the foundations of modern particle physics.

[1]We omit other important topics such as the relation between symmetry and conservation laws and degeneracy in the particle spectrum, and discrete symmetries (parity and time reversal invariance, etc.) and their violation, etc.

16.2 Gauge invariance in classical electromagnetism

We present in this chapter a pedagogical introduction to gauge theory. Since most students have their first exposure to gauge invariance in classical electrodynamics, this is where we will start—with a review of electromagnetic (EM) potentials and their gauge transformation. We then discuss gauge transformation in quantum mechanics. Because a quantum mechanical description is through a Hamiltonian (or through other energy quantities such as a Lagrangian), which can include a system's coupling to electromagnetism only through EM potentials, gauge symmetry plays an integral role in the QM description.

Classical electromagnetism[2] Any field theoretical description of the interaction between two particles involves a "two-step description". Call one the "source particle", giving rise to a field everywhere, which in turn acts locally on the "test particle". This two-step description can be represented schematically as follows:

[2]Here we repeat the essential elements of Maxwell's equations, first in familiar 3D vector notation (as already discussed in Sections A.1 and 3.1 as well as in Chapters 9 and 10) and then in the 4D spacetime formalism (in Section 16.4 as we have already done in Section 11.2.3).

$$\boxed{\text{Source particle}} \xrightarrow{\text{field eqns}} \boxed{\text{Field}} \xrightarrow{\text{eqns of motion}} \boxed{\text{Test particle}}$$

The "field equations" tell us how a source particle gives rise to the field everywhere. For the case of electromagnetism, they are **Maxwell's equations**. The "equations of motion" tell us the effects of the field on the motion of a test particle: how does the field cause the particle to accelerate. For the case of electromagnetism, they form the **Lorentz force law**.

- Equations of motion (Lorentz force law):

$$\mathbf{F} = e\left(\mathbf{E} + \frac{1}{c}\mathbf{v} \times \mathbf{B}\right) \tag{16.1}$$

we note that this equation has a "double duty": It gives the definition of the electric and magnetic fields as well as acting as the equation of motion for a test charge placed in the electromagnetic field.

- Field equations (Maxwell equations):
  - Inhomogeneous Maxwell equations:

$$\nabla \cdot \mathbf{E} = \rho \qquad \text{Gauss's law} \tag{16.2}$$

$$\nabla \times \mathbf{B} - \frac{1}{c}\frac{\partial \mathbf{E}}{\partial t} = \frac{1}{c}\mathbf{j} \qquad \text{Ampere's law} \tag{16.3}$$

  - Homogeneous Maxwell equations:

$$\nabla \cdot \mathbf{B} = 0 \quad \text{Gauss's law for magnetism} \tag{16.4}$$

$$\nabla \times \mathbf{E} + \frac{1}{c}\frac{\partial \mathbf{B}}{\partial t} = 0 \quad \text{Faraday's law.} \tag{16.5}$$

16.2.1   Electromagnetic potentials and gauge transformation

It is easy to solve the homogeneous Maxwell equations (16.4) and (16.5) by noting that the divergence of any curl, as well as the curl of any gradient, must vanish:[3] Eq. (16.4) can be solved if the **B** field is the curl of a **vector potential A**:

$$\mathbf{B} = \nabla \times \mathbf{A}. \tag{16.6}$$

Substituting this into (16.5), the vanishing curl $\nabla \times \left(\mathbf{E} + \frac{1}{c}\frac{\partial \mathbf{A}}{\partial t}\right) = 0$ implies that the term in parentheses can be written as the gradient of a scalar potential Φ:

$$\mathbf{E} = -\nabla\Phi - \frac{1}{c}\frac{\partial \mathbf{A}}{\partial t}. \tag{16.7}$$

Thus we can replace $(\mathbf{E}, \mathbf{B})$ fields by scalar and vector potentials $(\Phi, \mathbf{A})$ through the relations (16.6) and (16.7). Substituting these expressions into the inhomogeneous Maxwell equations of (16.2) and (16.3), we obtain the dynamics of the potentials once the source distribution $(\rho, \mathbf{j})$ is given. In other words, one can regard the homogeneous parts of Maxwell's equations as the "boundary conditions" telling us that fields can be expressed in terms of the potentials, and the true dynamics is contained in the inhomogeneous Maxwell equations.

Gauge invariance in classical electromagnetism

As outlined above we can simplify the description of the EM interactions by using four components of potentials $(\Phi, \mathbf{A})$ instead of six components of $(\mathbf{E}, \mathbf{B})$. However this replacement of $(\mathbf{E}, \mathbf{B})$ by $(\Phi, \mathbf{A})$ is not unique as the fields $(\mathbf{E}, \mathbf{B})$,

[3] See the discussion leading up to Eq. (A.20) in Appendix A1.

hence also Maxwells equations, are invariant under the following change of potentials (called a **gauge transformation**):

$$\Phi \longrightarrow \Phi' = \Phi - \frac{1}{c}\frac{\partial \chi}{\partial t} \tag{16.8}$$

$$\mathbf{A} \longrightarrow \mathbf{A}' = \mathbf{A} + \nabla \chi \tag{16.9}$$

where $\chi = \chi(t, \mathbf{r})$ (called a **gauge function**) is an arbitrary scalar function of position and time. This invariance statement (**gauge symmetry**) can be easily verified:

$$\mathbf{B} = \nabla \times \mathbf{A} \longrightarrow \mathbf{B}' = \nabla \times \mathbf{A}' = \nabla \times \mathbf{A} + \nabla \times (\nabla \chi) = \nabla \times \mathbf{A} = \mathbf{B}$$

because the curl of a gradient must vanish. Similarly,

$$\mathbf{E} = -\nabla\Phi - \frac{1}{c}\frac{\partial \mathbf{A}}{\partial t} \longrightarrow \mathbf{E}' = -\nabla\Phi' - \frac{1}{c}\frac{\partial \mathbf{A}'}{\partial t}$$

$$= -\nabla\Phi + \frac{1}{c}\frac{\partial \nabla\chi}{\partial t} - \frac{1}{c}\frac{\partial \mathbf{A}}{\partial t} - \frac{1}{c}\frac{\partial \nabla\chi}{\partial t} = -\nabla\Phi - \frac{1}{c}\frac{\partial \mathbf{A}}{\partial t} = \mathbf{E}.$$

Gauge symmetry in classical electromagnetism does not seem to be very profound. It is merely the freedom to choose potentials (Coulomb gauge, radiation gauge, Lorentz gauge, etc.) to simply calculations. One can in principle avoid using potentials and stick with the $(\mathbf{E}, \mathbf{B})$ fields throughout, with no arbitrariness. On the other hand, the situation in quantum mechanics is different. As we shall see, the QM description of the electromagnetic interaction must necessarily involve potentials. Gauge symmetry must be taken into account in the QM description. As a consequence, it acquires a deeper significance.

16.2.2 Hamiltonian of a charged particle in an electromagnetic field

Before moving on to a discussion of gauge symmetry in QM, we undertake an exercise in classical EM of writing the Lorentz force law (16.1) in terms of the EM potentials. This form of the force law will be needed in the subsequent QM description of a charged particle in an EM field.

Lorentz force in terms of potentials

The ith component of the force law (16.1) may be written out in terms of the potentials via (16.7) and (16.6):

$$m\frac{dv_i}{dt} = -e\nabla_i\Phi - \frac{e}{c}\frac{\partial A_i}{\partial t} + \frac{e}{c}\epsilon_{ijk}v_j\epsilon_{klm}\nabla_l A_m. \tag{16.10}$$

For the last term, we shall use the identity $\epsilon_{ijk}\epsilon_{klm} = \delta_{il}\delta_{jm} - \delta_{im}\delta_{jl}$:

$$\frac{e}{c}(\mathbf{v} \times \mathbf{B})_i = \frac{e}{c}\epsilon_{ijk}v_j\epsilon_{klm}\nabla_l A_m = \frac{e}{c}[\mathbf{v} \cdot (\nabla_i\mathbf{A}) - (\mathbf{v} \cdot \nabla)A_i]. \tag{16.11}$$

The above expressions involve the differentiation of the vector potential $\mathbf{A}(\mathbf{r}, t)$ which depends on the time variables in two ways: through its explicit dependence on t, as well as implicitly through its dependence on position $\mathbf{r} = \mathbf{r}(t)$.

(The particle is moving!) Thus its full time derivative has a more complicated structure:

$$\frac{dA_i}{dt} = \frac{\partial A_i}{\partial t} + \nabla_j A_i \frac{dr_j}{dt} = \frac{\partial A_i}{\partial t} + (\mathbf{v} \cdot \nabla) A_i. \tag{16.12}$$

The two factors on the RHS of this equation are just those that appear on the RHS's of Eqs. (16.10) and (16.11); thus they can be combined into a $-\frac{e}{c}\frac{dA_i}{dt}$ term:

$$m \frac{dv_i}{dt} = -e \nabla_i \Phi - \frac{e}{c} \frac{dA_i}{dt} + \frac{e}{c} \mathbf{v} \cdot (\nabla_i \mathbf{A}) \tag{16.13}$$

which is then the expression of the Lorentz force in terms of $(\Phi, \mathbf{A})$ that we shall use in the discussion below, cf. Eq. (16.19).

Hamiltonian of a charged particle in an Electromagnetic field

Recall in QM that we do not use the concept of force directly in our description of particle interactions. Instead, the dynamics is governed by the Schrödinger equation,[4] which involves the Hamiltonian of the system. How do we introduce EM interactions in the Hamiltonian formalism? What is the Hamiltonian that represents the Lorentz force?

Recall that the Hamiltonian $H(\mathbf{r}, \mathbf{p})$ is a function of the position coordinate $\mathbf{r}$ and the canonical momentum $\mathbf{p}$, and the classical equations of motion are **Hamilton's equations**

$$\frac{dr_i}{dt} = \frac{\partial H}{\partial p_i}, \qquad \frac{dp_i}{dt} = -\frac{\partial H}{\partial r_i}. \tag{16.14}$$

One can easily check that for $H = \frac{\mathbf{p}^2}{2m} + V(\mathbf{r})$, the first equation is just $\mathbf{p} = m\mathbf{v}$ (i.e. the canonical momentum is the same as the kinematic momentum) and the second, the usual $\mathbf{F} = m\mathbf{a}$. Now we claim that the Hamiltonian description of a charged particle (with mass m and charge e) in an EM field (represented by the potentials $\Phi, \mathbf{A}$), is given by

$$H = \frac{\left(\mathbf{p} - \frac{e}{c}\mathbf{A}\right)^2}{2m} + e\Phi. \tag{16.15}$$

To check this claim, let us work out the two Hamilton's equations:

1. The first equation in (16.14): Φ being a function of $\mathbf{r}$ only,

$$v_i \equiv \frac{dr_i}{dt} = \frac{\partial}{\partial p_i}\left[\frac{\left(\mathbf{p} - \frac{e}{c}\mathbf{A}\right)^2}{2m} + e\Phi\right] = \frac{1}{m}\left(p_i - \frac{e}{c}A_i\right).$$

Thus the canonical momentum ($\mathbf{p}$) differs from the kinematic momentum ($m\mathbf{v}$) by a factor related to the charge and vector potential,

$$\mathbf{p} = m\mathbf{v} + \frac{e}{c}\mathbf{A}, \tag{16.16}$$

and the first term in the Hamiltonian (16.15) remains the kinetic energy of $\frac{1}{2}mv^2$ (the second one, the electric potential energy).

[4] Here we start with elementary nonrelativistic quantum theory. However, all the results can be extended in a straightforward manner to relativistic Klein–Gordon and Dirac equations. For these cases, the simplest approach is (instead of the Hamiltonian) through the Lagrangian density as discussed in Section 16.4.2.

2. We expect the second equation in (16.14) to be the Lorentz force law. Let us verify this. Using the relation (16.16) between canonical and kinematic momenta, we have the LHS as

$$\frac{dp_i}{dt} = m\frac{dv_i}{dt} + \frac{e}{c}\frac{dA_i}{dt}. \tag{16.17}$$

The RHS may be written as

$$-\frac{\partial H}{\partial r_i} = -\nabla_i \left[\frac{\left(p_j - \frac{e}{c}A_j\right)\left(p_j - \frac{e}{c}A_j\right)}{2m} + e\Phi \right]$$

$$= \frac{\left(p_j - \frac{e}{c}A_j\right)}{m}\frac{e}{c}\nabla_i A_j - e\nabla_i\Phi$$

$$= v_j\frac{e}{c}\nabla_i A_j - e\nabla_i\Phi \tag{16.18}$$

where we have again used the relation (16.16). Equating (16.17) and (16.18) as in (16.14), we have

$$m\frac{dv_i}{dt} + \frac{e}{c}\frac{dA_i}{dt} = \frac{e}{c}\mathbf{v}\cdot(\nabla_i\mathbf{A}) - e\nabla_i\Phi, \tag{16.19}$$

which we recognize as the expression (16.13) of the Lorentz force in terms of the potentials, verifying our claim that Eq. (16.15) is the correct Hamiltonian for the Lorentz force law.

We have demonstrated that the Hamiltonian for a charged particle moving in an electromagnetic field can be compactly written in terms of the EM potentials $(\Phi, \mathbf{A})$ as (16.15). Perhaps the more important point is that there is no simple way to write the Hamiltonian, hence the QM description, in terms of the field strength $(\mathbf{E}, \mathbf{B})$ directly. As a consequence, we must study the invariance of the relevant QM equations under gauge transformation.

16.3 Gauge symmetry in quantum mechanics

Before launching into the study of gauge symmetry in QM, we shall take another look at the Hamiltonian (16.15) for a charged particle in the presence of an EM field. This prepares us for a new understanding of the theoretical significance of EM potentials $(\Phi, \mathbf{A})$.

16.3.1 The minimal substitution rule

Given the Hamiltonian (16.15), the Schrödinger equation $H\Psi = i\hbar\partial_t\Psi$, with the coordinate space representation of the canonical momentum $\mathbf{p} \doteq -i\hbar\nabla$, can be written out for a charged particle in an electromagnetic field as

$$\left[\frac{\left(\frac{\hbar}{i}\nabla - \frac{e}{c}\mathbf{A}\right)^2}{2m} + e\Phi \right]\Psi = i\hbar\frac{\partial\Psi}{\partial t}. \tag{16.20}$$

After a slight rearrangement of terms, it can be written as

$$-\frac{\hbar^2}{2m}\left(\nabla - \frac{ie}{\hbar c}\mathbf{A}\right)^2 \Psi = i\hbar\left(\partial_t + \frac{ie}{\hbar}\Phi\right)\Psi. \qquad (16.21)$$

When (16.21) is compared to the Schrödinger equation for **a free particle**,

$$-\frac{\hbar^2}{2m}\nabla^2\Psi = i\hbar\partial_t\Psi, \qquad (16.22)$$

we see that the EM interaction (also referred to as the "EM coupling") can be introduced via the following replacement:

$$\nabla \longrightarrow \left(\nabla - \frac{ie}{\hbar c}\mathbf{A}\right) \equiv \mathbf{D} \quad \text{and} \quad \partial_t \longrightarrow \left(\partial_t + \frac{ie}{\hbar}\Phi\right) \equiv D_t. \qquad (16.23)$$

This scheme of introducing the EM coupling is called **the minimal substitution rule**.[5] This procedure follows from the Hamiltonian (16.15) and is thus equivalent to the assumption of the Lorentz force law. While the procedure is simple, one is naturally curious for a deeper understanding: Is there a natural justification for this minimal coupling scheme? Namely, why does the EM coupling have the structure that it does?

Incidentally, the combinations $(D_t, \mathbf{D})$ of ordinary derivatives with EM potentials as defined in (16.23) are called **covariant derivatives**. As we shall discuss below they have the sane geometrical and physical significance as the covariant derivatives we encountered in our study of general relativity.

16.3.2   The gauge transformation of wavefunctions

Since QM must necessarily involve the EM potentials, one wonders how gauge invariance is implemented here. A direct inspection of the effects of the gauge transformations $(\Phi, \mathbf{A}) \longrightarrow (\Phi', \mathbf{A}')$ would show that the Schrödinger equation (16.21) is **not** invariant under the transformation (16.8) and (16.9):

$$\text{LHS} \longrightarrow -\frac{\hbar^2}{2m}\left(\nabla - \frac{ie}{\hbar c}\mathbf{A}'\right)^2\Psi = -\frac{\hbar^2}{2m}\left(\nabla - \frac{ie}{\hbar c}\mathbf{A} - \underbrace{\frac{ie}{\hbar c}\nabla\chi}\right)^2\Psi,$$

$$\text{RHS} \longrightarrow i\hbar\left(\partial_t + \frac{ie}{\hbar}\Phi'\right)\Psi = i\hbar\left(\partial_t + \frac{ie}{\hbar}\Phi - \underbrace{\frac{ie}{\hbar c}\partial_t\chi}\right)\Psi. \qquad (16.24)$$

Namely, there are these extra terms $\underbrace{\cdots}$ involving the gauge function χ that do not match on two sides of the transformed equation; hence gauge invariance is lost under (16.8) and (16.9). However, as observed by Fock (1926), the invariance could be obtained if we supplement the transformations of (16.8) and (16.9) by an appropriate spacetime-dependent phase change of the wavefunction $\Psi(\mathbf{r}, t)$,

$$\Psi(\mathbf{r}, t) \longrightarrow \Psi'(\mathbf{r}, t) = \exp\left[\frac{ie}{\hbar c}\chi(\mathbf{r}, t)\right]\Psi(\mathbf{r}, t), \qquad (16.25)$$

[5] This is "minimal", because of the absence of other possible, but more complicated, couplings, e.g. those involving the spin operator and magnetic field $\boldsymbol{\sigma} \cdot \mathbf{B}$, etc.

so that the above-mentioned extra terms can be cancelled. The function χ in the exponent is the same gauge function that appears in (16.8) and (16.9). Let us see how the various terms in the Schrödinger equation (16.21) change under this combined transformation (16.8), (16.9), and (16.25). First consider the RHS:

$$\left(\partial_t + \frac{ie}{\hbar}\Phi\right)\Psi \longrightarrow \left(\partial_t + \frac{ie}{\hbar}\Phi'\right)\Psi' = \left(\partial_t + \frac{ie}{\hbar}\Phi - \frac{ie}{\hbar c}\partial_t\chi\right)\exp\left(\frac{ie}{\hbar c}\chi\right)\Psi.$$

When the time derivative ∂_t acts on the product $\left[\exp\left(\frac{ie}{\hbar c}\chi\right)\Psi(\mathbf{r},t)\right]$, two terms result: $\exp\left(\frac{ie}{\hbar c}\chi\right)\partial_t\Psi(\mathbf{r},t) + \left(\frac{ie}{\hbar c}\partial_t\chi\right)\exp\left(\frac{ie}{\hbar c}\chi\right)\Psi(\mathbf{r},t)$, thus the effect of "pulling the phase factor $\exp\left(\frac{ie}{\hbar c}\chi\right)$ to the left of the ∂_t operator" will result in another extra term which just cancels the unwanted term in (16.24):

$$\left(\partial_t + \frac{ie}{\hbar}\Phi\right)\Psi \longrightarrow \exp\left(\frac{ie}{\hbar c}\chi\right)\left(\partial_t + \frac{ie}{\hbar}\Phi - \frac{ie}{\hbar c}\partial_t\chi + \frac{ie}{\hbar c}\partial_t\chi\right)\Psi$$

$$= \exp\left(\frac{ie}{\hbar c}\chi\right)\left(\partial_t + \frac{ie}{\hbar}\Phi\right)\Psi. \tag{16.26}$$

Similarly we have

$$\left(\nabla - \frac{ie}{\hbar c}\mathbf{A}\right)^2\Psi \longrightarrow \exp\left(\frac{ie}{\hbar c}\chi\right)\left(\nabla - \frac{ie}{\hbar c}\mathbf{A}\right)^2\Psi. \tag{16.27}$$

As a consequence, the transformed equation

$$-\frac{\hbar^2}{2m}\left(\nabla - \frac{ie}{\hbar c}\mathbf{A}'\right)^2\Psi' = i\hbar\left(\partial_t + \frac{ie}{\hbar}\Phi'\right)\Psi' \tag{16.28}$$

becomes

$$\exp\left(\frac{ie}{\hbar c}\chi\right)\left(\nabla - \frac{ie}{\hbar c}\mathbf{A}\right)^2\Psi = \exp\left(\frac{ie}{\hbar c}\chi\right)\left(\partial_t + \frac{ie}{\hbar}\Phi\right)\Psi.$$

The same exponential factor $\exp\left(\frac{ie}{\hbar c}\chi\right)$ appears on both sides of the transformed Schrödinger equation; they cancel, showing that the validity the transformed equation (16.28) follows from the original equation (16.21), and we have gauge invariance restored. From now on, whenever we refer to gauge transformation it is understood to be the combined transformations of (16.8), (16.9), and (16.25).

16.3.3 The gauge principle

We will now turn the argument around and regard the transformation (16.25) of the wavefunction as being more fundamental, and from this we can "derive" the gauge transformation of the EM potentials, (16.8) and (16.9). The rationale for doing it this way will become clear as we proceed. Our wish is to generalize gauge symmetry beyond electromagnetism, and to use this symmetry as a tool to discover new physics. In such an endeavor it is much easier to start with the generalization of the gauge transformation of the wavefunction rather than that for the potentials. More importantly, as we shall see, this reversed procedure will also explain why the electromagnetic couplings (16.23) have the structure that they have.

This is similar to what Einstein did when he elevated the empirically observed equality between gravitational and inertial masses to the principle of equivalence between gravitation and inertia. With this focus, he then applied EP to the physics beyond mechanics (cf. Sections 12.2 and 12.3). In the same way, once the approach is formulated as the gauge principle for electromagnetism, we can then apply it to the physics beyond, to strong and weak interactions, etc.

The Schrödinger equation for a free charged particle has global $U(1)$ symmetry

It is easier to start the generalization process by starting with the gauge transformation of the wavefunction because we can associate this part of the gauge transformation to a more familiar symmetry transformation. Consider the Schrödinger equation for a charged free particle,

$$-\frac{\hbar^2}{2m}\mathbf{\nabla}^2\Psi = i\hbar\partial_t\Psi. \tag{16.29}$$

This equation is unchanged under the **global** phase change:

$$\Psi(\mathbf{r},t) \longrightarrow \Psi'(\mathbf{r},t) = \exp\left(\frac{ie}{\hbar c}\chi\right)\Psi(\mathbf{r},t). \tag{16.30}$$

In contrast to the transformation as given in (16.25), here the phase factor is a constant $\chi \neq \chi(\mathbf{r},t)$. Namely, we make the **same** phase change for the wavefunction at all space-time points! This simple phase change is a "unitary transformation in one dimension";[6] hence called a "global $U(1)$ transformation". Clearly Eq. (16.29) is invariant, as every term acquires the same phase that can be cancelled out, and this theory possesses global $U(1)$ symmetry. This symmetry has the associated electric charge conservation law, as expressed by the continuity equation

$$\partial_t\rho_e + \mathbf{\nabla}\cdot\mathbf{j}_e = 0 \tag{16.31}$$

with $\rho_e = |\Psi|^2$ and $\mathbf{j}_e = \frac{-i\hbar}{2m}\left(\Psi^*\mathbf{\nabla}\Psi - \Psi\mathbf{\nabla}\Psi^*\right)$. We leave it as an elementary QM exercise to prove that this continuity equation follows from the free Schrödinger equation (16.29).

[6]The transformation $U = e^{i\chi}$ is "unitary" because it satisfies the condition $U^\dagger U = 1$; it is in one dimensional because it is specified by one parameter.

Gauging the symmetry

One may be dissatisfied with this **global** feature of the transformation: Why should the wavefunctions everywhere all undergo the same phase change? A more desirable form of symmetry would require the theory to be invariant under a **local** transformation. Namely, we replace the phase factor in the transformation (16.30) by a spacetime-dependent function

$$[\chi = \text{constant}] \quad \longrightarrow \quad [\chi = \chi(\mathbf{r},t)], \tag{16.32}$$

just as in (16.25). That is, we want the freedom of choosing the phase of the charge's wavefunction locally: a different one at each spacetime point.

Now the Schrödinger equation (16.29) is not invariant under such a local transformation as the derivative terms would bring down extra terms (because of the spacetime-dependent phase) that cannot be canceled, indicating that the equation is no longer symmetric. We can overcome this difficulty by replacing ordinary derivatives (∂_t, ∇) by covariant derivatives $(D_t, \mathbf{D})$ as defined in (16.23). They have the desired property that covariant ("change in the same way") derivatives of the wavefunction transform in the same way as the wavefunction itself. Namely, just as $\Psi' = \exp\left(\frac{ie}{\hbar c}\chi\right)\Psi$, we have

$$(D_t\Psi)' = \exp\left(\frac{ie}{\hbar c}\chi\right)(D_t\Psi) \tag{16.33}$$

and similarly,

$$(\mathbf{D}\Psi)' = \exp\left(\frac{ie}{\hbar c}\chi\right)(\mathbf{D}\Psi) \text{ and also } (\mathbf{D}^2\Psi)' = \exp\left(\frac{ie}{\hbar c}\chi\right)(\mathbf{D}^2\Psi). \tag{16.34}$$

This replacement of derivatives

$$(\partial_t, \nabla) \longrightarrow (D_t, \mathbf{D}) \tag{16.35}$$

calls to mind the principle of general covariance when going from special relativity to general relativity as discussed in Section 13.4. Similar to the situation here, when proceeding from SR to GR we go from a global symmetry to a local symmetry. This replacement (16.35) turns the Schrödinger equation (16.29) into

$$-\frac{\hbar^2}{2m}\mathbf{D}^2\Psi = i\hbar D_t\Psi. \tag{16.36}$$

Under the gauge transformations of (16.8), (16.9), and (16.25), we have

$$-\frac{\hbar^2}{2m}\left(\mathbf{D}^2\Psi\right)' = i\hbar(D_t\Psi)'. \tag{16.37}$$

The invariance of the equation can be checked because, through the relations in (16.33) and (16.34), it is

$$-\frac{\hbar^2}{2m}\left[\exp\left(\frac{ie}{\hbar c}\chi\right)\right](\mathbf{D}^2\Psi) = i\hbar\left[\exp\left(\frac{ie}{\hbar c}\chi\right)\right](D_t\Psi). \tag{16.38}$$

With the exponential factors [...] canceled, this is just the original equation (16.36). This completes the proof of the equation's invariance[7] under such a local transformation.

The covariant derivatives (16.23) are constructed by an artful combination of the ordinary derivative with a set of newly introduced "compensating fields" $(\Phi, \mathbf{A})$ which themselves transform in such a way to compensate, to cancel, the unwanted extra factors that spoil the invariance. The replacement of ordinary derivatives by covariant derivatives as in (16.35) justifies the "principle of minimal substitution", used to introduce the EM coupling as done in (16.23). Equation (16.36) is just the Schrödinger equation (16.21) for a charged particle in an EM field that we discussed earlier. Thus we can understand this

[7] Properly speaking we should say "covariance of the equation", as the terms in an equation are not invariant, but they transform "in the same way" so that their **relation is unchanged**, and the same equation is obtained for the transformed quantitites.

coupling scheme as resulting from requiring the theory to have a **local** $U(1)$ symmetry in the charge space (i.e. with respect to a change of the wavefunction phase associated with the particle's charge). The general practice is that such local symmetry is called **gauge symmetry** and the process of turning a global symmetry into a local one as in (16.32) has come to be called "**gauging a symmetry**".

One can then understand the "origin" of the electromagnetic potentials $(\Phi, \mathbf{A})$ as the **gauge fields**[8] required to implement such a gauge symmetry. This procedure of understanding the origins of some dynamics (e.g. electromagnetism) through the process of turning a global symmetry into a local one is now called **the gauge principle**.

We would like to emphasize the similarity of 'gauging a symmetry' and 'turning the global Lorentz symmetry in special relativity into the local coordinate symmetry of general relativity'. In both cases, we need to replace the usual derivatives by covariant derivatives. In the case of gauge theory, this introduces the gauge field; in the case of relativity this inserts a gravitational field intensity (in the form of Christoffel symbols) into the theory.

[8]From now on we shall often refer to $(\Phi, \mathbf{A})$ as "fields", rather than "potentials", and the equations they satisfy as field equations. This is compatible with the general practice in physics of calling any function of space and time "field".

16.4   Electromagnetism as a gauge interaction

The above discussion has allowed us to have a better understanding of the EM coupling as displayed in the Hamiltonian of (16.15), which is equivalent to the Lorentz force law. We will now show that gauge symmetry, together with special relativity, allows us to "derive" the electromagnetic field equations, the Maxwell equations. For this purpose we need a language that will simplify the expression of relativistic invariance. This is provided by Minkowski's four-dimensional spacetime formalism (cf. Chapter 11). We first provide a rapid review of this subject. Not only will this allow us to understand Maxwell's equations, it will also provide us, in a simple way, to infer the pattern of the $\pm$ signs and factors of c that have appeared in Eqs. (16.6)–(16.9), which were written in non relativistic notation.

16.4.1   The 4D spacetime formalism recalled

Gauging the $U(1)$ symmetry requires us to introduce the potentials $(\Phi, \mathbf{A})$ and thus the existence of electromagnetism. As we shall demonstrate, the requirements of gauge symmetry and Lorentz invariance (special relativity) can basically lead us to Maxwell's equations. For this we shall adopt the language of 4-vectors and 4-tensors in 4D Minkowski spacetime (as discussed in Chapter 11) so as to make it simpler to implement the condition of Lorentz symmetry.

The principal message of special relativity is that the arena for physics events is 4D Minkowski spacetime, with spatial and temporal coordinates being treated on an equal footing (cf. Section 11.3). In this 4D space, a position vector has four components x^{μ} with index $\mu = 0, 1, 2, 3$.

$$x^{\mu} = \left(x^0, x^1, x^2, x^3\right) = (ct, \mathbf{x}). \tag{16.39}$$

Noting that the contravariant and covariant tensor components are related, as shown in (11.19), by the Minkowski metric $\eta_{\mu\nu} = \eta^{\mu\nu} = \mathrm{diag}(-1, 1, 1, 1)$, we list some of the 4-position and 4-derivatives:

$$\text{4D del operator} \quad \partial_\mu = \left(\frac{1}{c}\partial_t, \mathbf{\nabla}\right) \quad \text{and} \quad \partial^\mu = \left(-\frac{1}{c}\partial_t, \mathbf{\nabla}\right) \quad (16.40)$$

$$\text{momentum 4-vector} \quad p^\mu = \left(\frac{E}{c}, \mathbf{p}\right) \quad \text{with} \quad p_\mu p^\mu = -\frac{E^2}{c^2} + \mathbf{p}^2 = -m^2 c^2. \quad (16.41)$$

Maxwell's equations

The six components of $(\mathbf{E}, \mathbf{B})$ are taken to be the elements of a 4×4 antisymmetric matrix: the "EM field tensor", $F_{\mu\nu} = -F_{\nu\mu}$ as displayed in Eq. (11.34). The inhomogeneous Maxwell equations (16.2)–(16.3) can then be written compactly as

$$\partial^\mu F_{\mu\nu} = -\frac{1}{c} j_\nu \quad \text{Gauss + Ampere (inhomogeneous Maxwell)} \quad (16.42)$$

where $j^\mu \equiv (c\rho, \mathbf{j})$ is the "4-current density", and the homogeneous (16.4)–(16.5) as

$$\partial^\mu \tilde{F}_{\mu\nu} = 0 \quad \text{Faraday + mag-Gauss (homogeneous Maxwell)} \quad (16.43)$$

where $\tilde{F}^{\mu\nu} = -\frac{1}{2}\varepsilon^{\mu\nu\lambda\rho}F_{\lambda\rho}$ is the dual field tensor.[9]

[9] The duality transformation (9.41) discussed in Section 9.5.1 corresponds to $F_{\mu\nu} \to \tilde{F}_{\mu\nu}$.

Electromagnetic potentials

It is easy to see that, with the "4-potential" being $A^\mu = (\Phi, \mathbf{A})$, namely, $\Phi = A^0 = -A_0$, the relation between potentials and the field tensor, (16.6) and (16.7), can be summarized as ($F_{\mu\nu}$ as the 4-curl of A_μ)

$$F_{\mu\nu} = \partial_\mu A_\nu - \partial_\nu A_\mu, \quad (16.44)$$

while the gauge transformations (16.8) and (16.9) can be compactly written in the Minkowski notation as

$$A_\mu \longrightarrow A'_\mu = A_\mu + \partial_\mu \chi. \quad (16.45)$$

The electromagnetic field strength tensor $F_{\mu\nu}$, with components of $(\mathbf{E}, \mathbf{B})$, being related to the potentials A^μ as in (16.44), is clearly unchanged[10] under this transformation (16.45).

[10] $F'_{\mu\nu} = \partial_\mu A'_\nu - \partial_\nu A'_\mu = (\partial_\mu A_\nu - \partial_\nu A_\mu) + (\partial_\mu \partial_\nu \chi - \partial_\nu \partial_\mu \chi) = F_{\mu\nu}$ because $\partial_\mu \partial_\nu = \partial_\nu \partial_\mu$.

Such a notation also simplifies the steps when showing that (16.44) solves the homogeneous Maxwell equation (16.43):

$$\partial^\mu \tilde{F}_{\mu\nu} = \frac{1}{2}\epsilon_{\mu\nu\lambda\rho}\partial^\mu F^{\lambda\rho} = \frac{1}{2}\epsilon_{\mu\nu\lambda\rho}\partial^\mu \left(\partial^\lambda A^\rho - \partial^\rho A^\lambda\right) = \epsilon_{\mu\nu\lambda\rho}\partial^\mu \partial^\lambda A^\rho = 0.$$

The two RHS terms are combined when the dummy indices λ and ρ are relabeled. The final result vanishes because the indices $\mu\lambda$ are antisymmetric in 4D Levi-Civita symbols $\epsilon_{\nu\mu\lambda\rho}$ but symmetric in the double derivative

$\partial^\mu \partial^\lambda$. When the relation (16.44) is plugged into the inhomogeneous Maxwell equation (16.42), we have

$$\partial^\mu \left(\partial_\mu A_\nu - \partial_\nu A_\mu \right) = \Box A_\nu - \partial_\nu \left(\partial^\mu A_\mu \right) = -\frac{1}{c} j_\nu$$

which, after imposing the Lorentz gauge condition $\partial^\mu A_\mu = 0$, reduces to the simple wave equation $\Box A_\nu = -\frac{1}{c} j_\nu$ that we displayed in Section 3.1.

Similarly, the covariant derivatives D_t and $\mathbf{D}$ defined in (16.23) can be combined into a "4-covariant derivative" as $D_\mu = \left(\frac{1}{c} D_t, \mathbf{D} \right)$ so that

$$D_\mu \equiv \left(\partial_\mu - \frac{ie}{\hbar c} A_\mu \right), \tag{16.46}$$

and the minimal substitution rule is simply the replacement of $\partial_\mu \longrightarrow D_\mu$.

We also take note of a useful relation between the commutator of covariant derivatives and the field strength tensor [cf. Eq. (14.17)]

$$\left[D_\mu, D_\nu \right] = -\frac{ie}{\hbar c} F_{\mu\nu}. \tag{16.47}$$

This operator equation is understood that each term is an operator that acts, from the left, on some spacetime-dependent test function (such as a wavefunction). This can be verified by explicit calculation:

$$\left[D_\mu, D_\nu \right] \psi$$

$$= \left(\partial_\mu - \frac{ie}{\hbar c} A_\mu \right) \left(\partial_\nu - \frac{ie}{\hbar c} A_\nu \right) \psi - \left(\partial_\nu - \frac{ie}{\hbar c} A_\nu \right) \left(\partial_\mu - \frac{ie}{\hbar c} g A_\mu \right) \psi$$

$$= -\frac{ie}{\hbar c} \left\{ \partial_\mu \left(A_\nu \psi \right) + A_\mu \left(\partial_\nu \psi \right) - \partial_\nu \left(A_\mu \psi \right) - A_\nu \left(\partial_\mu \psi \right) \right\} \tag{16.48}$$

$$= -\frac{ie}{\hbar c} \left(\partial_\mu A_\nu - \partial_\nu A_\mu \right) \psi = -\frac{ie}{\hbar c} F_{\mu\nu} \psi.$$

The relevance of this relation in a generalized gauge symmetry will be discussed below—see the displayed equation (16.68).

16.4.2   The Maxwell Lagrangian density

We now add further detail to the statement: "the electromagnetic interaction is a gauge interaction", or equivalently, "electrodynamics is a gauge theory". So far we have concentrated on the "equation of motion" part of the field description (the Lorentz force law). Now we discuss the "field equation" part. In the case of electromagnetism, it is Maxwell's equation. A field can be viewed as a system having an infinite number of degrees freedom with its generalized coordinate being the field itself $q = \phi(x)$, where $\phi(x)$ is some generic field. For such a continuum system, the field equation, as discussed in Section A.5.2, is the Euler–Lagrange equation (A.64) written in terms of the **Lagrangian density**

$\mathcal{L}$ (with a Lagrangian $L = \int d^3x\mathcal{L}$ and an action $S = \int d^4x\mathcal{L}(x)$) which is a function of the field and its derivatives $\mathcal{L} = \mathcal{L}(\phi, \partial_\mu\phi)$:

$$\partial_\mu \frac{\partial\mathcal{L}}{\partial(\partial_\mu\phi)} - \frac{\partial\mathcal{L}}{\partial\phi} = 0. \tag{16.49}$$

Knowledge of the (Lorentz invariant) Lagrangian density $\mathcal{L}$ is equivalent to knowing the (Lorentz covariant) field equation. Thus, knowledge of Maxwell's Lagrangian density:

$$\mathcal{L}(x) = -\frac{1}{4}F_{\mu\nu}F^{\mu\nu} + \frac{1}{c}j^\nu A_\nu \tag{16.50}$$

is tantamount to knowing Maxwell's field equations. The Euler–Lagrange equation for the $A_\mu(x)$ field is [namely, Eq. (16.49) with $\phi(x) = A_\nu(x)$]:

$$\partial_\mu \frac{\partial\mathcal{L}}{\partial(\partial_\mu A_\nu)} - \frac{\partial\mathcal{L}}{\partial A_\nu} = 0, \tag{16.51}$$

which is just the familiar Maxwell equation (16.42), as we have

$$\frac{\partial\mathcal{L}}{\partial A_\nu} = \frac{1}{c}j^\nu, \tag{16.52}$$

$$\frac{\partial\mathcal{L}}{\partial(\partial_\mu A_\nu)} = \frac{\partial}{\partial(\partial_\mu A_\nu)}\left(-\frac{1}{4}\left(\partial_\alpha A_\beta - \partial_\beta A_\alpha\right)^2\right)$$

$$= \frac{\partial}{\partial(\partial_\mu A_\nu)}\left(-\frac{1}{2}\partial_\alpha A_\beta\left(\partial^\alpha A^\beta - \partial^\beta A^\alpha\right)\right)$$

$$= -F^{\mu\nu}. \tag{16.53}$$

16.4.3 Maxwell equations from gauge and Lorentz symmetries

From the above discussion, we see that a derivation of Maxwell's Lagrangian density (16.50) is tantamount to a derivation of Maxwell's equations (16.42) themselves. Gauging a symmetry requires the introduction of the gauge field; in the case of $U(1)$ symmetry, it is the vector $A_\mu(x)$ field. To have a dynamical theory for the $A_\mu(x)$ field, we need to construct a Lagrangian density from this $A_\mu(x)$ field and its derivatives $\partial_\mu A_\nu$ (because the kinetic energy term must involve spacetime derivatives). The simplest gauge-invariant combination of $\partial_\mu A_\nu$ is

$$\partial_\mu A_\nu - \partial_\nu A_\mu = F_{\mu\nu}. \tag{16.54}$$

The Lagrangian density must also be a Lorentz scalar (i.e. all spacetime indices are contracted) so the resulting Euler–Lagrange equations are relativistic covariant. The simplest combination[11] is the expression

$$\mathcal{L}_A = -\frac{1}{4}F_{\mu\nu}F^{\mu\nu}. \tag{16.55}$$

[11] In principle, higher powers $(F_{\mu\nu}F^{\mu\nu})^n$ are also gauge and Lorentz symmetric. However such terms are "nonrenormalizable" and our current understanding of quantum field theory informs us that they should be highly suppressed (i.e. at the relevant energy scale we consider, they make negligible contribution).

One can then add $\frac{1}{c}j^\nu A_\nu$ as the source term to arrive at the result in (16.50). (Factors like $-\frac{1}{4}$ and c are unimportant—all a matter of system of units and convention.)

The Maxwell equations were discovered by experimentation and deep theoretical invention; but this derivation shows **why** the four equations,[12] (16.2)–(16.5), take on the form they take. Their essence is special relativity plus gauge invariance. With this insight of electromagnetism we can then generalize the approach to the investigation of other fundamental interactions.

[12]Of course these equations have already been greatly simplified with the vector notation embodying rotational symmetry, which is part of Lorentz invariance.

16.5   Gauge theories: A narrative history

The above discussion shows that we can understand electromagnetism as a "gauge interaction". From the requirement of a local $U(1)$ symmetry in charge space, the presence of a vector gauge field $A_\mu(x)$ is deduced. In the rather restrictive framework of special relativity, its dynamics can be fixed (to be that described by Maxwell's equation). This way of using symmetry to deduce the dynamics has been very fruitful in our attempts to understand (i.e. to construct theories of) other particle interactions as well.

One of the crowning achievement in the physics of the twentieth century is the establishment of the Standard Model (SM) of elementary particle interactions.[13] This gives a complete and correct description of all nongravitational physics. This theory is based on the principle of gauge symmetry. Strong, weak, and electromagnetic interactions are all gauge interactions. In this section we give a very brief account of this SM gauge theory of particle physics.

[13]The progress of physics depends both on theory and experiment. A proper account of the experimental accomplishments in the establishment of the Standard Model is however beyond the scope of this presentation. This omission should not be viewed in any way as the author's lack of appreciation of their importance.

16.5.1   Einstein's inspiration, Weyl's program, and Fock's discovery

The rich and interesting history of gauge invariance and electromagnetic potentials in classical electromagnetism is beyond the scope of our presentation; we refer the curious reader to the authoritative and accessible account given by Jackson and Okun (2001). Here we shall present a narrative of the development of the gauge symmetry idea[14] as rooted in Einstein's general theory of relativity.

[14]For a more detailed gauge theory survey with extensive references, see Cheng and Li (1988).

What is the origin of the name "gauge symmetry"? The term **eichinvarianz** (gauge invariance) was coined in 1919 by Hermann Weyl (1885–1955) in the context of his attempt to "geometrize" the electromagnetic interaction and to construct in this way a unified geometrical theory of gravity and electromagnetism (Weyl 1918, 1919). He invoked the invariance under a local change of the scale, the "gauge", of the metric field $g_{\mu\nu}(x)$:

$$g_{\mu\nu}(x) \longrightarrow g'_{\mu\nu}(x) = \lambda(x)g_{\mu\nu}(x), \tag{16.56}$$

where $\lambda(x)$ is an arbitrary function of space and time. Weyl was inspired by Einstein's geometric theory of gravity, general relativity, which was published in 1916 (cf. Chapters 13 and 14). This was, of course, before the emergence of modern quantum mechanics in 1925–26. A key QM concept was to identify the dynamical variables of energy and momentum with the operators $-i\hbar\partial_\mu$ and, in the presence of an electromagnetic field, with $-i\hbar\partial_\mu + \frac{e}{c}A_\mu$ as in the "minimal substitutional rule". In this context, Vladimir Fock (1898–1974) discovered in 1926 that the quantum mechanical wave equation was invariant under the combined transformation $A'_\mu = A_\mu + \partial_\mu\chi$ and

$$\Psi'(x) = \exp\left[\frac{ie}{\hbar c}\chi(x)\right]\Psi(x); \qquad (16.57)$$

he called it the **gradient transformation**.[15] Given the central role played by (16.57) in showing that here one is discussing transformations in charge space, one would say that it was Fock who first discovered the modern notion of gauge invariance in physics.

Fritz London observed in 1927 that, if the i was dropped from Fock's exponent in (16.57), this phase transformation becomes a scale change (16.56) and the transformation of (16.45) and (16.57) was equivalent to Weyl's old eich-transformation (London 1927). However, when Weyl finally worked out this approach later on he retained his original terminology of "gauge invariance" because he believed that a deep understanding of the local transformation of gauge invariance could come about only through the benefit of general relativity.[16] Most importantly it was Weyl who first declared [especially in his famous book: *Theory of Groups and Quantum Mechanics* (Weyl 1928, 1931)] gauge invariance as a fundamental principle—the requirement of the matter wave equation being symmetric under the gauge transformation leading to the introduction of the electromagnetic field. Subsequently this principle has become the key pathway in the discovery of modern theories of fundamental particle interactions; this calling a local symmetry (in charge space) a "gauge symmetry" has become the standard practice in physics.

[15]This designation originates from the transformation $A'_\mu = A_\mu + \partial_\mu\chi$. From the title of the Fock (1926) paper, it is clear what Fock wanted to emphasize is that this new symmetry, involving the transformation of $\Psi(x)$ and $A_\mu(x)$, is a symmetry of charge space. Thus in the first 40 years or so after the invention of the gauge principle, people often followed the practice of designating global symmetry in charge space as '**gauge symmetry of the first kind**' and local symmetry in charge space as '**gauge symmetry of the second kind**'. But nowadays by gauge symmetry we mean gauge symmetry of the second kind and distinguish it from the first kind simply by calling the latter global symmetry.

[16]The connection between gauge symmetry and general relativity would be deepened further with the advent of the nonabelian theory of Yang and Mills in the 1950s.

16.5.2 Quantum electrodynamics

We have so far discussed gauge symmetry in the quantum description of a nonrelativistic charged particle interacting with an electromagnetic field. The proper framework for particle interaction should be quantum field theory,[17] which is a union of quantum mechanics with special relativity. We first comment on the prototype quantum field theory of quantum electrodynamics (QED), which is the quantum description of relativistic electrons and photons. However in this introductory presentation of gauge symmetry, we shall by and large stay with a classical field description.

[17]Some elementary aspects of quantum field theory were presented in Section 6.4. For an introduction to the Standard Model in the proper quantum field theoretical framework, see, for example, Cheng and Li (1984).

Dirac equation

QED is the theory of electrons interacting through the electromagnetic field. While the EM field equation is already relativistic, we must replace the Schrödinger equation by the relativistic wave equation for the electron, first discovered by Paul Dirac. Namely, instead of the Schrödinger equation (16.29), we should use the Dirac equation for a free electron,

$$\left(i\hbar\gamma^\mu\partial_\mu - mc\right)\psi = 0 \qquad (16.58)$$

with ψ being the four-component electron spinor field and γ^μ is a set of four 4×4 "Dirac gamma matrices" obeying the anticommutation relation $\{\gamma^\mu, \gamma^\nu\} = -2g^{\mu\nu}$. In momentum space with $p^\mu = (E/c, \mathbf{p})$ being the 4-momentum vector this equation becomes

$$\left(\gamma^\mu p_\mu - mc\right)\psi = 0. \tag{16.59}$$

When operated on from the LHS by $(\gamma^\nu p_\nu + mc)$, this equation, after using the anticommutation relation, implies[18] $\left(p^\mu p_\mu + m^2 c^2\right)\psi = 0$, which we recognize as the relativistic energy momentum relation of (16.41). To couple it to an EM field, we replace the derivative ∂_μ in (16.58) by the covariant derivative D_μ of (16.46): $\left(i\hbar\gamma^\mu D_\mu - mc\right)\psi = 0$—in just the same way as we obtained Eq. (16.21) from the free Schrödinger equation (16.22). We can display the role of the gauge field $\left(A_\mu\right)$/electron field (ψ) cross-term as the source factor by separating out and moving it to the RHS:

$$\left(i\hbar\gamma^\mu \partial_\mu - mc\right)\psi = \frac{e}{c}\gamma^\mu A_\mu \psi. \tag{16.60}$$

Lagrangian density for QED

Instead of field equations, we can equivalently work with the Lagrangian density. Thus instead of Eq. (16.58) we can concentrate on the equivalent quantity

$$\mathcal{L}_\psi = \bar{\psi}\left(i\hbar\gamma^\mu \partial_\mu - mc\right)\psi, \tag{16.61}$$

which is manifestly Lorentz invariant, with $\bar{\psi}$ being the conjugate $\psi^\dagger \gamma_0$. As discussed above, EM coupling can be introduced through the covariant derivative and, after adding the density $\mathcal{L}_A$ of the EM field (16.55), we have the full QED Lagrangian density

$$\mathcal{L}_{\text{QED}} = \mathcal{L}_\psi + \mathcal{L}_A + \mathcal{L}_{\text{int}}. \tag{16.62}$$

The interaction density,[19] which comes from part of the covariant derivative, is just the source density in (16.50)

$$\mathcal{L}_{\text{int}} = \frac{1}{c}j^\mu A_\mu = \frac{e}{c}\bar{\psi}\gamma^\mu \psi A_\mu \tag{16.63}$$

where j^μ is shown now as the 4-current density of electron. A graphical representation of a gauge boson coupled to a current is shown in Fig. 16.1(a).

[19]$\mathcal{L}_\psi$ and $\mathcal{L}_A$ are the Lagrangians for free electrons and photons as they are quadratic (harmonic) in their respective fields, while $\mathcal{L}_{\text{int}}$ represents the interaction as it has more than two field powers (hence anharmonic).

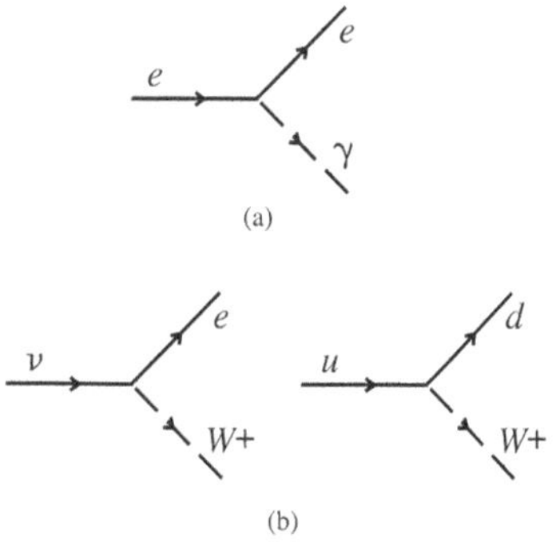

(a)

(b)

Fig. 16.1 (a) Trilinear coupling of a photon to an electron; (b) weak vector boson W coupled respectively to weak currents of leptons and quarks.

[18]We first find

$$\left(\gamma^\nu p_\nu + mc\right)\left(\gamma^\mu p_\mu - mc\right) = \gamma^\nu p_\nu \gamma^\mu p_\mu - m^2 c^2.$$

Since $p_\nu p_\mu = p_\mu p_\nu$, we should symmetrize the gamma matrices as well:

$$\frac{1}{2}\{\gamma^\nu, \gamma^\mu\}p_\mu p_\nu - m^2 c^2 = -p^\mu p_\mu - m^2 c^2 = 0.$$

To reach the last expression, we have used the anticommutation relation of gamma matrices.

QED as a $U(1)$ gauge theory

The discussion carried out in the previous sections of this chapter demonstrates that one can "derive" $\mathcal{L}_{QED}$ from the requirement of Lorentz and gauge symmetry. Namely, the theory can be understood as following from "gauging the $U(1)$ symmetry" of the free Dirac equation. The original global $U(1)$ symmetry is directly linked to the familiar electric charge conservation. The quantization (and renormalization) procedure based on $\mathcal{L}_{QED}$ is rather complicated and is beyond the scope of this presentation. Still, what we need to know is that the full QED theory can be worked out on the basis of this Lagrangian density (16.62). We also note that quanta of the electromagnetic field are photons. They can now be viewed as the gauge particles (spin-1 bosons) of QED theory. Parenthetically, the common practice in quantum field theory of describing electrons interacting through the electromagnetic field is "interaction through the exchange of photons". This language is particularly convenient when, as we shall see, describing the strong and weak interactions. An important feature of the QED Lagrangian is the absence of a term of the form of $A^\mu A_\mu$ because it is forbidden by gauge invariance. Such a term would correspond to a gauge boson (photon) mass. Thus gauge invariance automatically predicts a massless photon, which accounts for the long-range nature of the (electromagnetism) interaction that it transmits.[20]

Because of its many redundant degrees of freedom, quantization of gauge theory is rather intricate. The necessary renormalization program for QED was successfully formulated through the work of Julian Schwinger (1918–94), Richard Feynman (1918–88), Sin-Itiro Tomonaga (1906–79), and Freeman Dyson (1923–). The close interplay of high-precision experimental measurement and theoretical prediction brought about this notable milestone in the history of physics.

[20] The relation between the range of interaction and the mass of the mediating particle is discussed in Section 6.4.2.

16.5.3   QCD as a prototype Yang–Mills theory

Here we shall discuss a highly nontrivial extension of the gauge symmetry of electromagnetism. This extension makes it even clearer that the transformation in the charge space involves the change of particle/field labels of the theory.

Abelian versus nonabelian gauge symmetries

The gauge symmetry for electromagnetism is based on the $U(1)$ symmetry group; its transformation involves the multiplications of phase factors to the wavefunction $\Psi \to \Psi' = U\Psi$ with $U(x) = e^{i\theta(x)}$ and the wavefunction $\Psi(x)$ is itself a simple function. A $U(1)$ phase transformation is equivalent to a rotation around a fixed axis in a 2D plane (in charge space). Hence $U(1)$ is isomorphic to the 2D rotation group (also called the 2D special orthogonal group): $U(1) = SO(2)$. Clearly such transformations are commutative $U_1 U_2 = U_2 U_1$, and the symmetry is said to be an **abelian symmetry**. On the other hand, general rotations in 3D space are represented by noncommutative matrices. The symmetry based on such rotations, called $SO(3) = SU(2)$, is **nonabelian symmetry**. The corresponding wavefunction (i.e. field) Ψ is a multiplet in some multidimensional charge space; its components would correspond to different particle states. As it turns out, we can understand other elementary particle interactions due to strong and weak forces as gauge interactions also, but their gauge symmetries are nonabelian—their respective symmetry transformations are noncommutative. Gauge symmetry with such noncommutative transformations was first studied in 1954 in particle physics by C.N. Yang (1922–) and Robert L. Mills (1927–99), hence nonabelian gauge theory is often referred to as **Yang–Mills theory**.

Quarks and gluons   From the heroic experimental and deep phenomenological studies of the strong and weak interactions, it was discovered that strongly interacting particles (called **hadrons**[21]) are composed of even more elementary constituents. These spin-1/2 particles were invented and named **quarks** by Murray Gell-Mann (1929–). There are six 'quark flavors' (*up, down, strange, charm, bottom* and *top*); each has three hidden degrees of freedom called 'color'.[22] Namely, each quark flavor can be in three different color states—they form an $SU(3)$ triplet representation in the color charge space:

$$q(x) = \begin{pmatrix} q_1(x) \\ q_2(x) \\ q_3(x) \end{pmatrix}. \tag{16.64}$$

If it is an 'up-quark', we can call them, for example, 'red', 'blue', and 'white' up-quarks. This triplet undergoes the transformation, $q' = Uq$, with U being a 3×3 unitary matrix having unit determinant (hence called special unitary). Namely, the particle fields can not only change their phases but the particle labels as well. When this symmetry is "gauged" (i.e. turned into a local symmetry) we have the $SU(3)$ gauge theory, called **quantum chromodynamics** (QCD). Just as QED is the theory of electrons interacting through the exchange of abelian gauge fields of photons, QCD is the fundamental strong interaction of quarks through the exchange of a set (8) of nonabelian gauge particles called **gluons**.

Yang–Mills gauge particles   To implement such a local symmetry, we need to introduce a covariant derivative involving gauge fields:

$$D_\mu = \partial_\mu - ig_s G_\mu. \tag{16.65}$$

This is the same as (16.46) of the abelian case (for simplicity of notation we have suppressed, or absorbed, the factor $\hbar c$). In the strong interacting QCD case, in place of the electromagnetic coupling strength e we have the strong coupling g_s. Instead of the $U(1)$ gauge field A_μ, we have Yang–Mills fields G_μ. But now the D_μ and G_μ are matrices in the color charge space; the gluon field matrix G_μ has eight independent components (being a traceless 3×3 Hermitian matrix) corresponding to the eight gluons of the strong interaction.

The basic property for covariant derivatives in Yang–Mills theory is still the same as that for the abelian case: the covariant derivative of a wavefunction $(D_\mu q)$ transforms in the same way as the wavefunction $q(x)$ itself:

$$q' = Uq \quad \text{and} \quad (D'_\mu q') = U(D_\mu q). \tag{16.66}$$

[21]Examples of hadrons are the proton, neutron, pion and omega, etc.

[22]'Color' is the whimsical name given to the strong interaction charge and has nothing to do with the common understanding of different frequencies of visible EM waves. Here we give an example of the type of phenomenology from which the color degrees of freedom were deduced. The omega baryon is composed of three strange quarks. Being a system of identical fermions, its wavefunction should be antisymmetric with respect to the interchange of any two quarks. Yet both its spin (3/2) and orbital angular momentum (S-wave) wavefunctions are symmetric. Spin-statistics is restored only when its antisymmetric color (singlet) wavefunction is taken into account.

We then have $D'_\mu = UD_\mu U^{-1}$ and more explicitly

$$\partial_\mu - ig_s G'_\mu = U\partial_\mu U^{-1} - ig_s UG_\mu U^{-1}.$$

This means that the gauge fields must transform as

$$G'_\mu = UG_\mu U^{-1} - \frac{1}{ig_s}U(\partial_\mu U^{-1}). \tag{16.67}$$

One can easily check that in the abelian case with $U = \exp(ig_s\chi)$ this reduces to Eq. (16.45). The factor $UG_\mu U^{-1}$ indicates that the gauge field itself transform nontrivially under the gauge group. Namely, the gauge fields (or gauge particles) themselves carry gauge charges.

The QCD Lagrangian   Another place where one can see that nonabelian gauge fields themselves carry gauge charges is in a property of the nonabelian field tensor $F_{\mu\nu}$, which is similarly related to the covariant derivatives as (16.47), $[D_\mu, D_\nu] = -ig_s F_{\mu\nu}$. Working it out as in (16.48), we find

$$F_{\mu\nu} = \partial_\mu G_\nu - \partial_\nu G_\mu - \frac{1}{ig_s}[G_\mu, G_\nu] \tag{16.68}$$

showing a nonvanishing commutator because now G_μ is also a matrix in the charge space. This nonabelian $F_{\mu\nu}$ is now quadratic in G_μ. Thus when we construct[23] the QCD Lagrangian density for the gauge field, like we did for the abelian case of (16.55),

$$\mathcal{L}_A = -\frac{1}{4}\mathrm{tr}F_{\mu\nu}F^{\mu\nu}, \tag{16.69}$$

we get, besides quadratic (G^2) terms, also cubic (G^3) and quartic (G^4) gauge field terms. While quadratic terms in $\mathcal{L}$ correspond to the free-particle Lagrangian, higher powers represent interactions. Again, these cubic and quartic couplings reflect the fact that nonabelian gauge fields, the gluons, must now be charged fields. See Fig. 16.2. Very much like Eq. (16.62) for QED, the Lagrangian density for QCD can be written down:

$$\mathcal{L}_{QCD} = \mathcal{L}_q + \mathcal{L}_A + \mathcal{L}_{qA}. \tag{16.70}$$

$\mathcal{L}_q = \bar{q}(i\hbar\gamma^\mu\partial_\mu - mc)q$ is the Lagrangian density for free quarks, much like $\mathcal{L}_\psi$ in Eq. (16.61) for free electrons. Since the quark field is a triplet, a sum of three terms (one for each color) is understood in $\mathcal{L}_q$. The Euler Lagrange equation for the gluon field based on $\mathcal{L}_A$ is nonlinear,[24] reflecting the fact that gluons carry color charges themselves. The quark (q)/gauge field (G_μ) coupling

[23]The symbol "tr" in (16.69) stands for the operation "trace", taken in charge space, (i.e. all charge space indices of the two $F_{\mu\nu}s$ are to be summed over in order to get a gauge symmetric quantity).

[24]This is entirely similar to the nonlinearity of the Einstein field equation in GR. A gravitational field carries energy ("gravity charge"), thus is itself a source of a gravitational field.

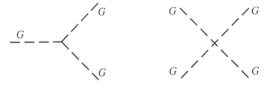

Fig. 16.2 Cubic and quartic self-couplings of charged gauge bosons. The similarity to the trilinear couplings shown in Fig.16.1 should be noted.

$\mathcal{L}_{qA}$ comes from the covariant derivative and is of a form entirely similar to the QED interaction term (16.63)

$$\mathcal{L}_{qA} = \frac{1}{c} j^\mu G_\mu = \frac{g_s}{c} \bar{q} \gamma^\mu G_\mu q. \qquad (16.71)$$

However now the quark field q is a triplet and G_μ is a 3×3 Hermitian matrix in the color charge space. With nonvanishing off-diagonal terms in the color matrix G_μ, a quark's color charges can be changed by such a quark/gluon coupling.

Asymptotic freedom and quark confinement The Yang–Mills gauge particles as transmitters of interactions being charged, this feature leads to the important physical consequence that the effective interaction strength[24a] (the so-called "running coupling") grows, logarithmically, as the distance between quarks increases. Namely, we have an antiscreen effect on the color charge— as the color charge is probed further away from the source, the effective charge is seen to increase! The increase in coupling strength means that it would take more and more energy to separate colored charges. Thus a colored particle must be confined to short subnuclear distances. This explains why no free quarks have ever been seen. All the observed strong-interaction particles (hadrons) are colorless compounds of quarks and gluons. This short-range confinement effect explains why even though gluons, like photons, are massless, the strong interaction they transmit, unlike the EM interaction, is nevertheless short-ranged. The other side of the same property (called **asymptotic freedom**[25]) is that the effective coupling becomes small at short distances and a perturbation approach can be used to solve the QCD equations in the high-energy and large-momentum-transfer regime, leading to precise QCD predictions that have been verified to high accuracy by experiments.

16.5.4   Hidden gauge symmetry and the electroweak interaction

The Standard Model of particle interactions describes the strong, weak, and electroweak interactions. We have already discussed the gauge theories of the electromagnetic and strong interactions. We now discuss the gauge theory of the weak interaction.

The electroweak $SU(2) \times U(1)$ gauge symmetry

In the early 1930s Enrico Fermi proposed a quantum field description of weak interactions. It was modeled on QED. His proposal[26] can be translated and updated in the language of quarks and weak vector bosons as follows. Just as electron–electron scattering $e + e \to e + e$ is described in quantum field theory as due to the exchange of a photon, the weak process of neutrino scattering off a down-quark producing an electron and an up-quark $v + d \to e + u$ is due to the exchange of a heavy vector boson W (see Fig. 16.3). Thus instead of the trilinear coupling of $e\bar{e}A_\mu$ we need $\bar{v}eW_\mu^+$ and $\bar{u}dW_\mu^+$, etc. (Here, except for the photon, we use particle names for their respective fields.) Namely, unlike

[24a]In quantum field theory the interaction strength is always modified by the quantum fluctuations represented by the production and reabsorption of virtual particles. The effect of such a quantum cloud depends how closely in distance is the coupling being probed.

[25]This fundamental property of Yang–Mills theory was discovered in 1973 by David Gross (1941–), Frank Wilczek (1951–), and David Politzer (1949–).

[26]Fermi's theory was invented to describe the then only known weak interaction—the neutron's beta decay: $n \to p + e + \bar{v}$. The neutron/proton transition can be interpreted at the quark level as a down/up transition because a neutron has valence quarks of (ddu) and a proton has (udu). The quark decay of $d \to u + e + \bar{v}$ is directly related to the scattering $v + d \to e + u$ when the final antineutrino is turned into a neutrino in the initial state as depicted in Fig. 16.3(b).

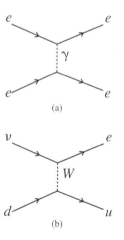

Fig. 16.3 (a) QED description of $e + e \to e + e$ through the exchange of a vector photon; (b) weak scattering $v + d \to e + u$ as due to the exchange of a heavy vector boson W. In this sense Fermi's weak interaction theory was based on the analog of QED.

electrodynamics, weak interaction couplings change particle labels, or we can say, "changes the weak interaction charge of a particle" (i.e. we can regard, for example, a neutrino and an electron as different weak interaction states of a leptonic particle).

This feature can be easily accommodated by a nonabelian gauge symmetry. For example, in the weak charge space the electron and electron–neutrino can be placed in an $SU(2)$ doublet of leptons, and in the same way, the up- and down-quarks form a weak doublet:[27]

$$l = \begin{pmatrix} \nu_e \\ e \end{pmatrix} \quad \text{and} \quad q = \begin{pmatrix} u \\ d \end{pmatrix}. \qquad (16.72)$$

The lepton/gauge-boson coupling, much like the electron/photon coupling shown in (16.71), can have the weak charge structure:[28]

$$g_2 \bar{l} \gamma^\mu W_\mu l = g_2 (\bar{\nu}_e, \bar{e}) \gamma^\mu \begin{pmatrix} W_\mu^0 & W_\mu^+ \\ W_\mu^- & -W_\mu^0 \end{pmatrix} \begin{pmatrix} \nu_e \\ e \end{pmatrix} \qquad (16.73)$$

which (when only particle labels are displayed) contains a trilinear vertex $\bar{\nu}eW_\mu^+$ of an (electrically) charged gauge boson W^+ coupled to an electron and an antineutrino. Similarly, if we replace the lepton by the quark doublet we can have a flavor-changing quark and gauge boson coupling like $\bar{u}dW_\mu^+$. See Fig. 16.1(b).

One would naturally try to identify the neutral gauge boson W_μ^0 with the photon. However, this would not be feasible because, as can be seen in (16.73), W_μ^0 must couple oppositely to the neutrino and to the electron: $\left(\bar{\nu}\nu W_\mu^0 \right.$ and $-\bar{e}eW_\mu^0)$; similarly, oppositely to up- and to down-quarks $\left(\bar{u}uW_\mu^0 \right.$ and $-\bar{d}dW_\mu^0)$, but all these fermions do not have electric charges opposite to their doublet partners. While such a unification of the weak and electromagnetic interactions, involving only a symmetry group of $SU(2)$ with gauge bosons W_μ^0 and $W_\mu^\pm$, does not work out, we can nevertheless achieve a partial unification by the simple addition of another $U(1)$ gauge factor, having an abelian gauge boson B_μ. Namely, we have a unified theory of electromagnetic and weak interactions (electroweak theory) based on the gauge symmetry of $SU(2) \times U(1)$. While neither W_μ^0 nor B_μ can be the photon field, we can assign leptons and quarks with new $U(1)$ charges (called "weak hypercharges") so that one of their linear combination has just the correct coupling property for a photon field A_μ:

$$A_\mu = \cos\theta_{\rm w} B_\mu + \sin\theta_{\rm w} W_\mu^0 \qquad (16.74)$$

$$Z_\mu^0 = -\sin\theta_{\rm w} B_\mu + \cos\theta_{\rm w} W_\mu^0$$

where the mixing angle $\theta_{\rm w}$ is called the Weinberg angle. The combination Z_μ^0, orthogonal to A_μ, is another physical neutral vector boson mediating yet another set of weak interaction processes.[29] We have only a "partial unification" because, to describe two interactions, we still have two independent coupling strengths as each gauge factor comes with an independent gauge coupling:[30] g_1 for $U(1)$ and g_2 for $SU(2)$.

[27] An $SU(3)$ quark color triplet has components of quarks with different '**colors**', while a quark weak doublet has different '**flavors**'.

[28] The discovery of parity violation in weak interactions, through the work of T.D. Lee (1926–), C.N. Yang, and many others in the mid-1950s, stimulated a great deal of progress in particle physics. This symmetry violation can be accommodated elegantly by the stipulation that the above displayed weak doublets involve only the left-handed helicity state of each particle. Parity violation comes about because the left-handed states and right-handed states have different weak charges (i.e. they belong to different types of weak multiplets). See footnote 36 in this chapter for further comments.

[29] Historically the weak interactions that were first studied are those mediated by the charged vector bosons $W_\mu^\pm$; they are called charged current reactions. Thus one of the firm predictions of this electroweak unification is the existence of Z_μ^0-mediated "neutral current processes".

[30] These gauge coupling constants are directly related to the experimentally more accessible constants of electric charge and Weinberg angle

$$e = \frac{g_1 g_2}{\left(g_1^2 + g_2^2\right)^{1/2}} \quad \text{and} \quad \cos\theta_{\rm w} = \frac{g_2}{\left(g_1^2 + g_2^2\right)^{1/2}}.$$

Spontaneous symmetry breaking

Mass problems in electroweak gauge theory   Gauge boson mass terms are forbidden by gauge invariance. While gluons are massless, the QCD gauge interaction is still effectively short-ranged because color particles are confined within short distances. Now if the weak interaction is to be formulated as a gauge interaction with the weak vector bosons $W^{\pm}$ and Z^0 identified as gauge bosons, they would also be required to be massless. But observationally the weak interaction is very short ranged (even shorter than the strong interaction range) hence the interaction transmitters must be massive.[31] This had been a major obstacle in the formulation of weak interaction as a gauge force. If we simply insert vector-boson mass terms (hence breaking the weak gauge symmetry), one would end up with uncontrollable ultraviolet divergences (technically speaking, making the theory unrenormalizable). This is the gauge boson mass problem. There is also a fermion mass problem. Symmetry (whether global or local) implies mass degeneracy of particles belonging to the same symmetry multiplet. Thus all three color states of the quark triplet (16.64) have identical masses. But to have a gauge theory of the weak interaction with the weak doublets as shown in (16.72), such fermion mass degeneracy would contradict observation, as the electron and the neutrino have different masses $m_e \neq m_v$, so have the up- and down-quarks $m_u \neq m_d$.

Symmetry is hidden   The mass problems discussed above were solved eventually by spontaneous symmetry breaking (SSB).[32] This is the possibility that physics equations with symmetry may have asymmetric solutions. A ferromagnet is a familiar example: above the critical temperature ($T > T_c$) it is a system of randomly oriented magnetic dipoles, reflecting the rotational symmetry of the physics equation describing such a system. But, below the critical temperature, all the dipoles are aligned in one particular direction—breaking the rotational symmetry even though the underlying physics equation is rotational invariant. This comes about because in a certain parameter space the theory would yield a ground state, instead being symmetric (i.e. a symmetry singlet state as shown in Fig. 16.4a), being a set of degenerate states related to each other through the symmetry transformation (as shown in Fig. 16.4b). Since the physical ground state (the vacuum state in a quantum field system) has to be unique, its selection, out of the degenerate set, must necessarily break the symmetry. Thus, the ground state, a solution to the symmetrical equation, is itself asymmetric. The rest of the physics (built on this vacuum state) will also have asymmetric features such as nondegenerate masses in a multiplet, etc. Since the underlying equations are symmetric while the outward appearance is not, SSB can best be described as the case of a "symmetry being hidden".

The Higgs sector   A hidden symmetry scenario can take place in both global and local symmetries. For global symmetry, one has the interesting consequence that such a hidden symmetry scheme leads to the existence of massless scalar bosons (called Nambu–Goldstone bosons). One would then be concerned with the following unpalatable prospect of a theory with local

[31] The relation between the interaction range and the mass of the mediating particle is discussed in Section 6.4.2.

[32] Important contribtutions were made by P.W. Anderson (1923–), Y. Nambu (1921–), J. Goldstone (1933–), S. Weinberg (1933–), J. Schwinger, P.W. Higgs (1929–), and many others.

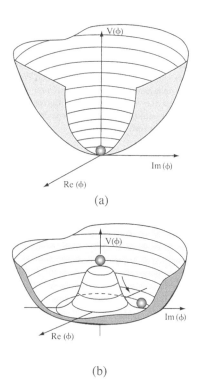

Fig. 16.4   The potential energy function $V(\phi)$ illustrates the occurrence of spontaneous symmetry breaking. (a) Normal symmetry realization: the ground state is a symmetry singlet. (b) A case of hidden symmetry: the ground state is a set of degenerate states— the circle at the bottom of the wine-bottle shaped energy surface; the selection of the true vacuum as one point in this circle breaks the symmetry. The small ball indicates the location of the physical ground state.

symmetry: not only do we have the unwanted massless vector gauge bosons, we also have these unwanted massless scalar bosons. As it turns out, in the realization of SSB in gauge theories (**the Higgs mechanism**) each of these two ills is the cure of the other. The massless Goldstone scalar combines with the (two states of) a massless vector boson to form (the three states of) a massive vector boson. In the end we have a hidden gauge symmetry without any unwanted massless states. The $SU(2) \times U(1)$ electroweak gauge theory starts out in the symmetric limit with all particles (gauge bosons, leptons, and quarks) being massless. The explicit realization of the Higgs mechanism involves the introduction of a doublet of elementary (complex) scalar boson fields (ϕ^+, ϕ^0), and their dynamics is such that the ground state value of ϕ^0 is a nonvanishing constant. The vacuum is permeated with this constant scalar field. All the particles, originally massless, gain their respective masses while propagating in this vacuum. The electroweak theory has a structure such that the photon gauge particle (as well as the neutrino states) remains massless. However this scalar sector, often referred to as the Higgs sector, is less constrained by the symmetry of the theory. In particular we are free to adjust the couplings of the scalars to leptons and quarks in order to obtain their respective observed masses. (Namely the Standard Model does not predict the lepton and quark masses.) Of the complex doublet (ϕ^+, ϕ^0) we have four independent scalar bosons; while $\phi^\pm$ and one of $\phi^{0\prime}s$ are "eaten" by the gauge bosons to make three massive vector bosons of $W^\pm$ and Z^0, the remaining scalar boson is a real massive spin-0 particle. This 'Higgs boson' with characteristic couplings (to leptons, quarks, photons, and other intermediate vector bosons) should be an observable signature of the SSB feature of the electroweak theory.

The development of electroweak gauge theory

The Glashow–Weinberg–Salam model Many have contributed to the development of the gauge theory of electroweak interactions. We mention some milestones. Sheldon Glashow (1932–) was the first one in 1957 to write down an $SU(2) \times U(1)$ gauge theory and also made major contributions later on in building a consistent quark sector of the theory. However in the original Glashow model, the vector boson masses were introduced by hand, hence it was not a self-consistent quantum field theory. In 1967 Steven Weinberg formulated an electroweak gauge theory of leptons with gauge bosons and electron masses generated by the Higgs mechanism. At about the same time Abdus Salam (1926–96) presented an electroweak gauge theory with spontaneous symmetry breaking as well, although not in a formal journal publication. Their results did not generate great enthusiasm right away in the physics community because the quantization[33] and renormalization[34] of nonabelian gauge theories were still being worked out in those years.

Yang–Mill theories are renormalizable with or without SSB For almost two decades (1950s and 1960s), there was in fact a great deal of pessimism in the physics community that quantum field theory could be the proper framework for the study of strong and weak interactions. The strong interaction did not appear to have a small coupling and its field equation could not be solved

[33]The quantization of Yang–Mills theory, because of its many redundant degrees of freedom, is highly nontrivial. Its consistent program was finally achieved through the work of many, by Bryce DeWitt (1923–2004), R.P. Feynman, Ludvig Faddeev (1934–) and Victor Popov (1937–94), *et al.*

[34]One of the important steps in the renormalization is the implementation of the regularization procedure that renders the divergent integrals finite so that the calculation is well defined for further mathematical manipulations. The renormalizability of a theory with symmetry depends critically on the cancellation of divergences as enforced by symmetry relations. The **dimensional regularization scheme**, by going to a lower spacetime dimension, makes theory finite without violating its symmetry properties. This elegant procedure was invented independently by several groups, among them G. 't Hooft (1946–) and M.J.G. Veltman (1931–).

by the only known method: perturbation theory. Without knowing their solutions, one did not know how to test such theories. While the weak interaction had features of a gauge interaction and the perturbation should be applicable, it was generally thought that quantum theory with massive vector bosons was not renormalizable. It is in this light that one must appreciate the result obtained in 1971 by Gerard 't Hooft, a student of Martinus Veltman, proving that Yang–Mills theory was renormalizable, with or without spontaneous symmetry breaking. The significance of this achievement was appreciated very quickly by their worldwide physics colleagues. This transformed the whole field of theoretical particle physics and brought about the renaissance of quantum field theory in the 1970s.

We discussed QCD before electroweak theory, because QCD, without the need of a hidden symmetry, is a simpler gauge theory to present. Historically the nonabelian gauge theory for the weak interaction was successfully developed first. Politzer, Gross, and Wilczek then proved that Yang–Mills theory, and only Yang–Mills theory, has the property of asymptotic freedom. That allowed the QCD quantum field theory of strong interactions at the short-distance regime to be solved perturbatively and tested experimentally.

16.5.5   The Standard Model and beyond

The Standard Model of particle interactions (Table 16.1) is a gauge theory based on the symmetry group of $SU(3) \times SU(2) \times U(1)$. QCD is the $SU(3)$ gauge theory for the strong interaction. The $SU(2) \times U(1)$ gauge theory with spontaneous symmetry breaking describes the electroweak interaction. Even though their coupling strengths are the same, the weak interaction appears to be much weaker than the electromagnetic force because its effects are usually suppressed by the large masses of the $W^{\pm}$ and Z bosons.

Grand unified gauge theories

The Standard Model has been remarkably successful in its confrontation with experiment tests. Nevertheless it does not explain why the three generations of leptons and quarks have the same charge and representation assignments.[35] Furthermore, the theory must be specified by 18 parameters: three gauge couplings, the SSB energy scale (which fixes the vector boson masses), three lepton masses, as well as six masses and four angles of a complex mixing matrix of the quarks, and, finally, the Higgs boson mass. The consensus is that the Standard

[35]The theoretical structures for each of the three generations (e, ν_e, u, d), (μ, ν_μ, c, s), and (τ, ν_τ, t, b) in the Standard Model are identical.

Table 16.1 Gauge symmetry, gauge bosons, and gauge couplings of the Standard Model.

interactions	sym group	vector gauge fields	partial unification
Electromagnetic	$U(1)$	photon A_μ	$SU(2) \times U(1)$
Weak	$SU(2)$	weak vector bosons $W_\mu^{\pm}$, Z_μ	electroweak, e and θ_w
Strong	$SU(3)$	gluons G_μ^a, $a = 1, 2, .. 8$	g_s

Model is only a low-energy effective theory of some more fundamental theory with an intrinsic energy scale much higher than the electroweak scale.

As a first step going beyond the Standard Model, people have explored the possibility of 'grand unification' of strong, weak, and electromagnetic interactions in the framework of larger groups that are 'simple' (with only one gauge coupling) that contain $SU(3) \times SU(2) \times U(1)$ as their subgroup. Namely, in the Standard Model the three interactions are still described by three separate gauge groups with distinctive coupling strengths. In a more unified simple gauge group there is only one coupling strength—truly one gauge interaction. This unified strength at some very high 'grand unified' energy scale Λ_{GU} can evolve into the distinctive couplings of strong, weak, and electromagnetic couplings at a lower energy scale if there is another spontaneous symmetry breaking taking place at Λ_{GU} with all gauge bosons other than those belonging to the $SU(3) \times SU(2) \times U(1)$ group gaining masses $O(\Lambda_{\text{GU}})$. The decoupling of these heavy particles implies that the subgroup couplings g_1, g_2, and g_3 will evolve differently below the Λ_{GU} scale, giving rise to the observed different interaction strengths for the strong, weak, and electromagnetic forces observed in our more familiar low-energy scales (see Fig. 16.5).

Successful GUTs such as the gauge theory based on $SU(5)$ have been constructed; they can explain (as the coupling unification discussed above) why the strong interaction is strong and why the weak interaction is weak. Moreover all the quark lepton gauge charges can be understood based on a simple assignment of GUT charges for these fermions.[36] Their description can in fact be improved upon with the introduction of supersymmetry; in particular the precise coupling unification discussed above can come about only by the inclusion of supersymmetric particles. This program is still a work in progress; it very much needs guidance from experimental discoveries. In this connection, we comment below on the distinction that Einstein made between constructive theories versus theories of principle.

The Standard Model as a constructive theory and as a theory of principle

Abraham Pais in his Einstein biography wrote[37]

... a distinction that Einstein liked to make between two kinds of physical theories. Most theories, according to Einstein, are constructive, they interpret complex phenomena in terms of relatively simple propositions. An example is the kinetic theory of gases, in which the mechanical, thermal, and diffusional properties of gases are reduced to molecular interactions and motions ... then there are the theories of principle, which use the analytic rather than the synthetic method ... An example is the impossibility of a perpetuum mobile in thermodynamics. Then Einstein went on to say, 'The theory of relativity is a theory of principle'.

We would like to suggest that the Standard Model of elementary particle interactions is a good example of a theory that is **both** a constructive theory **and** a theory of principle.

The discovery of the quark and lepton as the basic constituents of matter, and that of the symmetry groups of $SU(3)$ and $SU(2) \times U(1)$, followed the

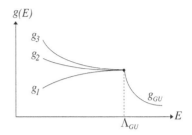

Fig. 16.5 Running coupling strengths as a function of energy. Coupling constant unification occurs at some super-high energy ($\Lambda_{\text{GU}} = 10^{16}$ GeV?). Spontaneous symmetry breaking of the unification gauge group with a single gauge coupling g_{GU} will cause the gauge couplings of subgroup $SU(3) \times SU(2) \times U(1)$ to evolve differently towards lower-energy regimes, giving rise to the different interaction strengths as observed of the strong, weak, and electromagnetic forces.

[36]That two-component fermions form the fundamental representations of the Lorentz group provides us with a natural explanation of parity violation by fundamental interactions. That QCD turns out to be parity conserving is explained by the GUT charge assignment which just leads to the same $SU(3)$ color charges for the left-handed and right-handed quarks.

[37]See Pais (1982, p. 27), based on Einstein's letter to his gymnasium teacher H. Friedmann, March 18, 1929.

practice of a constructive theory with its trial-and-error theoretical propositions followed by the experimental checks.

Einstein and the Standard Model   Einstein did not participate directly in the construction of the Standard Model as described above. Also, the Standard Model is an example of a quantum field theory, which Einstein never accepted as an acceptable theoretical framework. However the influence of his idea has been of paramount importance in the successful creation of the Standard Model. Besides the fundamental importance of special relativity, photons, and Bose–Einstein statistics to particle physics, the use of local symmetry to generate dynamics (the gauge principle) is very much in the spirit of Einstein's theory of principle as represented in particular by his masterful deployment of the invariance principle, and the equivalence principle. This approach of utilizing an overarching principle in the search of the new patterns in Nature will become even more relevant as we explore physical realms that are ever more inaccessible to direct experimentation.

The Kaluza–Klein theory and extra dimensions

17

- Einstein famously spent the latter half of his physics working life on his program of unified field theories. His conviction was that a unification program not only could combine his GR gravitational field theory with Maxwell's equations but also shed light on the quantum mystery. In this chapter, we discuss the Kaluza–Klein (KK) theory which, in many ways, is a shining example of Einstein's unification program. It has stayed relevant even for physics research in the twenty-first century. However, the Kaluza–Klein theory uses the more conventional quantum idea and does not illuminate its origin as Einstein had envisioned for a unified theory.

- The Kaluza–Klein theory is a GR field theory in a spacetime with an extra spatial dimension. The nonobservation of the extra fifth dimension is assumed to result from its compact size. (The extra dimension is curled up.) The theory not only achieves a unification of gravitation with electrodynamics but also suggests a possible interpretation of the charge space and gauge symmetry as reflecting the existence of this compactified extra dimension.

- By way of a long calculation, Theodor Kaluza has shown that the 5D general relativity field equation with a particular geometry is composed of two parts, one being the Einstein field equation and the other the Maxwell equation. This remarkable discovery has been called "the Kaluza–Klein miracle". The details of this calculation are provided in the SuppMat Section 17.5. We also discuss the motivation and meaning of the assumed geometry for this 5D spacetime.

- As gauge symmetry was being developed, Oskar Klein showed that a gauge transformation could be identified with a displacement in the extra dimension coordinate in the KK theory. This went a long way in explaining the KK "miracle". Also it was demonstrated that the relativistic Klein–Gordon wave equation in KK spacetime is equivalent to a set of decoupled 4D Klein–Gordon equations for a tower of particles with increasing masses. Thus the signature of an extra dimension is the existence of a tower of KK particles, having identical spin and gauge quantum numbers, with increasing masses controlled by the compactification scale.

- In the last section brief comments are offered of the more recent efforts in the construction of unified theories with extra dimensions.

17.1   Unification of electrodynamics and gravity

17.1.1   Einstein and unified field theory

Unifying different realms of physics has always led to fresh insight into our physical world. Maxwell's and Faraday's fusion of electricity and magnetism brought new understanding of light and radiation. Einstein's motivation for new physics was often prompted by the promise of wider comprehension that a new synthesis would bring. Recall his motivation for the principle of relativity. His special relativity brought about the deep cognizance that space and time were interchangeable. The resultant insight that spacetime was the arena in which physical events took place ultimately brought about his geometric theory of gravitation—in the form of a dynamical spacetime. Concurrently extending his atomic hypothesis of matter (doctoral thesis and Brownian motion) as well to radiation, his light quantum idea, he found that the electromagnetic field could have the puzzling feature of being both wave and particle at the same time. As we have discussed in Chapter 8, while Einstein appreciated the specific successes of the new quantum mechanics, he could not believe it as an acceptable description of reality. It was in this context that Einstein had hopes of finding a unification of electrodynamics and general relativity that would also shed light on the quantum mystery. This was the driving force behind his 20-year effort in the unified field theory program.

17.1.2   A geometric unification

Einstein's accomplishment in formulating a geometric theory of gravitation naturally led him, and others, in efforts to find a geometric formulation of Maxwell's theory. As mentioned in Section 16.5.1, this was the original motivation of Hermann Weyl and it eventually led to fruition in the form of interpreting electrodynamics as a gauge interaction. If a more direct geometric formulation of electromagnetism is possible, it would perhaps make the unification with gravity more likely. As we shall see in this chapter, a geometric theory of electrodynamics was actually obtained through a unification attempt, but it is the geometry of a spacetime with an extra dimension.

While gauge symmetry bears a resemblance to general relativity of being also a local symmetry, the result still does not seem like much help in finding a unified theory. Gauge invariance being a local symmetry, not in ordinary spacetime, but in the internal charge space, the question naturally arises: what exactly is this charge space? A possible answer was found[1] in 1919 by the Prussian mathematician Theodor Kaluza (1885–1954). He suggested extending the general principle of relativity to a hypothetical 5D spacetime—the usual

[1] Kaluza's paper was sent to Einstein in 1919, but did not come out in print until two years later (Kaluza 1921).

4D spacetime augmented by an extra spatial dimension.[2] It was discovered that a particular restricted 5D space geometry would lead to a 5D general theory of relativity with a field equation composed of two parts, one being the Einstein equation and another being the Maxwell equation.

The possibility of such a 'miraculous' unification was explained by the Swedish physicist Oskar Klein (1894–1977) in the late 1920s when quantum mechanics and gauge theory were being developed. Klein showed that, in Kaluza's 5D theory, a gauge transformation had the geometric significance of being a displacement in the extra dimension. Thus charge space has the physical meaning of being the extra space dimension. If this extra dimension is compactified (so as to have avoided direct detection), quantum theory predicts the existence of a tower of particles with ever increasing masses with the mass difference controlled by the compactification length size. This unified field theory has come to be called the Kaluza–Klein (KK) theory. However, the KK theory made use of the more conventional quantum idea and did not shed light on its origin as Einstein had envisioned for the unified theory.

The Kaluza–Klein theory postulates the existence of an extra spatial dimension. Our spacetime is actually five-dimensional; it was demonstrated that electromagnetism can be viewed as part of 5D general relativity. That is, one can "derive" electrodynamics by postulating the principle of general relativity in a 5D spacetime. To prepare for the study of this embedding of electromagnetic gauge theory in a 5D general relativity, we first recall the relevant parts of gauge symmetry as well as of the GR theory.

17.1.3 A rapid review of electromagnetic gauge theory

Under a $U(1)$ gauge transformation, the 4-vector potential (regarded as the fundamental electromagnetic field) transforms as

$$A_\mu \longrightarrow A'_\mu = A_\mu + \partial_\mu \theta \qquad (17.1)$$

where $\theta(x)$ is the gauge function, cf. Eq. (16.45). The EM field intensity, being the 4-curl of the gauge field

$$F_{\mu\nu} = \partial_\mu A_\nu - \partial_\nu A_\mu, \qquad (17.2)$$

is clearly invariant under the gauge transformation of (17.1). Requiring the Lagrangian density to be a relativistic and gauge invariant scalar leads to a free Maxwell density of

$$\mathcal{L}_{EM} = -\frac{1}{4} F_{\mu\nu} F^{\mu\nu}. \qquad (17.3)$$

As discussed in Sections 16.4.2 and 16.4.3, the Euler–Lagrange equation based on this $\mathcal{L}_{EM}$ is the Maxwell equation. In this sense we say Maxwell's theory is essentially determined by (special) relativity and $U(1)$ gauge symmetry.

[2] Just about all attempts of extending spacetime dimensions involve extra spatial dimensions. The need to have the extra dimensions compactified would have to, in the case of an extra time dimension, overcome the serious causality difficulties associated with looped times.

17.1.4　A rapid review of general relativistic gravitational theory

GR equations are covariant under the general coordinate transformation that leaves invariant the spacetime interval

$$ds^2 = g_{\mu\nu} dx^\mu dx^\nu,$$

where $g_{\mu\nu}$ is the metric tensor of the 4D spacetime. Just as the metric can be interpreted as a relativistic gravitational potential, the Christoffel symbols

$$\Gamma^\lambda_{\mu\nu} = \frac{1}{2} g^{\lambda\rho} \left[\partial_\nu g_{\mu\rho} + \partial_\mu g_{\nu\rho} - \partial_\rho g_{\mu\nu} \right], \tag{17.4}$$

being the first derivative of the potential [cf. Eq. (13.6)] can be thought as the gravitational field intensities. The Riemann–Christoffel curvature tensor [cf. Eq. (14.9)]

$$R^\mu_{\lambda\alpha\beta} = \partial_\alpha \Gamma^\mu_{\lambda\beta} - \partial_\beta \Gamma^\mu_{\lambda\alpha} + \Gamma^\mu_{\nu\alpha} \Gamma^\nu_{\lambda\beta} - \Gamma^\mu_{\nu\beta} \Gamma^\nu_{\lambda\alpha}, \tag{17.5}$$

being the nonlinear second derivatives of the metric, is the relativistic tidal forces, its contracted version enters into the GR field equation (14.35)

$$R_{\mu\nu} - \frac{1}{2} R g_{\mu\nu} = \kappa T_{\mu\nu} \tag{17.6}$$

where the Ricci tensor $R_{\mu\nu}$ and scalar R are contractions of the Riemann curvature

$$R_{\mu\nu} \equiv g^{\alpha\beta} R_{\alpha\mu\beta\nu} \quad \text{and} \quad R \equiv g^{\alpha\beta} R_{\alpha\beta} \tag{17.7}$$

and $T_{\mu\nu}$ is the energy–momentum tensor for an external source. κ is proportional to Newton's constant.

The Einstein–Hilbert action

Just as Maxwell's equation can be compactly presented as the Euler–Lagrangian equation resulting from the variation of the Maxwell action, the Einstein equation (17.6) is similarly related to the GR Lagrangian density, the Ricci scalar:

$$\mathcal{L}_g = R, \tag{17.8}$$

for the source-free case (Hilbert 1915, Einstein 1916d). Since only the product $\sqrt{-g} d^4 x$ (where g is the determinant of the metric tensor $g_{\mu\nu}$) is invariant under the general coordinate transformation,[3] the relevant action, called the **Einstein–Hilbert action**, is the 4D integral

$$I_g = \int \sqrt{-g} d^4 x \mathcal{L}_g = \int \sqrt{-g} d^4 x g^{\mu\nu} R_{\mu\nu}. \tag{17.9}$$

We can then derive Eq. (17.6) as the Euler–Lagrange equation from the minimization of this action. The variation of the action δI_g has three parts involving $\delta R_{\mu\nu}$, $\delta g^{\mu\nu}$, and $\delta \sqrt{-g}$. The integral containing the $\delta R_{\mu\nu}$ factor after an integration by parts turns into a vanishing surface term; the metric matrix being symmetric[4] hence obeys the general relation $\ln(\det g_{\mu\nu}) = Tr(\ln g_{\mu\nu})$ leading

[3] For such more advanced GR topics, see, for example, Carroll (2004).

[4] A symmetric matrix M can always be diagonalized by a similarity transformation $SMS^\mathsf{T} = M_d$.

to the variation of $\delta\sqrt{-g} = -\frac{1}{2}\sqrt{-g}g_{\mu\nu}\delta g^{\mu\nu}$. Consequently the variation principle requires

$$\delta I_g = \int \sqrt{-g}d^4x \left(R_{\mu\nu} - \frac{1}{2}Rg_{\mu\nu}\right)\delta g^{\mu\nu} = 0,$$

which implies the Einstein equation of (17.6). In this sense we can interpret the general coordinate-invariant Ricci scalar R as representing the (source-free) 4D gravitational theory.

17.2 General relativity in 5D spacetime

In this section we show how a particular version of the 5D spacetime metric leads to the Ricci scalar being the sum of Lagrangian densities of Einstein's gravity theory and Maxwell's electromagnetism.

17.2.1 Extra spatial dimension and the Kaluza–Klein metric

One can motivate the geometric unification by the observation that while we have the 4-potential A_μ in electromagnetism, the spacetime metric $g_{\mu\nu}$ is the relativistic gravitational potential. One would like to combine these two types of potentials into one mathematical entity.[5]

Kaluza starts out by postulating a spacetime with an extra spatial dimension[6]

$$\hat{x}^M = \left(x^0, x^1, x^2, x^3, x^5\right). \tag{17.10}$$

However, the metric $\hat{g}_{MN}$ for this 5D spacetime is assumed to have a particular structure $\hat{g}_{MN} = \hat{g}_{MN}^{(kk)}$ having its elements related to the 4D $g_{\mu\nu}$ and A^μ as,

$$\hat{g}_{\mu\nu}^{(kk)} = g_{\mu\nu} + A_\mu A_\nu, \quad \hat{g}_{\mu 5}^{(kk)} = \hat{g}_{5\mu}^{(kk)} = A_\mu, \quad \hat{g}_{55}^{(kk)} = 1. \tag{17.11}$$

When displayed in 5×5 matrix form, we have

$$\hat{g}_{MN}^{(kk)} \equiv \begin{pmatrix} \hat{g}_{\mu\nu}^{(kk)} & \hat{g}_{\mu 5}^{(kk)} \\ \hat{g}_{5\nu}^{(kk)} & \hat{g}_{55}^{(kk)} \end{pmatrix} = \begin{pmatrix} g_{\mu\nu} + A_\mu A_\nu & A_\mu \\ A_\nu & 1 \end{pmatrix}. \tag{17.12}$$

Equivalently, the corresponding invariant interval in this 5D spacetime can be written as

$$ds_{(kk)}^2 = \hat{g}_{MN}^{(kk)}d\hat{x}^M d\hat{x}^N = g_{\mu\nu}dx^\mu dx^\nu + (dx^5 + A_\lambda dx^\lambda)^2. \tag{17.13}$$

We should note the particular feature that, with $g_{\mu\nu}$ and A^μ being functions of the 4D coordinate x^μ, all the elements of the 5D metric $\hat{g}_{MN}^{(kk)}$ have no dependence on the extra dimensional coordinate $\hat{x}^5$ and we also set $\hat{g}_{55}^{(kk)} = 1$. From this point on we shall drop the cumbersome superscript label (kk) and the relation $\hat{g}_{MN} = \hat{g}_{MN}^{(kk)}$ is always understood.

One can also check that the 5D inverse metric $\hat{g}^{MN}$ must have the components of

$$\hat{g}^{\mu\nu} = g^{\mu\nu}, \quad \hat{g}^{\mu 5} = \hat{g}^{5\mu} = -A^\mu, \quad \hat{g}^{55} = 1 + A^\nu A_\nu, \tag{17.14}$$

[5]Further discussion can be found in Section 17.3.1.

[6]5D quantities will be denoted with a caret symbol ^. The capital Latin index $M = (\mu, 5) = (0, 1, 2, 3, 5)$ is for a 5D spacetime, with the Greek index μ for the usual 4D spacetime and the index 5 for the extra spatial dimension. Our system skips the index 4, so as not to be confused with another common practice of labeling the 4D spacetime by the indices $(1, 2, 3, 4)$ with the fourth index being the time coordinate. In our system the time component continues to be denoted by the zeroth index.

or,

$$\hat{g}^{MN} = \begin{pmatrix} g^{\mu\nu} & -A^{\mu} \\ -A^{\nu} & 1 + A^{\lambda}A_{\lambda} \end{pmatrix} \tag{17.15}$$

so that a simple matrix multiplication can check out the correct metric relation $\hat{g}^{MN}\hat{g}_{MN} = \delta^{M}_{K}$.

17.2.2 "The Kaluza–Klein miracle"

From the 5D metric, we can obtain the other curved spacetime quantities by the usual relations. The **5D Christoffel symbols**, cf. Eq. (17.4), are first-order derivatives of the 5D metric

$$\hat{\Gamma}^{M}_{NL} = \frac{1}{2}\hat{g}^{MK}\left(\partial_N \hat{g}_{LK} + \partial_L \hat{g}_{NK} - \partial_K \hat{g}_{NL}\right). \tag{17.16}$$

The 5D **Riemann curvature tensor** is a nonlinear derivative of the 5D **Christoffel symbols**, cf. Eq. (17.5):

$$\hat{R}^{L}_{MSN} = \partial_S \hat{\Gamma}^{L}_{MN} - \partial_N \hat{\Gamma}^{L}_{MS} + \hat{\Gamma}^{L}_{ST}\hat{\Gamma}^{T}_{MN} - \hat{\Gamma}^{L}_{NT}\hat{\Gamma}^{T}_{MS}. \tag{17.17}$$

Contracting the pair of indices (L, S), we obtain the 5D **Ricci tensor**, cf. Eq. (14.23),

$$\hat{R}_{MN} = \hat{R}^{L}_{MLN} = \partial_L \hat{\Gamma}^{L}_{MN} - \partial_N \hat{\Gamma}^{L}_{ML} + \hat{\Gamma}^{L}_{LT}\hat{\Gamma}^{T}_{MN} - \hat{\Gamma}^{L}_{NT}\hat{\Gamma}^{T}_{ML}. \tag{17.18}$$

Contracting one more time, we obtain the **5D Ricci scalar**, cf. Eq. (14.25),

$$\hat{R} = \hat{g}^{MN}\hat{R}_{MN}. \tag{17.19}$$

Given Kaluza's stipulation of the 5D metric, it is a straightforward, but rather tedious, task to calculate the 5D Christoffel symbols, in terms of the familiar 4D metric tensor and 4D electromagnetic potential, and then all the other 5D geometric quantities as listed in Eqs. (17.16)–(17.19). After an enormous calculation,[7] Kaluza obtained the remarkable result that the 5D Ricci scalar $\hat{R}$, which should be the Lagrangian density of a 5D general theory of relativity, is simply the sum of the Lagrangian densities of the 4D general relativity R and Maxwell's electromagnetism, $-\frac{1}{4}F_{\mu\nu}F^{\mu\nu}$ where $F_{\mu\nu}$ is the Maxwell field tensor (17.2):

[7] The details of calculating the 5D Ricci scalar $\hat{R}$ in term of $g_{\mu\nu}$ and A^{μ} are provided in SuppMat Section 17.5.

$$\hat{R} = R - \frac{1}{4}F_{\mu\nu}F^{\mu\nu}, \tag{17.20}$$

namely,

$$\mathcal{L}^{(5)}_{g} = \mathcal{L}^{(4)}_{g} + \mathcal{L}^{(4)}_{EM}. \tag{17.21}$$

[8] It can be similarly shown that the 4D Lorentz force law follows from the 5D geodesic equation (the GR equation of motion).

The Einstein and Maxwell equations are all components of the 5D GR field equation.[8] In this rather "miraculous" way a geometric unification of gravitation and electromagnetism is indicated.

17.3 The physics of the Kaluza–Klein spacetime

As indicated above, once we have the KK metric (17.11), the unification of gravitation and electromagnetism follows by a straightforward calculation. Thus the whole unification program relies on the structure of the KK metric. What is the physics behind the KK metric ansatz?

17.3.1 Motivating the Kaluza–Klein metric ansatz

The amazing unification results having their origin in the KK prescription (17.11) for the metric $\hat{g}_{MN}^{(kk)}$; it may be worthwhile to motivate the algebra that can lead one to this metric ansatz.

The metric $g_{\mu\nu}$ being the gravitational potential and comparable to the EM potential A_μ, one would like to fit both of them into the 5D $\hat{g}_{MN}$. What should be the precise identification of the metric elements? Similarly, can one fit the Christoffel symbols $\Gamma_{\nu\lambda}^\mu$ and EM field tensor $F_{\mu\nu}$ (both being first derivatives of the potentials) into $\hat{\Gamma}_{NL}^M$? Or, after lowering the upper index in (17.16),

$$\hat{\Gamma}_{M\,NL} \equiv \hat{g}_{MJ}\hat{\Gamma}_{NL}^J = \frac{1}{2}\left(\partial_N \hat{g}_{LM} + \partial_L \hat{g}_{NM} - \partial_M \hat{g}_{NL}\right). \tag{17.22}$$

Out of the 50 elements, we will concentrate on the set with the indices $M = \mu$, $L = 5$, and $N = \nu$ in (17.22) as a possible match for the EM field intensity $F_{\mu\nu}$:

$$\hat{\Gamma}_{\mu\,5\nu} = \frac{1}{2}\left(\partial_\nu \hat{g}_{5\mu} + \partial_5 \hat{g}_{\mu\nu} - \partial_\mu \hat{g}_{\nu 5}\right). \tag{17.23}$$

This suggests the identification with $-\frac{1}{2}F_{\mu\nu}$ if the 5D metric elements actually do not depend on the x^5 coordinate so that the middle term vanishes, $\partial_5 \hat{g}_{\mu\nu} = 0$, and if $\hat{g}_{\mu 5} = \hat{g}_{5\mu} = A_\mu$. With the further simplifying assumption of $\hat{g}_{55} = 1$ and ds^2, Kaluza ends up trying the ansatz of (17.11), hence (17.13).

17.3.2 Gauge transformation as a 5D coordinate change

The invariant interval $ds_{(kk)}^2 = \hat{g}_{MN}^{(kk)}d\hat{x}^M d\hat{x}^N$, with the metric $\hat{g}_{MN}^{(kk)}$ not being the most general 5D metric tensor, will not be invariant under the most general coordinate transformation in the 5D spacetime $\hat{x}^M \to \hat{x}'^M$. However, $ds_{(kk)}^2$, as we shall show, is unchanged under a subset of coordinate transformations holding the 4D coordinates fixed and a local displacement of the extra dimensional coordinate:

$$x^\mu \to x'^\mu = x^\mu \quad \text{and} \quad x^5 \to x'^5 = x^5 + \theta(x). \tag{17.24}$$

Since we have $g_{\mu\nu}(x)$ depending on the 4D coordinate only, this leads to

$$g'_{\mu\nu}(x) = g_{\mu\nu}(x), \quad \text{and, of course,} \quad \hat{g}'_{55} = \hat{g}_{55} = 1. \tag{17.25}$$

Most interestingly, according to the general transformation rule, cf. Eq. (13.29), with $\partial \hat{x}^M / \partial x'^5 = \delta_5^M$, we have

$$\hat{g}'_{5\mu} = \frac{\partial \hat{x}^M}{\partial x'^5}\frac{\partial \hat{x}^N}{\partial x'^\mu}\hat{g}_{MN} = \frac{\partial \hat{x}^N}{\partial x'^\mu}\hat{g}_{5N}$$

$$= \frac{\partial x^\nu}{\partial x'^\mu}\hat{g}_{5\nu} + \frac{\partial x^5}{\partial x'^\mu}\hat{g}_{55} = \delta_\mu^\nu \hat{g}_{5\nu} - \frac{\partial \theta}{\partial x^\mu}. \tag{17.26}$$

With the identification of $\hat{g}_{5\mu} = A_\mu$, this is just the gauge transformation of Eq. (17.1). It is then an easy exercise to check that the KK interval of (17.13) is unchanged under this restricted 5D coordinate transformation (17.25) and (17.26).

Recall the discussion in Chapter 16 and reviewed in Section 17.1 that electromagnetism can to a large extent be determined by the $U(1)$ gauge symmetry. As the $U(1)$ transformation $e^{i\theta(x)}$ is equivalent to a rotation by an angle θ, we can have the literal realization of the gauge transformation as a displacement along a circle if the x^5 coordinate is compactified into a circle. (See further discussion below.) That we can interpret gauge transformations as coordinate transformations in an extra dimension goes a long way in explaining why Maxwell's theory is embedded in this higher dimensional GR theory.

17.3.3   Compactified extra dimension

We have explained that the 5D coordinate transformation that leaves the KK interval invariant is a gauge transformation. Still, why does one restrict the metric to this $\hat{g}_{MN}^{(kk)}$ form?

The key feature of the KK metric is that its elements are independent of the extra dimensional coordinate x^5. This is rather strange: one postulates a spacetime with extra dimension $(x^0, x^1, x^2, x^3, x^5)$ yet the fields are not allowed to depend on the extra coordinate x^5!

Kaluza, the mathematician, is silent on the physical reality of the extra dimension; but Klein, the physicist, proposes that the fifth dimension is real. It has not been observed because it is extremely small; the extra dimension is curled up. Just like a garden hose is viewed at a distance as a 1D line, upon closer inspection one finds a surface composed of a series of circles (Fig. 17.1). Here one of the two dimensions of the surface is compactified into circles. In the same manner, Klein proposes that one spatial dimension of our 5D spacetime is compactified: every point in the observed 3D spatial space is actually a circle.

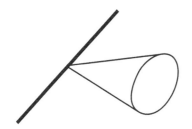

Fig. 17.1 Compactified dimension. A 1D line is revealed to be a 2D surface with one dimension compactified into a circle.

17.3.4   Quantum fields in a compactified space

The Kaluza–Klein "miracle" was discovered by Kaluza, and was explained by Klein, who used the then new quantum mechanics to deduce the consequence of a compactified dimension. This also justifies the restrictions imposed on the 5D metric (Klein 1926).

Consider, as the simplest case, a scalar field $\phi(\hat{x}^M) = \phi(x^\mu, x^5)$ satisfying a 5D relativistic wave equation (the Klein–Gordon equation):

$$\left(\Box^{(5)} - \frac{m_0^2 c^2}{\hbar^2} \right) \phi(\hat{x}^N) = 0 \qquad (17.27)$$

[9]Cf. Eq. (11.25).

where $\Box^{(5)}$ is the five-dimensional D'Alembertian operator[9] $\Box^{(5)} = \Box^{(4)} + \partial^2/\partial x_5^2$. Since the extra dimension is a circle (with compactification radius a), we have the identification of x^5 and $x^5 + 2\pi a$. Thus the wavefunction $\phi(\hat{x}^M)$ must satisfying the boundary condition of

$$\phi(x^\mu, x^5) = \phi(x^\mu, x^5 + 2\pi a). \tag{17.28}$$

The field must have a sinusoidal dependence on the x^5 coordinate, and has the harmonic expansion

$$\phi(x^\mu, x^5) = \sum_n \phi_n(x^\mu) e^{ip_n x^5/\hbar}. \tag{17.29}$$

In order that the boundary condition (17.28) is satisfied, the momentum in the extra dimension must be quantized[10]

$$p_n = n\frac{\hbar}{a} \quad \text{with} \quad n = 0, 1, 2, \ldots \tag{17.30}$$

[10]Recall the similar problem of "a particle in a box" in quantum mechanics.

A tower of Kaluza–Klein particles in 4D spacetime

To see the implications in the familiar 4D spacetime, we can write out the 5D Klein–Gordon equation in the 4D spacetime

$$\left(\Box^{(5)} - \frac{m_0^2 c^2}{\hbar^2}\right)\phi(\hat{x}^N) = \left(\Box^{(4)} + \partial_5^2 - \frac{m_0^2 c^2}{\hbar^2}\right)\phi(\hat{x}^N) = 0,$$

which, after substituting in the series expansion, becomes

$$\sum_n \left[\left(\Box^{(4)} - \frac{n^2}{a^2} - \frac{m_0^2 c^2}{\hbar^2}\right)\phi(x^\mu)\right] e^{ip_n x^5/\hbar} = 0.$$

Namely, we have an infinite number of decoupled 4D Klein–Gordon equations

$$\left(\Box^{(4)} - \frac{m_n^2 c^2}{\hbar^2}\right)\phi(x^\mu) = 0, \tag{17.31}$$

with a tower of "Kaluza–Klein states" having masses

$$m_n^2 = m_0^2 + n^2\frac{\hbar^2}{a^2 c^2}. \tag{17.32}$$

Thus the signature of the extra dimension in 4D spacetime is a tower of KK particles, having identical spin and gauge quantum numbers, with increasing masses controlled by the compactification scale of a.

Compactification by quantum gravity?

The natural expectation is that the compactification is brought about by the dynamics of quantum gravity. In this way the compactification radius should be the order of the Planck length, discussed in Section 3.3.2. That the first KK state has a mass of at least 10^{19} GeV would mean that such particles would not be detectable in the foreseeable future. Nevertheless, the "decoupling" of the large KK state masses do explain the basic structure of the KK metric ansatz—the x^5 independence of the metric elements.

Since only the $n = 0$ state is physically relevant, we have from Eqs. (17.30) and (17.29) the approximation

$$\phi(x^\mu, x^5) = \phi_0(x^\mu) \tag{17.33}$$

and the x^5 dependence of the theory disappears.

17.4   Further theoretical developments

We have presented the Kaluza–Klein theory as an illustration of Einstein's unified field theory. While it is an example of unification of fundamental forces, it certainly does not have a bearing on Einstein's loftier goal of a unified theory that would explain the mystery of quantum physics. In fact KK theory makes use of conventional quantum mechanics in extracting physical consequences of a compactified spatial dimension.

Extending the original Kaluza–Klein theory

Scalar–tensor gravity theory   Even restricted to the original theory, there is no strong theoretical argument for setting the metric element $g_{55} = 1$. A more natural alternative is to replace it by a field. This would lead to a scalar–tensor theory of gravity, for which there is no experimental support.

From a circle to Calabi–Yau space   From a modern perspective, the KK theory cannot be a complete unified theory because the list of fundamental forces must be expanded beyond the gravitational and electromagnetic interactions: it must at least include the strong and weak particle interactions.[11] Nevertheless, the modern development of particle physics has led to the discovery of superstring theory as a possible quantum gravity theory that has the potential to unify all fundamental forces. What is most relevant for our discussion here is the finding that the self-consistency requirement of superstring theory requires a spacetime to have 10 dimensions. Thus what is needed is not just one extra dimension curled into a very small circle but six extra dimensions into a more complicated geometric entity. A much discussed compactification scheme is the Calabi–Yau space.[12] In short, the spirit of Einstein's unification program, especially in the form of Kaluza–Klein extra dimensions, is being carried on in the foremost theoretical physics research of the twenty-first century.

Speculations of a large extra dimension   As the current thinking of unification theory is directly related to the quest for quantum gravity, the natural unification distance scale is thought to be the Planck length, which is something like nine orders of magnitude smaller than the highest accelerator can probe. This makes an experimental test of such theories extremely difficult. Yet researchers have found the whole idea of extra dimensions so attractive that there have been serious speculations on the possibility that the compactification scale is much larger than the Planck size. Maybe the extra dimension is on the electroweak scale that can possibly be revealed in experiments being performed at the Large Hadron Collider. These are intriguing speculations that are being actively pursued.

17.4.1   Lessons from Maxwell's equations

In this book we have repeatedly discussed the importance of Maxwell's equations. We will conclude our presentation by recalling the important lessons that we have learnt from the structure of these equations:

[11] In this connection, we note that Oskar Klein in the late 1930s constructed a 5D theory that attempted to include not only gravity and electromagnetism, but also Yukawa-meson-mediated nuclear forces. Although he did not explicitly consider any nonabelian gauge symmetry, his theory had foreshadowed the later development of Yang–Mills theory, in particular, charged gauge bosons, etc. He presented these results in a 1938 conference held in Warsaw, but never published them formally. For an appreciation of this 1938 contribution see Gross (1995).

[12] For a comprehensive and nontechnical discussion, see Yau and Nadis (2010).

- The idea of the photon and quantum theory, brought forth through a deep statistical thermodynamic study of electromagnetic radiation.
- Einstein's principle of relativity, that taught us that the arena of physics is 4D spacetime; this paved the way for a geometric understanding of gravitation.
- The idea of local symmetry in the charge space, leading eventually to the viewpoint that all fundamental interactions have a connection with gauge symmetry.
- Finally, the possibility that spacetime has extra dimensions. This may be the origin of the charge space.

17.4.2 Einstein and mathematics

"Mathematics is the language of physics." Such a statement implies a rather passive role for mathematics. In fact this language has often led the way in opening up new understanding in physics. This is one of the powerful lessons that one gathers from the history of theoretical physics; this account of Einstein's physics, we believe, confirms this opinion too. Clearly Einstein's discovery that Riemannian geometry offers a truer description of nature is a brilliant example of such a role. One aspect of Einstein's scientific biography can be viewed as the story of his growing appreciation of the role of mathematics: starting from his skepticism of higher mathematics, doubting the usefulness of Minkowski's geometric formulation of special relativity to the role that Riemannian geometry played in the implementation of the general principle of relativity—first learning the new mathematics with the help of Marcel Grossmann and the eventual discovery of the GR field equation after much struggle on his own. In the process, Einstein became greatly appreciative of the role of mathematics as fundamental in setting up new physical theory. Finding the correct mathematical structure to describe the physical concepts, and postulating the simplest equation compatible with that structure, are all key elements in the invention of new theories.

That Einstein had not made more advances in his unified field theory program may also imply that this great physicist was after all not a comparably great mathematician (like the case of Newton). One probably needs to invent new mathematics in order to forge progress in such a pursuit.[13] It is also interesting to observe the important contributions that mathematicians have made in furthering Einstein's vision: Hermann Weyl's gauge symmetry program and Theodor Kaluza's 5D GR leading to the 'Kaluza–Klein miracle'.

[13] In this connection see the comments made by Roger Penrose in the new *Forward* he wrote in the 2005 re-issue of Pais (1982).

17.5 SuppMat: Calculating the 5D tensors

Here we provide the details of calculating the 5D tensors in terms of the gravitational potential $g_{\mu\nu}$ and electromagnetic potential A_μ—all based on the KK metric prescription given by (17.11).

17.5.1   The 5D Christoffel symbols

The 5D Christoffel symbols $\hat{\Gamma}^M_{NL}$ are first-order derivatives of the 5D metric (17.16). They have six distinctive types of terms:

$$\hat{\Gamma}^\mu_{\nu\lambda}, \quad \hat{\Gamma}^5_{\nu\lambda}, \quad \hat{\Gamma}^\mu_{5\lambda}, \quad \hat{\Gamma}^5_{5\lambda}, \quad \hat{\Gamma}^\mu_{55}, \quad \hat{\Gamma}^5_{55}.$$

We shall calculate them in terms of the 4D metric $g_{\mu\nu}$, the electromagnetic potential A^μ, and the field tensor $F_{\mu\nu} = \partial_\mu A_\nu - \partial_\nu A_\mu$.

The components $\hat{\Gamma}^\mu_{\nu\lambda}$

According to (17.16), with $M = \mu, N = \nu, L = \lambda$, we have

$$\hat{\Gamma}^\mu_{\nu\lambda} = \frac{1}{2}\hat{g}^{\mu K}\left(\partial_\nu \hat{g}_{\lambda K} + \partial_\lambda \hat{g}_{\nu K} - \partial_K \hat{g}_{\nu\lambda}\right)$$

$$= \frac{1}{2}\hat{g}^{\mu\rho}\left(\partial_\nu \hat{g}_{\lambda\rho} + \partial_\lambda \hat{g}_{\nu\rho} - \partial_\rho \hat{g}_{\nu\lambda}\right)$$

$$+ \frac{1}{2}\hat{g}^{\mu 5}\left(\partial_\nu \hat{g}_{\lambda 5} + \partial_\lambda \hat{g}_{\nu 5} - \partial_5 \hat{g}_{\nu\lambda}\right). \tag{17.34}$$

Plugging in the Kaluza metric components of (17.11) and (17.14), we find the first term on the RHS is just the 4D Christoffel symbols $\Gamma^\mu_{\nu\lambda}$ with an extra term coming from that fact that $\hat{g}_{\mu\nu} - g_{\mu\nu} = A_\mu A_\nu$, while in the second term we have $\partial_5 \hat{g}_{\nu\lambda} = 0$ because the Kaluza metric elements are independent of the extra coordinate x_5.

$$\hat{\Gamma}^\mu_{\nu\lambda} = \Gamma^\mu_{\nu\lambda} + \frac{1}{2}g^{\mu\rho}\left[\partial_\nu\left(A_\lambda A_\rho\right) + \partial_\lambda\left(A_\nu A_\rho\right) - \partial_\rho\left(A_\nu A_\lambda\right)\right]$$

$$- \frac{1}{2}A^\mu\left(\partial_\nu A_\lambda + \partial_\lambda A_\nu\right). \tag{17.35}$$

The first two terms in the square bracket on the RHS can be written out as

$$\frac{1}{2}g^{\mu\rho}\left[\partial_\nu\left(A_\lambda A_\rho\right) + \partial_\lambda\left(A_\nu A_\rho\right)\right]$$

$$= \frac{1}{2}g^{\mu\rho}\left[\left(\partial_\nu A_\lambda\right)A_\rho + A_\lambda\left(\partial_\nu A_\rho\right) + \left(\partial_\lambda A_\nu\right)A_\rho + A_\nu\left(\partial_\lambda A_\rho\right)\right]$$

$$= \frac{1}{2}\left[\left(\partial_\nu A_\lambda\right)A^\mu + g^{\mu\rho}A_\lambda\left(\partial_\nu A_\rho\right) + \left(\partial_\lambda A_\nu\right)A^\mu + g^{\mu\rho}A_\nu\left(\partial_\lambda A_\rho\right)\right]. \tag{17.36}$$

The first and third terms in this last square bracket just cancel the last term on the RHS of (17.35). On the other hand the last term in the square bracket on the RHS of (17.35), when expanded,

$$-\frac{1}{2}g^{\mu\rho}\partial_\rho\left(A_\nu A_\lambda\right) = -\frac{1}{2}g^{\mu\rho}\left[\left(\partial_\rho A_\nu\right)A_\lambda + A_\nu\left(\partial_\rho A_\lambda\right)\right] \tag{17.37}$$

can be combined with the second and fourth terms on the RHS of (17.36) to yield

$$\frac{1}{2}g^{\mu\rho}\left[\left(\partial_\nu A_\rho - \partial_\rho A_\nu\right)A_\lambda + \left(\partial_\lambda A_\rho - \partial_\rho A_\lambda\right)A_\nu\right] = -\frac{1}{2}g^{\mu\rho}\left(F_{\rho\nu}A_\lambda + F_{\rho\lambda}A_\nu\right).$$

All this leads to

$$\hat{\Gamma}^{\mu}_{\nu\lambda} = \Gamma^{\mu}_{\nu\lambda} - \frac{1}{2}g^{\mu\rho}\left(F_{\rho\nu}A_{\lambda} + F_{\rho\lambda}A_{\nu}\right). \tag{17.38}$$

Later on we shall also need the Christoffel symbols $\hat{\Gamma}^{\mu}_{\nu\lambda}$ with a pair of indices summed over:

$$\hat{\Gamma}^{\mu}_{\nu\mu} = \Gamma^{\mu}_{\nu\mu} - \frac{1}{2}g^{\mu\rho}\left(F_{\rho\nu}A_{\mu} + F_{\mu\rho}A_{\nu}\right).$$

The very last factor vanishes because of the opposite symmetry properties of the two tensors: $g^{\mu\rho}F_{\mu\rho} = 0$. Expanding out $\Gamma^{\mu}_{\nu\mu}$, we then have

$$\hat{\Gamma}^{\mu}_{\nu\mu} = \frac{1}{2}g^{\mu\rho}\left(\partial_{\nu}g_{\mu\rho} + \partial_{\mu}g_{\nu\rho} - \partial_{\rho}g_{\nu\mu} - F_{\rho\nu}A_{\mu}\right).$$

The two middle terms cancel, $g^{\mu\rho}\left(\partial_{\mu}g_{\nu\rho} - \partial_{\rho}g_{\nu\mu}\right) = 0$, and this leads to

$$\hat{\Gamma}^{\mu}_{\nu\mu} = \frac{1}{2}g^{\mu\rho}\partial_{\nu}g_{\mu\rho} - \frac{1}{2}A_{\mu}F^{\mu}_{\nu}. \tag{17.39}$$

The components $\hat{\Gamma}^{5}_{\nu\lambda}$

According to (17.16), with $M = 5, N = \nu, L = \lambda$,

$$\hat{\Gamma}^{5}_{\nu\lambda} = \frac{1}{2}\hat{g}^{5K}\left(\partial_{\nu}\hat{g}_{\lambda K} + \partial_{\lambda}\hat{g}_{\nu K} - \partial_{K}\hat{g}_{\nu\lambda}\right)$$

$$= \frac{1}{2}\hat{g}^{5\rho}\left(\partial_{\nu}\hat{g}_{\lambda\rho} + \partial_{\lambda}\hat{g}_{\nu\rho} - \partial_{\rho}\hat{g}_{\nu\lambda}\right)$$

$$+ \frac{1}{2}\hat{g}^{55}\left(\partial_{\nu}\hat{g}_{\lambda 5} + \partial_{\lambda}\hat{g}_{\nu 5} - \partial_{5}\hat{g}_{\nu\lambda}\right). \tag{17.40}$$

Plugging in the Kaluza metric components of (17.11) and (17.14), and noting $\partial_{5}\hat{g}_{\nu\lambda} = 0$, we have

$$\hat{\Gamma}^{5}_{\nu\lambda} = -A^{\rho}\Gamma_{\rho\nu\lambda} - \frac{1}{2}A^{\rho}\left[\partial_{\nu}\left(A_{\lambda}A_{\rho}\right) + \partial_{\lambda}\left(A_{\nu}A_{\rho}\right) - \partial_{\rho}\left(A_{\lambda}A_{\nu}\right)\right]$$

$$+ \frac{1}{2}\left(1 + A^{\sigma}A_{\sigma}\right)\left(\partial_{\nu}A_{\lambda} + \partial_{\lambda}A_{\nu}\right). \tag{17.41}$$

The first two terms in the square bracket can be written out as

$$-\frac{1}{2}A^{\rho}\left[\partial_{\nu}\left(A_{\lambda}A_{\rho}\right) + \partial_{\lambda}\left(A_{\nu}A_{\rho}\right)\right]$$

$$= \frac{1}{2}\left[-A^{\rho}A_{\rho}\left(\partial_{\nu}A_{\lambda}\right) - A^{\rho}\left(\partial_{\nu}A_{\rho}\right)A_{\lambda} - A^{\rho}A_{\rho}\left(\partial_{\lambda}A_{\nu}\right) - A^{\rho}\left(\partial_{\lambda}A_{\rho}\right)A_{\nu}\right].$$

We note that the first and third terms on the RHS just cancel the last two terms on the RHS of (17.41). The remaining second and fourth terms combine with the last term in the square bracket of (17.41)

$$-\frac{1}{2}A^\rho \left[(\partial_\nu A_\rho)A_\lambda + (\partial_\lambda A_\rho)A_\nu - \partial_\rho (A_\lambda A_\nu) \right]$$

$$-\frac{1}{2}A^\rho \left[(\partial_\nu A_\rho)A_\lambda - (\partial_\rho A_\nu)A_\lambda + (\partial_\lambda A_\rho)A_\nu - (\partial_\rho A_\lambda)A_\nu \right]$$

$$= -\frac{1}{2}A^\rho \left[F_{\nu\rho}A_\lambda + F_{\lambda\rho}A_\nu \right]. \tag{17.42}$$

Equation (17.41) then becomes

$$\hat{\Gamma}^5_{\nu\lambda} = -A^\rho \Gamma_{\rho\nu\lambda} - \frac{1}{2}A^\rho \left[F_{\nu\rho}A_\lambda + F_{\lambda\rho}A_\nu \right] + \frac{1}{2}B_{\nu\lambda} \tag{17.43}$$

where

$$B_{\nu\lambda} = \partial_\nu A_\lambda + \partial_\lambda A_\nu. \tag{17.44}$$

The components $\hat{\Gamma}^\mu_{5\lambda}$

According to (17.16), with $M = \mu, N = 5, L = \lambda$,

$$\hat{\Gamma}^\mu_{5\lambda} = \frac{1}{2}\hat{g}^{\mu K} \left(\partial_5 \hat{g}_{\lambda K} + \partial_\lambda \hat{g}_{5K} - \partial_K \hat{g}_{5\lambda} \right). \tag{17.45}$$

With $\partial_5 \hat{g}_{\lambda K} = 0$ and separating the K-index summation into a $K = \rho$ sum and a $K = 5$ sum:

$$\hat{\Gamma}^\mu_{5\lambda} = \frac{1}{2}\hat{g}^{\mu\rho} \left(\partial_\lambda \hat{g}_{5\rho} - \partial_\rho \hat{g}_{5\lambda} \right) + \frac{1}{2}\hat{g}^{\mu 5} \left(\partial_\lambda \hat{g}_{55} - \partial_5 \hat{g}_{5\lambda} \right). \tag{17.46}$$

Again we have $\partial_5 \hat{g}_{5\lambda} = 0$ and as $\hat{g}_{55} = 1$ so that $\partial_\lambda \hat{g}_{55} = 0$;

$$\hat{\Gamma}^\mu_{5\lambda} = \frac{1}{2}g^{\mu\rho} \left(\partial_\lambda A_\rho - \partial_\rho A_\lambda \right) = \frac{1}{2}g^{\mu\rho} F_{\lambda\rho} = -\frac{1}{2}F^\mu_\lambda. \tag{17.47}$$

Recall from Section 17.3.1 that it is this simple relation that has motivated the original ansatz for the KK metric. Furthermore we note that the sum over the indices μ and λ in $\hat{\Gamma}^\mu_{5\lambda}$ vanishes because $g^{\mu\rho}$ is symmetric and $F_{\mu\rho}$ is antisymmetric:

$$\hat{\Gamma}^\mu_{5\mu} = \frac{1}{2}g^{\mu\rho} F_{\mu\rho} = 0. \tag{17.48}$$

The components $\hat{\Gamma}^5_{5\lambda}$

Following the same steps leading to (17.47), we now calculate the $M = 5$, $L = \lambda$, and $\mu = 5$ element:

$$\hat{\Gamma}^5_{5\lambda} = \frac{1}{2}g^{5\rho} \left(\partial_\lambda A_\rho - \partial_\rho A_\lambda \right) = -\frac{1}{2}F_{\lambda\rho}A^\rho. \tag{17.49}$$

The components $\hat{\Gamma}^\mu_{55}$ and $\hat{\Gamma}^5_{55}$

According to (17.46), with $\lambda = 5$, we have

$$\hat{\Gamma}^\mu_{55} = \frac{1}{2}\hat{g}^{\mu\rho} \left(\partial_5 \hat{g}_{5\rho} - \partial_\rho \hat{g}_{55} \right) = 0 \tag{17.50}$$

for the same reason that the last factor in (17.46) vanishes.

Similarly, with $\lambda = 5$, Eq. (17.49) becomes

$$\hat{\Gamma}^5_{55} = \frac{1}{2} g^{5\rho} \left(\partial_5 A_\rho - \partial_\rho \hat{g}_{55} \right) = 0. \tag{17.51}$$

Collecting all the $\hat{\Gamma}^M_{NL}$ components in one place, we have from (17.38), (17.43), (17.47), (17.49), (17.50), and (17.51):

$$\hat{\Gamma}^\mu_{\nu\lambda} = \Gamma^\mu_{\nu\lambda} - \frac{1}{2} \left(F^\mu_\nu A_\lambda + F^\mu_\lambda A_\nu \right)$$

$$\hat{\Gamma}^5_{\nu\lambda} = -A_\rho \Gamma^\rho_{\nu\lambda} - \frac{1}{2} A^\rho \left(F_{\nu\rho} A_\lambda + F_{\lambda\rho} A_\nu \right) + \frac{1}{2} B_{\nu\lambda}$$

$$\hat{\Gamma}^\mu_{5\lambda} = -\frac{1}{2} F^\mu_\lambda \tag{17.52}$$

$$\hat{\Gamma}^5_{5\lambda} = -\frac{1}{2} F_{\lambda\rho} A^\rho$$

$$\hat{\Gamma}^\mu_{55} = \hat{\Gamma}^5_{55} = 0.$$

17.5.2 The 5D Ricci tensor components

Knowing the Christoffel symbols, we are ready to calculate the Ricci tensor $\hat{R}_{MN}$ according to Eq (17.18).

The 5D Ricci tensor components $\hat{R}_{\mu\nu}$

There are two pairs of repeated indices, L and T in (17.18). We will now consider the separate cases when they take on 4D values of $L = \lambda$ and $T = \tau$, or the extra dimensional value of 5.

1. $L = \lambda$ and $T = \tau$:

$$\left(\hat{R}_{\mu\nu} \right)_1 = \underbrace{\partial_\lambda \hat{\Gamma}^\lambda_{\mu\nu}}_{(1)} - \underbrace{\partial_\nu \hat{\Gamma}^\lambda_{\mu\lambda}}_{(2)} + \underbrace{\hat{\Gamma}^\lambda_{\lambda\tau} \hat{\Gamma}^\tau_{\mu\nu}}_{(3)} - \underbrace{\hat{\Gamma}^\lambda_{\nu\tau} \hat{\Gamma}^\tau_{\mu\lambda}}_{(4)}. \tag{17.53}$$

2. $L = 5$ and $T = \tau$:

$$\left(\hat{R}_{\mu\nu} \right)_2 = \underbrace{\partial_5 \hat{\Gamma}^5_{\mu\nu} - \partial_\nu \hat{\Gamma}^5_{\mu5}}_{(5)} + \underbrace{\hat{\Gamma}^5_{5\tau} \hat{\Gamma}^\tau_{\mu\nu}}_{(6)} - \underbrace{\hat{\Gamma}^5_{\nu\tau} \hat{\Gamma}^\tau_{\mu5}}_{(7)}. \tag{17.54}$$

3. $L = \lambda$ and $T = 5$:

$$\left(\hat{R}_{\mu\nu} \right)_3 = \hat{\Gamma}^\lambda_{\lambda5} \hat{\Gamma}^5_{\mu\nu} - \underbrace{\hat{\Gamma}^\lambda_{\nu5} \hat{\Gamma}^5_{\mu\lambda}}_{(7)}. \tag{17.55}$$

4. $L = T = 5$:

$$\left(\hat{R}_{\mu\nu} \right)_4 = \hat{\Gamma}^5_{55} \hat{\Gamma}^5_{\mu\nu} - \underbrace{\hat{\Gamma}^5_{\nu5} \hat{\Gamma}^5_{\mu5}}_{(8)}. \tag{17.56}$$

Out of the 12 terms on the RHS, three terms vanish: besides the absence of the x^5 dependence $\partial_5 \hat{\Gamma}^5_{\mu\nu} = \hat{\Gamma}^5_{55} \hat{\Gamma}^5_{\mu\nu} = 0$, we also have $\hat{\Gamma}^\lambda_{\lambda5} \hat{\Gamma}^5_{\mu\nu} = 0$ because, according to (17.52), $\hat{\Gamma}^\lambda_{\lambda 5} = -\frac{1}{2}F^\lambda_\lambda = 0$. Furthermore, the last term in (17.55) may be written as

$$-\hat{\Gamma}^\lambda_{\nu 5}\hat{\Gamma}^5_{\mu\lambda} = -\hat{\Gamma}^5_{\mu\lambda}\hat{\Gamma}^\lambda_{\nu 5} = -\hat{\Gamma}^5_{\mu\tau}\hat{\Gamma}^\tau_{\nu 5}$$

where we reach the last expression by changing the labels of the dummy indices from λ to τ. This makes it clear that this term is really the same as the last term in (17.54); they are symmetrized with respect to the indices (μ, ν) which is justified because the Ricci tensor $\hat{R}_{\mu\nu}$ must be symmetric. As a result, altogether we have eight distinctive terms—four from (17.53), three from (17.54), and one from (17.56).

In the following we shall substitute into these eight terms the expression of $\hat{\Gamma}$ as given in (17.52). Keep in mind that, in the first four terms, term 1 to term 4, the 4D Γs in the 5D $\hat{\Gamma}$ just combine to form the 4D Ricci tensor $R_{\mu\nu}$. For the remaining terms, we have

(1) The $\partial_\lambda\hat{\Gamma}^\lambda_{\mu\nu}$ term:

$$\partial_\lambda\hat{\Gamma}^\lambda_{\mu\nu} = -\frac{1}{2}\partial_\lambda\left(F^\lambda_\mu A_\nu + F^\lambda_\nu A_\mu\right)$$

$$= -\frac{1}{2}\left(A_\nu\partial_\lambda F^\lambda_\mu + A_\mu\partial_\lambda F^\lambda_\nu + F^\lambda_\mu\partial_\lambda A_\nu + F^\lambda_\nu\partial_\lambda A_\mu\right). \quad (17.57)$$

(2) The $-\partial_\nu\hat{\Gamma}^\lambda_{\mu\lambda}$ term:

$$-\partial_\nu\hat{\Gamma}^\lambda_{\mu\lambda} = \frac{1}{2}\partial_\nu\left(F^\lambda_\mu A_\lambda + F^\lambda_\lambda A_\mu\right) = \frac{1}{2}\left(A_\lambda\partial_\nu F^\lambda_\mu + F^\lambda_\mu\partial_\nu A_\lambda\right). \quad (17.58)$$

(3) The $\hat{\Gamma}^\lambda_{\lambda\tau}\hat{\Gamma}^\tau_{\mu\nu}$ term:

$$\hat{\Gamma}^\lambda_{\lambda\tau}\hat{\Gamma}^\tau_{\mu\nu} = -\frac{1}{2}\left(F^\lambda_\lambda A_\tau + F^\lambda_\tau A_\lambda\right)\Gamma^\tau_{\mu\nu} - \frac{1}{2}\Gamma^\lambda_{\lambda\tau}\left(F^\tau_\mu A_\nu + F^\tau_\nu A_\mu\right)$$

$$+ \frac{1}{4}\left(F^\lambda_\lambda A_\tau + F^\lambda_\tau A_\lambda\right)\left(F^\tau_\mu A_\nu + F^\tau_\nu A_\mu\right)$$

$$= -\frac{1}{2}A_\lambda\Gamma^\tau_{\mu\nu}F^\lambda_\tau - \frac{1}{2}A_\nu\Gamma^\lambda_{\lambda\tau}F^\tau_\mu - \frac{1}{2}A_\mu\Gamma^\lambda_{\lambda\tau}F^\tau_\nu$$

$$+ \frac{1}{4}F^\lambda_\tau F^\tau_\mu A_\lambda A_\nu + \frac{1}{4}F^\lambda_\tau F^\tau_\nu A_\lambda A_\mu. \quad (17.59)$$

(4) The $-\hat{\Gamma}^\lambda_{\nu\tau}\hat{\Gamma}^\tau_{\mu\lambda}$ term:

$$-\hat{\Gamma}^\lambda_{\nu\tau}\hat{\Gamma}^\tau_{\mu\lambda} = \frac{1}{2}\left(F^\lambda_\nu A_\tau + F^\lambda_\tau A_\nu\right)\Gamma^\tau_{\mu\lambda} + \frac{1}{2}\Gamma^\lambda_{\nu\tau}\left(F^\tau_\mu A_\lambda + F^\tau_\lambda A_\mu\right)$$

$$- \frac{1}{4}\left(F^\lambda_\nu A_\tau + F^\lambda_\tau A_\nu\right)\left(F^\tau_\mu A_\lambda + F^\tau_\lambda A_\mu\right)$$

$$= \frac{1}{2}\left(A_\tau\Gamma^\tau_{\mu\lambda}F^\lambda_\nu + A_\nu\Gamma^\tau_{\mu\lambda}F^\tau_\lambda + A_\lambda\Gamma^\lambda_{\nu\tau}F^\tau_\mu + A_\mu\Gamma^\lambda_{\nu\tau}F^\tau_\lambda\right)$$

$$- \frac{1}{4}\left(F^\lambda_\nu F^\tau_\mu A_\tau A_\lambda + F^\lambda_\tau F^\tau_\mu A_\nu A_\lambda + F^\lambda_\tau F^\tau_\lambda A_\nu A_\mu + F^\lambda_\nu F^\tau_\lambda A_\tau A_\mu\right).$$

$$(17.60)$$

(5) The $-\partial_\nu \hat{\Gamma}^5_{\mu 5}$ term:

$$-\partial_\nu \hat{\Gamma}^5_{\mu 5} = \frac{1}{2}\partial_\nu\left(F_{\mu\tau}A^\tau\right) = \frac{1}{2}A^\tau\partial_\nu F_{\mu\tau} + \frac{1}{2}F_{\mu\tau}\partial_\nu A^\tau. \qquad (17.61)$$

(6) The $\hat{\Gamma}^5_{5\tau}\hat{\Gamma}^\tau_{\mu\nu}$ term:

$$\hat{\Gamma}^5_{5\tau}\hat{\Gamma}^\tau_{\mu\nu} = -\frac{1}{2}F_{\tau\rho}A^\rho\Gamma^\tau_{\mu\nu} + \frac{1}{4}F_{\tau\rho}A^\rho\left(F^\tau_\mu A_\nu + F^\tau_\nu A_\mu\right)$$

$$= -\frac{1}{2}A^\lambda\Gamma^\tau_{\mu\nu}F_{\tau\lambda} - \frac{1}{4}F_{\lambda\tau}F^\tau_\mu A^\lambda A_\nu - \frac{1}{4}F_{\lambda\tau}F^\tau_\nu A^\lambda A_\mu. \qquad (17.62)$$

(7) The $-2\hat{\Gamma}^5_{\nu\tau}\hat{\Gamma}^\tau_{\mu 5}$ term:

$$-2\hat{\Gamma}^5_{\nu\tau}\hat{\Gamma}^\tau_{\mu 5} = -A_\rho\Gamma^\rho_{\nu\tau}F^\tau_\mu - \frac{1}{2}A^\rho\left(F_{\nu\rho}A_\tau + F_{\tau\rho}A_\nu\right)F^\tau_\mu + \frac{1}{2}B_{\nu\tau}F^\tau_\mu$$

$$= -A_\rho\Gamma^\rho_{\nu\tau}F^\tau_\mu - \frac{1}{2}F_{\nu\rho}F^\tau_\mu A^\rho A_\tau$$

$$- \frac{1}{2}F_{\tau\rho}F^\tau_\mu A^\rho A_\nu + \frac{1}{2}F^\tau_\mu B_{\nu\tau}. \qquad (17.63)$$

(8) The $-\hat{\Gamma}^5_{\nu 5}\hat{\Gamma}^5_{\mu 5}$ term:

$$-\hat{\Gamma}^5_{\nu 5}\hat{\Gamma}^5_{\mu 5} = -\frac{1}{4}F_{\nu\rho}F_{\mu\lambda}A^\rho A^\lambda. \qquad (17.64)$$

Collecting all terms of the type $F\partial A$: the last two terms in (17.57), one from (17.58), one from (17.61), and one from (17.63), we have

$$-\frac{1}{2}F^\lambda_\mu\partial_\lambda A_\nu - \frac{1}{2}F^\lambda_\nu\partial_\lambda A_\mu + \frac{1}{2}F^\lambda_\mu\partial_\nu A_\lambda + \frac{1}{2}F_{\mu\tau}\partial_\nu A^\tau + \frac{1}{2}F^\tau_\mu B_{\nu\tau}. \qquad (17.65)$$

The fourth term may be rewritten as follows

$$\frac{1}{2}F_{\mu\tau}\partial_\nu A^\tau = -\frac{1}{2}F_{\tau\mu}\partial_\nu A^\tau = -\frac{1}{2}F^\lambda_\mu\partial_\nu A_\lambda$$

which just cancels the third term in (17.65). For the remaining terms, after using (17.44), we have

$$-\frac{1}{2}F^\lambda_\mu\partial_\lambda A_\nu - \frac{1}{2}F^\lambda_\nu\partial_\lambda A_\mu + \frac{1}{2}F^\lambda_\mu\partial_\nu A_\lambda + \frac{1}{2}F^\lambda_\mu\partial_\lambda A_\nu$$

$$= -\frac{1}{2}F^\lambda_\nu\partial_\lambda A_\mu + \frac{1}{2}F^\lambda_\mu\partial_\nu A_\lambda$$

$$= -\frac{1}{2}F^\lambda_\nu\partial_\lambda A_\mu + \frac{1}{2}F^\lambda_\nu\partial_\mu A_\lambda - \frac{1}{2}F^\lambda_\nu\partial_\mu A_\lambda + \frac{1}{2}F^\lambda_\mu\partial_\nu A_\lambda$$

$$= -\frac{1}{2}F^\lambda_\nu F_{\lambda\mu} \qquad (17.66)$$

where the final result comes from the combination of the first two terms; we drop the last two terms because they are antisymmetric with respect to (μ, ν) as the Ricci tensor must be symmetric.

Collecting all terms of the type *FFAA*: the last two terms in (17.59), the last four from (17.60), the last two from (17.62), two from (17.63), and one from (17.64), we have (not displaying the common 1/4 coefficients)

$$F_\tau^\lambda F_\mu^\tau A_\lambda A_\nu + F_\tau^\lambda F_\nu^\tau A_\lambda A_\mu - F_\nu^\lambda F_\mu^\tau A_\tau A_\lambda - F_\tau^\lambda F_\mu^\tau A_\nu A_\lambda$$

$$-F_\tau^\lambda F_\lambda^\tau A_\nu A_\mu - F_\nu^\lambda F_\lambda^\tau A_\tau A_\mu - F_{\lambda\tau} F_\mu^\tau A^\lambda A_\nu - F_{\lambda\tau} F_\nu^\tau A^\lambda A_\mu$$

$$-2F_{\nu\rho} F_\mu^\tau A^\rho A_\tau - 2F_{\tau\rho} F_\mu^\tau A^\rho A_\nu - F_{\nu\rho} F_{\mu\lambda} A^\rho A^\lambda. \tag{17.67}$$

Grouping similar terms

$$F_\tau^\lambda F_\mu^\tau A_\lambda A_\nu - F_\tau^\lambda F_\mu^\tau A_\nu A_\lambda - F_{\lambda\tau} F_\mu^\tau A^\lambda A_\nu - 2F_{\tau\rho} F_\mu^\tau A^\rho A_\nu$$

$$= -F_{\lambda\tau} F_\mu^\tau A^\lambda A_\nu + 2F_{\lambda\tau} F_\mu^\tau A^\lambda A_\nu = F_{\lambda\tau} F_\mu^\tau A^\lambda A_\nu, \tag{17.68}$$

and

$$F_\tau^\lambda F_\nu^\tau A_\lambda A_\mu - F_\nu^\lambda F_\lambda^\tau A_\tau A_\mu - F_{\lambda\tau} F_\nu^\tau A^\lambda A_\mu = -F_{\lambda\tau} F_\nu^\tau A^\lambda A_\mu, \tag{17.69}$$

again we note that the results from (17.68) and (17.69) form an antisymmetric combination in (μ, ν), hence can be dropped. After eliminating seven out of the 11 terms in (17.67) we have four left, with three of them mutually canceling:

$$-F_\nu^\lambda F_\mu^\tau A_\tau A_\lambda - 2F_{\nu\rho} F_\mu^\tau A^\rho A_\tau - F_{\nu\rho} F_{\mu\lambda} A^\rho A^\lambda$$

$$= -F_\nu^\lambda F_\mu^\tau A_\tau A_\lambda + 2F_\nu^\tau F_\mu^\lambda A_\tau A_\lambda - F_\nu^\tau F_\mu^\lambda A_\tau A_\lambda = 0.$$

The only nonvanishing term from (17.67) is (and putting back the 1/4 factor)

$$-\frac{1}{4} F_\tau^\lambda F_\lambda^\tau A_\nu A_\mu = \frac{1}{4} F_{\lambda\tau} F^{\lambda\tau} A_\nu A_\mu. \tag{17.70}$$

Collecting the remaining terms from (17.57)–(17.64), we have (not displaying the common $-1/2$ coefficients)

$$A_\nu \partial_\lambda F_\mu^\lambda + A_\mu \partial_\lambda F_\nu^\lambda \overleftarrow{- A_\lambda \partial_\nu F_\mu^\lambda} + \underbrace{A_\lambda \Gamma_{\mu\nu}^\tau F_\tau^\lambda} + A_\nu \Gamma_{\lambda\tau}^\lambda F_\mu^\tau$$

$$+ \underline{A_\mu \Gamma_{\lambda\tau}^\lambda F_\nu^\tau} - A_\tau \Gamma_{\mu\lambda}^\tau F_\nu^\lambda - A_\nu \Gamma_{\mu\lambda}^\tau F_\tau^\lambda - \underline{A_\lambda \Gamma_{\nu\tau}^\lambda F_\mu^\tau}$$

$$- A_\mu \Gamma_{\nu\tau}^\lambda F_\lambda^\tau \overleftarrow{- A^\tau \partial_\nu F_{\mu\tau}} + \underbrace{A^\lambda \Gamma_{\mu\nu}^\tau F_{\tau\lambda}} + 2A_\rho \Gamma_{\nu\tau}^\rho F_\mu^\tau. \tag{17.71}$$

The terms with the same underlines in (17.71) mutually cancel; we are left with six terms that can be grouped into two combinations (putting back the $-1/2$ coefficients):

$$-\frac{1}{2}\left(A_\nu \partial_\lambda F_\mu^\lambda - A_\nu \Gamma_{\mu\lambda}^\tau F_\tau^\lambda + A_\nu \Gamma_{\lambda\tau}^\lambda F_\mu^\tau \right) = -\frac{1}{2} A_\nu D_\lambda F_\mu^\lambda \tag{17.72}$$

and

$$-\frac{1}{2}\left(A_\mu \partial_\lambda F_\nu^\lambda - A_\mu \Gamma_{\nu\tau}^\lambda F_\lambda^\tau + A_\mu \Gamma_{\lambda\tau}^\lambda F_\nu^\tau \right) = -\frac{1}{2} A_\mu D_\lambda F_\nu^\lambda \tag{17.73}$$

where D_λ is the covariant derivative.[14]

[14]Recall from Eq. (13.43) that the covariant derivative of a tensor with multiple indices (i.e. tensor of higher rank) has a Christoffel factor for each index.

Collecting the results from (17.66), (17.70), (17.72), and (17.73) we have, after putting back the $R_{\mu\nu}$ discussed just before our calculation of the factors (1)–(8),

$$\hat{R}_{\mu\nu} = R_{\mu\nu} - \frac{1}{2}F^\lambda_\nu F_{\lambda\mu} + \frac{1}{4}F_{\lambda\tau}F^{\lambda\tau}A_\nu A_\mu - \frac{1}{2}\left(A_\nu D_\lambda F^\lambda_\mu + A_\mu D_\lambda F^\lambda_\nu\right). \quad (17.74)$$

The 5D Ricci tensor components $\hat{R}_{5\nu}$ and $\hat{R}_{55}$

Equation (17.18) with $M = 5$ and $N = \nu$ can now be written as

$$\hat{R}_{5\nu} = \partial_L \hat{\Gamma}^L_{5\nu} - \partial_\nu \hat{\Gamma}^L_{5L} + \hat{\Gamma}^L_{LT}\hat{\Gamma}^T_{5\nu} - \hat{\Gamma}^L_{\nu T}\hat{\Gamma}^T_{5L}. \quad (17.75)$$

There are two pairs of repeated indices, L and T. We will now consider the separate cases when they take on 4D values of $L = \lambda$ and $T = \tau$, or the extra dimensional index value of 5.

1. $L = \lambda$ and $T = \tau$:

$$\left(\hat{R}_{5\nu}\right)_1 = \partial_\lambda \hat{\Gamma}^\lambda_{5\nu} - \underline{\partial_\nu \hat{\Gamma}^\lambda_{5\lambda}} + \hat{\Gamma}^\lambda_{\lambda\tau}\hat{\Gamma}^\tau_{5\nu} - \hat{\Gamma}^\lambda_{\nu\tau}\hat{\Gamma}^\tau_{5\lambda}. \quad (17.76)$$

2. $L = 5$ and $T = \tau$:

$$\left(\hat{R}_{5\nu}\right)_2 = \underline{\partial_5 \hat{\Gamma}^5_{5\nu}} - \underline{\partial_\nu \hat{\Gamma}^5_{55}} + \underbrace{\hat{\Gamma}^5_{5\tau}\hat{\Gamma}^\tau_{5\nu}} - \underline{\hat{\Gamma}^5_{\nu\tau}\hat{\Gamma}^\tau_{55}}. \quad (17.77)$$

3. $L = \lambda$ and $T = 5$:

$$\left(\hat{R}_{5\nu}\right)_3 = \underline{\hat{\Gamma}^\lambda_{\lambda 5}\hat{\Gamma}^5_{5\nu}} - \underbrace{\hat{\Gamma}^\lambda_{\nu 5}\hat{\Gamma}^5_{5\lambda}}. \quad (17.78)$$

4. $L = T = 5$:

$$\left(\hat{R}_{5\nu}\right)_4 = \underline{\hat{\Gamma}^5_{55}\hat{\Gamma}^5_{5\nu}} - \underline{\hat{\Gamma}^5_{\nu 5}\hat{\Gamma}^5_{55}}. \quad (17.79)$$

Those with straight underlines will individually vanish by themselves, while the two with braces under them cancel each other. We are left with

$$\hat{R}_{5\nu} = \partial_\lambda \hat{\Gamma}^\lambda_{5\nu} + \hat{\Gamma}^\lambda_{\lambda\tau}\hat{\Gamma}^\tau_{5\nu} - \hat{\Gamma}^\lambda_{\nu\tau}\hat{\Gamma}^\tau_{5\lambda}. \quad (17.80)$$

According to (17.52), $\hat{\Gamma}^\lambda_{5\nu} = -\frac{1}{2}F^\lambda_\nu$ and $\hat{\Gamma}^\lambda_{\nu\tau} = \Gamma^\lambda_{\nu\tau} - \frac{1}{2}\left(F^\lambda_\nu A_\tau + F^\lambda_\tau A_\nu\right)$ so that $\hat{\Gamma}^\lambda_{\nu\lambda} = \Gamma^\lambda_{\nu\lambda} - \frac{1}{2}\left(F^\lambda_\nu A_\lambda + F^\lambda_\lambda A_\nu\right) = \Gamma^\lambda_{\nu\lambda} - \frac{1}{2}F^\lambda_\nu A_\lambda$, and we have, when adding up all the nonvanishing terms from (17.76) to (17.79),

$$\hat{R}_{5\nu} = -\frac{1}{2}\left(\partial_\lambda F^\lambda_\nu + \Gamma^\lambda_{\lambda\tau}F^\tau_\nu - \Gamma^\lambda_{\nu\tau}F^\tau_\lambda\right)$$

$$+ \frac{1}{4}A_\lambda F^\lambda_\tau F^\tau_\nu - \frac{1}{4}\left(F^\lambda_\nu A_\tau + F^\lambda_\tau A_\nu\right)F^\tau_\lambda$$

$$= -\frac{1}{2}D_\lambda F^\lambda_\nu - \frac{1}{4}A_\nu F^\lambda_\tau F^\tau_\lambda \quad (17.81)$$

because two middle terms cancel each other.

Finally, we need to calculate $\hat{R}_{55}$. Just setting $\nu = 5$ in (17.80) we have

$$\hat{R}_{55} = \partial_\lambda \hat{\Gamma}^\lambda_{55} + \hat{\Gamma}^\lambda_{\lambda\tau}\hat{\Gamma}^\tau_{55} - \hat{\Gamma}^\lambda_{5\tau}\hat{\Gamma}^\tau_{5\lambda}. \tag{17.82}$$

Since $\hat{\Gamma}^\lambda_{55} = 0$, we obtain the simple expression

$$\hat{R}_{55} = -\frac{1}{4}F^\lambda_\tau F^\tau_\lambda = +\frac{1}{4}F_{\mu\nu}F^{\mu\nu}. \tag{17.83}$$

17.5.3   From 5D Ricci tensor to 5D Ricci scalar

Separating out the 5D spacetime index $M = (\mu, 5)$ into the 4D plus the extra dimensional indices, the Ricci scalar of (17.19) is seen to be composed of three terms:

$$\hat{R} = \hat{g}^{MN}\hat{R}_{MN} = \hat{g}^{\mu\nu}\hat{R}_{\mu\nu} + 2\hat{g}^{5\nu}\hat{R}_{5\nu} + \hat{g}^{55}\hat{R}_{55}. \tag{17.84}$$

Collecting all the $\hat{R}_{MN}$ components in one place, we have from (17.74), (17.81), and (17.83):

$$\hat{R}_{\mu\nu} = R_{\mu\nu} - \frac{1}{2}F^\lambda_\nu F_{\lambda\mu} + \frac{1}{4}F_{\lambda\tau}F^{\lambda\tau}A_\nu A_\mu - \frac{1}{2}\left(A_\nu D_\lambda F^\lambda_\mu + A_\mu D_\lambda F^\lambda_\nu\right)$$

$$\hat{R}_{5\nu} = -\frac{1}{2}D_\lambda F^\lambda_\nu - \frac{1}{4}A_\nu F^\lambda_\tau F^\tau_\lambda \tag{17.85}$$

$$\hat{R}_{55} = +\frac{1}{4}F_{\mu\nu}F^{\mu\nu}.$$

From Eqs. (17.84) and (17.85) and the inverse metric elements of (17.14), we have for the 5D Ricci scalar

$$\hat{R} = \hat{g}^{\mu\nu}\hat{R}_{\mu\nu} + 2\hat{g}^{5\nu}\hat{R}_{5\nu} + \hat{g}^{55}\hat{R}_{55}$$

$$= g^{\mu\nu}\left[R_{\mu\nu} - \frac{1}{2}F^\lambda_\nu F_{\lambda\mu} + \frac{1}{4}F_{\lambda\tau}F^{\lambda\tau}A_\nu A_\mu - \frac{1}{2}\left(A_\nu D_\lambda F^\lambda_\mu + A_\mu D_\lambda F^\lambda_\nu\right)\right]$$

$$+ A^\nu D_\lambda F^\lambda_\nu + \frac{1}{2}A^\nu A_\nu F^\lambda_\tau F^\tau_\lambda + \frac{1}{4}\left(1 + A^\lambda A_\lambda\right)F_{\mu\nu}F^{\mu\nu}. \tag{17.86}$$

The $F^2 A^2$ terms cancel

$$\frac{1}{4}F_{\lambda\tau}F^{\lambda\tau}A_\nu A^\nu + \frac{1}{2}A^\nu A_\nu F_{\lambda\tau}F^{\tau\lambda} + \frac{1}{4}A^\lambda A_\lambda F_{\mu\nu}F^{\mu\nu} = 0. \tag{17.87}$$

The ADF terms cancel because

$$-\frac{1}{2}g^{\mu\nu}\left(A_\nu D_\lambda F^\lambda_\mu + A_\mu D_\lambda F^\lambda_\nu\right) + A^\nu D_\lambda F^\lambda_\nu$$

$$= -\frac{1}{2}A^\mu D_\lambda F^\lambda_\mu - \frac{1}{2}A^\nu D_\lambda F^\lambda_\nu + A^\nu D_\lambda F^\lambda_\nu = 0. \tag{17.88}$$

Remarkably, with the Kaluza postulate for the 5D metric $\hat{g}_{MN}$ of (17.11), the resultant 5D Ricci scalar (17.86) reduces to the simple expression of (17.20):

$$\hat{R} = R - \frac{1}{4}F_{\mu\nu}F^{\mu\nu}. \tag{17.89}$$

Some people call this the **Kaluza–Klein miracle**.

APPENDICES

Mathematics supplements

A.1 Vector calculus

A vector is a quantity with a magnitude and direction. In three-dimensional Euclidean space a vector $\mathbf{A}$ can be expanded in terms of the Cartesian basis unit vectors $(\mathbf{i}, \mathbf{j}, \mathbf{k})$

$$\mathbf{A} = A_1\mathbf{i} + A_2\mathbf{j} + A_3\mathbf{k} \tag{A.1}$$

with $\{A_i\}$, $i = 1, 2, 3$, being the coefficients of expansion. The vector magnitude is related to these components as

$$A = \sqrt{A_1^2 + A_2^2 + A_3^2}. \tag{A.2}$$

Thus a vector can be represented by its components as

$$\mathbf{A} \doteq \begin{pmatrix} A_1 \\ A_2 \\ A_3 \end{pmatrix}. \tag{A.3}$$

A vector field $\mathbf{A}(\mathbf{r}, t)$ is simply a spacetime-dependent vector. Namely, at every spacetime point it has a definite magnitude and direction. Or, in terms of components, a vector field may be represented as $A_i(\mathbf{r}, t)$. Our discussion of the differential and integral calculus of vector fields in this appendix will be particularly relevant as the electric and magnetic fields, $\mathbf{E}(\mathbf{r}, t)$ and $\mathbf{B}(\mathbf{r}, t)$, are vector fields.

A.1.1 The Kronecker delta and Levi-Civita symbols

The dot product and the Kronecker delta

The product of two vectors can be a scalar or a vector:[1] The scalar product is also called the **dot product**

$$\mathbf{A} \cdot \mathbf{B} = A_1B_1 + A_2B_2 + A_3B_3 = A_iB_i. \tag{A.4}$$

In the last expression we have used the **Einstein summation convention**, which states that when there is a pair of repeated indices (in this case the indices i) a summation is understood:

[1] We do not discuss the possibility of the product being a tensor with components T_{ij}, which, in term of vector components, are:

$$T_{ij} = A_iB_j.$$

$$A_i B_i \equiv \sum_i A_i B_i. \tag{A.5}$$

The summation can also be written in terms of the invariant tensor called the Kronecker delta δ_{ij} which has the properties that $\delta_{ij} = 0$ whenever $i \neq j$ and $\delta_{ij} = 1$ when the two indices are equal, such as $\delta_{11} = \delta_{22} = 1$.

$$\mathbf{A} \cdot \mathbf{B} = A_i B_i = \delta_{ij} A_i B_j. \tag{A.6}$$

Again, following the Einstein summation convention, we have omitted the summation signs for the index i as well for the index j.

The cross-product and Levi-Civita symbols

When the product of two vectors results in a vector, it is called a **cross-product**. Its components are related to the vector components by the Levi-Civita symbols:

$$(\mathbf{A} \times \mathbf{B})_i = \varepsilon_{ijk} A_j B_k. \tag{A.7}$$

Levi-Civita symbols are antisymmetric with respect to the interchange of any pair indices. Hence they vanish if any two indices have the same value (for example $\varepsilon_{113} = \varepsilon_{111} = 0$) while $\varepsilon_{123} = 1$ and remain as unity after any even exchanges of the indices such as $\varepsilon_{231} = \varepsilon_{312} = 1$; they involve a sign change after any odd exchanges such as $\varepsilon_{213} = \varepsilon_{132} = -1$. Thus from (A.7) we have the familiar component expression of the cross-product

$$(\mathbf{A} \times \mathbf{B})_1 = \varepsilon_{1jk} A_j B_k = A_2 B_3 - A_3 B_2$$
$$(\mathbf{A} \times \mathbf{B})_2 = \varepsilon_{2jk} A_j B_k = A_3 B_1 - A_1 B_3$$
$$(\mathbf{A} \times \mathbf{B})_3 = \varepsilon_{3jk} A_j B_k = A_1 B_2 - A_2 B_1. \tag{A.8}$$

We also have quantities involving both scalar and vector products, such as

$$\mathbf{C} \cdot (\mathbf{A} \times \mathbf{B}) = C_i (\mathbf{A} \times \mathbf{B})_i = \varepsilon_{ijk} C_i A_j B_k. \tag{A.9}$$

From this we can easily prove identities such as

$$\mathbf{C} \cdot (\mathbf{A} \times \mathbf{B}) = (\mathbf{C} \times \mathbf{A}) \cdot \mathbf{B} = -\mathbf{A} \cdot (\mathbf{C} \times \mathbf{B}). \tag{A.10}$$

A useful identity of Levi-Civita symbols

Since the only invariant constant tensors are the Levi-Civita symbols and the Kronecker delta, the product of two Levi-Civita symbols with a pair indices summed over must be in the form of the Kronecker deltas, constructed with the correct (anti)symmetry properties; the overall constant coefficient, in this case a simple unity, can easily be deduced by an examination of some particular values of the free indices:

$$\varepsilon_{ijk} \varepsilon_{lmk} = \delta_{il} \delta_{jm} - \delta_{im} \delta_{jl}, \tag{A.11}$$

leading to vector identities such as

$$(\mathbf{A} \times \mathbf{B}) \cdot (\mathbf{C} \times \mathbf{D}) = (\varepsilon_{ijk} A_i B_j)(\varepsilon_{lmk} C_l D_m)$$
$$= (\mathbf{A} \cdot \mathbf{C})(\mathbf{B} \cdot \mathbf{D}) - (\mathbf{A} \cdot \mathbf{D})(\mathbf{B} \cdot \mathbf{C}) \tag{A.12}$$

and

$$[\mathbf{C}\times(\mathbf{A} \times \mathbf{B})]_i = \varepsilon_{ijk}C_j(\mathbf{A} \times \mathbf{B})_k = \varepsilon_{ijk}\varepsilon_{lmk}C_jA_lB_m$$
$$= (\mathbf{B} \cdot \mathbf{C})A_i - (\mathbf{A} \cdot \mathbf{C})B_i. \tag{A.13}$$

A.1.2 Differential calculus of a vector field

The del operator and the gradient of a scalar field

The position operator

$$\mathbf{r} = x\mathbf{i} + y\mathbf{j} + z\mathbf{k} = x_1\mathbf{i} + x_2\mathbf{j} + x_3\mathbf{k} \tag{A.14}$$

has components x_i. It can be shown that

$$\frac{\partial}{\partial x_1}\mathbf{i} + \frac{\partial}{\partial x_2}\mathbf{j} + \frac{\partial}{\partial x_3}\mathbf{k} \equiv \nabla \tag{A.15}$$

is a vector operator, called the **del operator**.[2] An operator must act on some field. It can be shown that if the field $\Phi(\mathbf{r})$ is a scalar field (e.g. the temperature field) the resultant operation by the del operator is a vector field, called the **gradient** of $\Phi(\mathbf{r})$:

$$\text{grad } \Phi \equiv \nabla\Phi = \frac{\partial\Phi}{\partial x_1}\mathbf{i} + \frac{\partial\Phi}{\partial x_2}\mathbf{j} + \frac{\partial\Phi}{\partial x_3}\mathbf{k}. \tag{A.16}$$

[2] It transforms under rotations as a vector. In our presentation we shall use a variety of notations for the del operator components: $\partial/\partial x_i = \partial_i = \nabla_i$, etc.

What is the significance of the gradient? It represents the spatial rate of change of the field Φ. The x component of grad Φ shows how fast Φ changes in the x direction, etc. Thus, the direction of the vector field grad Φ gives the direction in which Φ changes fastest.

The divergence and curl of a vector field

What about the del operator acting on a vector field? Just as two vectors can have a product being a scalar (the dot product) or a vector (the cross product), there are two types of differentiations when the del operator acts on a vector field. The scalar version is the **divergence** of a vector field $\mathbf{A}$:

$$\text{div } \mathbf{A} \equiv \nabla \cdot \mathbf{A} = \frac{\partial A_1}{\partial x_1} + \frac{\partial A_2}{\partial x_2} + \frac{\partial A_3}{\partial x_3}. \tag{A.17}$$

The **curl** of a vector field $\mathbf{A}$ is a vector field:

$$\text{curl } \mathbf{A} \equiv \nabla \times \mathbf{A} \tag{A.18}$$

with components that can be read off from (A.8):

$$(\nabla \times \mathbf{A})_1 = \frac{\partial A_3}{\partial x_2} - \frac{\partial A_2}{\partial x_3}$$

$$(\nabla \times \mathbf{A})_2 = \frac{\partial A_1}{\partial x_3} - \frac{\partial A_3}{\partial x_1}$$

$$(\nabla \times \mathbf{A})_3 = \frac{\partial A_2}{\partial x_1} - \frac{\partial A_1}{\partial x_2}. \tag{A.19}$$

The significance of the divergence and curl of a vector field will be discussed when we study the vector integral calculus in the next subsection.

One way that shows the advantage of denoting the divergence and curl as dot and cross-products involving the del operator is that one can immediately understand identities such as curl grad $\Phi = 0$ and div curl $\mathbf{A} = 0$:

$$[\text{curl grad }\Phi]_i = [\nabla \times \nabla\Phi]_i = \varepsilon_{ijk}\nabla_j\nabla_k\Phi = -\varepsilon_{ijk}\nabla_j\nabla_k\Phi = 0 \qquad (A.20)$$

and

$$\text{div curl }\mathbf{A} = \nabla \cdot (\nabla \times \mathbf{A}) = \varepsilon_{ijk}\nabla_i\nabla_j A_k = -\varepsilon_{ijk}\nabla_i\nabla_j A_k = 0. \qquad (A.21)$$

In both cases, the differential equals the negative of itself after a relabeling of dummy indices $j \leftrightarrow k$ because $\nabla_j\nabla_k = +\nabla_k\nabla_j$ while $\varepsilon_{ijk} = -\varepsilon_{ikj}$.

A.1.3   Vector integral calculus

The physical meaning of the del operator acting on scalar and vector fields can be clarified by consideration of the following integral theorems.

The line integral of a gradient

Consider the projection of the gradient of a scalar field (A.16) along the direction of $\Delta\mathbf{r}$:

$$\nabla\Phi \cdot \Delta\mathbf{r} = \frac{\partial\Phi}{\partial x_1}\Delta x_1 + \frac{\partial\Phi}{\partial x_2}\Delta x_2 + \frac{\partial\Phi}{\partial x_3}\Delta x_3. \qquad (A.22)$$

It is clear that if $\Delta\mathbf{r} = d\mathbf{l}$ is a segment of a path, then $\nabla\Phi \cdot d\mathbf{l}$ is a perfect differential $d\Phi$ along the path. This leads to the following integral theorem for a line integral along any path from some initial point i to some final point f:

$$\int_i^f \nabla\Phi \cdot d\mathbf{l} = \Phi(f) - \Phi(i). \qquad (A.23)$$

Being path-independent, such a line integral over any closed path C must vanish:

$$\oint_C \nabla\Phi \cdot d\mathbf{l} = 0. \qquad (A.24)$$

The divergence theorem (Gauss' theorem)

The theorem states that the volume integral of the divergence of a vector field $\mathbf{J}(\mathbf{r})$ is equal to the surface integral of the flux passing through the surface S enclosing the volume V—the "outward flow" of the vector field across S:

$$\int_V \text{div }\mathbf{J}dV = \oint_S \mathbf{J} \cdot d\mathbf{S}. \qquad (A.25)$$

One can prove this by dividing the volume V into infinitesimally small cubic volumes. Taking one such volume $dV = dxdydz$ and concentrating on the x component, we can evaluate the dot product $\mathbf{J} \cdot d\mathbf{S}$ on the RHS to find it to be $[-J_x + J_x + (\partial J_x/\partial x)dx]dydz = (\partial J_x/\partial x)dV$. The minus sign results from the negative x direction of the left-most cubic face (perpendicular to the y-z plane, pointing away from the volume), and the following two terms are the leading expansion of $J_x(x + dx)$ at the right-most face. See Fig. A.1(a). Clearly the result matches that on the LHS in (A.25). It is then straightforward to see that the same result holds for a finite volume built from such cubes. In this way we can understand div $\mathbf{J} = \nabla \cdot \mathbf{J}$ as the measure of the outward flow of the vector field (per unit volume) at every spatial point.

Conservation law in a field system and the divergence theorem

The divergence theorem is closely related to the statement of conservation in a field system, which is expressed by the **equation of continuity**. For definiteness, let us consider the case of charge (q) conservation:[3]

$$\frac{\partial \rho}{\partial t} + \nabla \cdot \mathbf{j} = 0 \tag{A.26}$$

where ρ is the charge density $\int \rho dV = q$ and $\mathbf{j} = \rho\mathbf{v}$ is the current density. This is a conservation statement as can be seen by performing a volume integration of the equation

$$0 = \frac{d}{dt}\int \rho dV + \int \nabla \cdot \mathbf{j} dV = \frac{dq}{dt} + \oint_S \mathbf{j} \cdot d\mathbf{S}. \tag{A.27}$$

To reach the RHS we have used Gauss' theorem (A.25). Charge is conserved as the equation shows that the increase of the charge in the volume equals the inward flow ($-\mathbf{j} \cdot d\mathbf{S}$) of the charge into the volume.

Circulation around a closed path: Stokes' theorem

While the divergence $\nabla \cdot \mathbf{J}$ expresses the local outward flow of the vector field $\mathbf{J}(\mathbf{r})$, the vector differentiation curl $\nabla \times \mathbf{J}$ is the local "circulation" (per unit area) of the field. This is the physical content of **Stokes' theorem:**

$$\int_S \text{curl}\,\mathbf{J} \cdot d\mathbf{S} = \oint_C \mathbf{J} \cdot d\mathbf{l} \tag{A.28}$$

where S is an open surface bounded by a closed curve C. We can prove this theorem by dividing the surface into infinitesimal squares For the circulation around such a square on the RHS (see Fig. A.1b),

$$[\mathbf{J} \cdot d\mathbf{l}]_\square = J_x(1)\,dx + J_y(2)\,dy - J_x(3)\,dx - J_y(4)dy.$$

We have minus signs at segments 3 and 4, because the circulating vector field is in the opposite direction the coordinate increments. For the dx terms, we have

$$[J_x(1) - J_x(3)]\,dx = \left[J_x(1) - J_x(1) - \frac{dJ_x}{dy}dy\right]dx = -\frac{dJ_x}{dy}dxdy;$$

similarly, for the dy terms,

$$\left[J_y(2) - J_y(4)\right]dy = \left[J_y(4) + \frac{dJ_y}{dx}dx - J_y(4)\right]dy = \frac{dJ_y}{dx}dxdy.$$

In this way we find

$$[\mathbf{J} \cdot d\mathbf{l}]_\square = \left(\frac{dJ_y}{dx} - \frac{dJ_x}{dy}\right)dxdy = (\text{curl}\,\mathbf{J})_z (dS)_z$$

where we have used the expression of the z component of curl $\nabla \times \mathbf{J}$ as given in (A.19) and the fact that the normal to the square surface $dxdy$ is in the z direction.

[3] See the related discussion in Section 1.4.1.

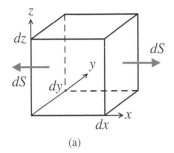

(a)

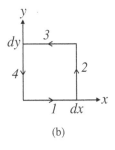

(b)

Fig. A.1 (a) A cube with sides (dx, dy, dz). Through its surfaces a field flows. The surface vectors $d\mathbf{S}$ are normal to their respective surfaces and pointing outward. (b) A square closed path with sides (dx, dy). The circulation is the (integral) sum of the projections of the field along each segment of the path.

A.1.4 Differential equations of Maxwell electrodynamics

The integral equations

Most students first learn Maxwell's theory in the form of integral equations of the electric and magnetic fields $(\mathbf{E}, \mathbf{B})$:

Gauss' law for $\mathbf{E}$
$$\oint_S \mathbf{E} \cdot d\mathbf{S} = q_{in}$$

Gauss' law for $\mathbf{B}$
$$\oint_S \mathbf{B} \cdot d\mathbf{S} = 0$$

Faraday's law
$$\oint_C \mathbf{E} \cdot d\mathbf{l} + \frac{1}{c}\frac{d}{dt}\Phi_B = 0$$

Ampere's law with displacement current
$$\oint_C \mathbf{B} \cdot d\mathbf{l} - \frac{1}{c}\frac{d}{dt}\Phi_E = \frac{1}{c}I.$$

(A.29)

We comment on this set of equations:

- These equations in (A.29) are written in the **Heaviside–Lorentz unit system**. In this system the measured parameter is the velocity of light c instead of the (measured) permittivity and the (defined) permeability of free space, with the relation $c = 1/\sqrt{\epsilon_0\mu_0}$. To go from the more familiar **SI unit system**, one has to scale the fields by $\sqrt{\epsilon_0}E_i \to E_i$, $\sqrt{1/\mu_0}B_i \to B_i$, and the charge and current densities by $\sqrt{1/\epsilon_0}(\rho, j_i) \to (\rho, j_i)$.
- Gauss's law is valid for any closed surface S with any charge distribution. The total charge enclosed inside the surface S is q_{in}. We recall that Gauss's law is equivalent to Coulomb's law, with its basic form written in terms of a point charge. Thus for a general source charge distribution, one has to divide up the source into point charges and then integrate over the resultant field due to each point in the source.
- Gauss's law for a magnetic field states that there is no magnetic monopole (magnetic charge).
- Faraday's law states that a change in magnetic flux $\Phi_B = \int_S \mathbf{B} \cdot d\mathbf{S}$ induces an electromotive force

$$\mathcal{E} = -\frac{d\Phi_B}{dt}. \tag{A.30}$$

The minus sign (the Lenz law) indicates that the induced field would always oppose the original change. The electromotive force is given by the line integral of the induced electric field over a closed path:

$$\mathcal{E} = \oint_C \mathbf{E} \cdot d\mathbf{l}. \tag{A.31}$$

- Ampere's law

$$\oint_C \mathbf{B} \cdot d\mathbf{l} = \frac{1}{c}I \tag{A.32}$$

states that the source of a magnetic field is a current distribution. James Maxwell showed that in order to have charge conservation in this theory, besides the usual conduction current I given by the surface integral through which the current density $\mathbf{j}$ passes

$$I = \int \mathbf{j} \cdot d\mathbf{S}, \tag{A.33}$$

there must also be a "displacement current"

$$I_d = \frac{1}{c} \frac{d\Phi_E}{dt} = \frac{1}{c} \frac{d}{dt} \int \mathbf{E} \cdot d\mathbf{S}. \tag{A.34}$$

This completes the set of field equations for electromagnetism. With the presence of the displacement current term, it becomes clear that a changing electric field would induce a magnetic field, which in turn induces an electric field (Faraday's law). In this way, a self-sustaining electromagnetic wave can be generated.

The differential equations

Many detailed features of the theory (the existence of electromagnetic waves and the validity of charge conservation, etc.) can be best understood when the theory is written in the form of differential equations. In this section, we shall convert the integral equations (A.29) into differential equations:

$$
\begin{array}{lll}
\text{Gauss' law for } \mathbf{E} & \nabla \cdot \mathbf{E} = \rho & \\[2mm]
\text{Gauss' law for } \mathbf{B} & \nabla \cdot \mathbf{B} = 0 & \\[2mm]
\text{Faraday's law} & \nabla \times \mathbf{E} + \dfrac{1}{c}\dfrac{\partial \mathbf{B}}{\partial t} = 0 & \text{(A.35)} \\[3mm]
\begin{array}{l}\text{Ampere's law with} \\ \text{displacement current}\end{array} & \nabla \times \mathbf{B} - \dfrac{1}{c}\dfrac{\partial \mathbf{E}}{\partial t} = \dfrac{1}{c}\mathbf{j}. &
\end{array}
$$

Their derivation from Eqs. (A.29) is discussed below:

- For Gauss' law for $\mathbf{E}$, we convert both sides to volume integrals

$$\int_V \nabla \cdot \mathbf{E} dV = \int_V \rho dV. \tag{A.36}$$

To obtain the LHS we have used the divergence theorem of (A.25). Since the volume is arbitrary, the equality must hold for the integrands as well. In this way we obtain the differential equation for Gauss' law.

$$\nabla \cdot \mathbf{E} = \rho. \tag{A.37}$$

- For Faraday's law, we write each term as a surface integral by applying Stokes' theorem to the line integral of the field

$$\int_S \nabla \times \mathbf{E} \cdot d\mathbf{S} + \frac{1}{c} \frac{d}{dt} \int_S \mathbf{B} \cdot d\mathbf{S} = 0. \tag{A.38}$$

Again the surface being arbitrary the integrands must obey the same relation, and we have the differential equation for Faraday's law as shown above in (A.35).

- In a similar manner and with the relation (A.33), Ampere's law with the displacement current can be written out as a differential equation

$$\nabla \times \mathbf{B} - \frac{1}{c}\frac{\partial \mathbf{E}}{\partial t} = \frac{1}{c}\mathbf{j}. \tag{A.39}$$

It is easy to show from these differential equations that the presence of the displacement current leads to charge conservation in the theory. Taking the time derivative of Gauss's law (A.37), we have

$$\frac{\partial \rho}{\partial t} = \nabla \cdot \frac{\partial \mathbf{E}}{\partial t}.$$

Since the RHS is related to the displacement current, we can use Ampere's law of (A.39) to replace it by the current density $\mathbf{j}$. Since the divergence of a curl vanishes $\nabla \cdot (\nabla \times \mathbf{B}) = 0$ as shown in (A.21), we have

$$\frac{\partial \rho}{\partial t} + \nabla \cdot \mathbf{j} = 0,$$

which is the equation of continuity (A.26) expressing the conservation of electric charge. That Maxwell's equations predict electromagnetic waves is demonstrated in Section 9.2.1.

A.2   The Gaussian integral

[3a] A notable example is the equipartition theorem of statistical mechanics, as discussed in Section 4.4.

Throughout the text[3a], we need the result of various Gaussian integrals.

We start with the basic result for a Gaussian integral:

$$I_1 \equiv \int_{-\infty}^{\infty} e^{-x^2}dx = \sqrt{\pi}. \tag{A.40}$$

Proof I_1 can first be converted into a two-dimensional integral by writing it as the product of two square root factors $\sqrt{I_1}\sqrt{I_1}$:

$$I_1 = \sqrt{\left(\int_{-\infty}^{\infty} e^{-x^2}dx\right)\left(\int_{-\infty}^{\infty} e^{-y^2}dy\right)}$$

$$= \sqrt{\int_{-\infty}^{\infty} e^{-(x^2+y^2)}dxdy}.$$

This 2D integral can then be evaluated when we change the coordinate system from a Cartesian to a polar system $dxdy = rd\theta dr$ and a change of integration variable $r \to z(=r^2)$:

$$I_1 = \sqrt{\int_0^{2\pi} d\theta \int_0^{\infty} e^{-r^2}rdr} = \sqrt{\pi \int_0^{\infty} e^{-z}dz}$$

$$= \sqrt{\pi\left[-e^{-z}\right]_0^{\infty}} = \sqrt{\pi}.$$

■

- By a straightforward change of integration variable, the above result can be written as

$$I_2 \equiv \int_{-\infty}^{\infty} e^{-ax^2} dx = \sqrt{\frac{\pi}{a}}. \tag{A.41}$$

- By completing the exponent as a perfect square, we have

$$I_3 \equiv \int_{-\infty}^{\infty} e^{-(ax^2+bx)} dx = \sqrt{\frac{\pi}{a}} e^{b^2/4a}. \tag{A.42}$$

- By reflection symmetry $x \leftrightarrow -x$, we have

$$I_4 \equiv \int_{-\infty}^{\infty} xe^{-ax^2} dx = 0. \tag{A.43}$$

- By differentiation of $\int_{-\infty}^{\infty} e^{-ax^2} dx = \sqrt{\frac{\pi}{a}}$ with respect to a, we obtain

$$I_5 \equiv \int_{-\infty}^{\infty} x^2 e^{-ax^2} dx = \frac{1}{2a}\sqrt{\frac{\pi}{a}}. \tag{A.44}$$

A.3   Stirling's approximation

In our discussion we shall repeatedly use Stirling's approximation

$$\ln n! \simeq n \ln n - n \qquad \text{for large } n. \tag{A.45}$$

We shall derive this formula through the integral representation of $n!$.

A.3.1   The integral representation for $n!$

A very useful mathematical relation is the integral representation of a factorial:

$$n! = \int_0^{\infty} x^n e^{-x} dx, \tag{A.46}$$

which can be proven by induction. We first check this equation for the simplest case when $n = 0$:

$$\int_0^{\infty} x^0 e^{-x} dx = -\left[e^{-x}\right]_0^{\infty} = 1 = 0!$$

and for $n = 1$ we have, through an integration-by-parts,

$$\int_0^{\infty} xe^{-x} dx = \left[-xe^{-x}\right]_0^{\infty} + \int_0^{\infty} e^{-x} dx = \left[-e^{-x}\right]_0^{\infty} = 1!$$

Assuming that this relation holds for $n = k$:

$$k! = \int_0^{\infty} x^k e^{-x} dx \tag{A.47}$$

we need to prove that the validity of the $n = k + 1$ case then follows; namely,

$$(k+1)! = \int_0^{\infty} x^{k+1} e^{-x} dx.$$

This can be demonstrated through an integration-by-parts:

$$\int_0^\infty x^{k+1} e^{-x} dx = \int_0^\infty x^{k+1} d\left[-e^{-x}\right]$$

$$= \left[-x^{k+1} e^{-x}\right]_0^\infty + \int_0^\infty e^{-x}(k+1)x^k dx$$

$$= 0 + (k+1)\int_0^\infty x^k e^{-x} dx = (k+1)!.$$

To reach the final expression we have used Eq. (A.47). This completes our proof of (A.46). ∎

A.3.2   Derivation of Stirling's formula

An inspection of the integrand in (A.46) shows that, as x increases, the factor x^n would enhance while the factor e^{-x} would suppress the integrand. Thus we expect the integrand to have a maximum (a bulge) and the integral would receive most of its contribution from this maximal region.

Since we will eventually be interested in $\ln n!$, let us take the logarithm of the integrand now:

$$f(x) \equiv \ln x^n e^{-x} = n \ln x - x. \tag{A.48}$$

Its maximum can be found by taking the derivative

$$\frac{df}{dx} = \frac{n}{x} - 1 = 0 \qquad \text{namely, } x = n,$$

which is a maximum as the second derivative is negative at this point:

$$\frac{d^2 f}{dx^2} = -\frac{n}{x^2}. \tag{A.49}$$

We now Taylor expand the function $f(x)$ around this maximum $x = n$:

$$f(x) = f(n) + \frac{1}{2!}(x-n)^2 \left.\frac{d^2 f}{dx^2}\right|_n + \cdots = (n \ln n - n) - \frac{(x-n)^2}{2n} + \cdots$$

Using this approximate expression for the integrand in (A.46), we have

$$n! = \int_0^\infty x^n e^{-x} dx = \int_0^\infty e^f dx = e^{n \ln n - n} \int_0^\infty e^{-\frac{(x-n)^2}{2n}} dx.$$

Since the main contribution to the last integral comes from the positive x region around $x \approx n$, the tail end of the integrand in the $x < 0$ region would be negligible, especially for large n. This allows us to approximate the last term by a standard Gaussian integral as discussed in Section A.2:

$$\int_0^\infty e^{-\frac{(x-n)^2}{2n}} dx \simeq \int_{-\infty}^\infty e^{-\frac{(x-n)^2}{2n}} dx = \sqrt{2\pi n} \tag{A.50}$$

where we have used Eq. (A.41). In this way we find

$$n! \simeq e^{n \ln n - n} \sqrt{2\pi n} \tag{A.51}$$

or

$$\ln n! \simeq n \ln n - n + \frac{1}{2} \ln 2\pi n \simeq n \ln n - n. \tag{A.52}$$

This is the claimed result of (A.45).

A.4   Lagrangian multipliers

The method of Lagrangian multipliers is a strategy to find the local maximum or minium of a function subject to some constraint conditions.

The extremum of a function $f(x, y, z)$ corresponds to $df = 0$. Namely,

$$df = \frac{\partial f}{\partial x} dx + \frac{\partial f}{\partial y} dy + \frac{\partial f}{\partial z} dz = 0. \tag{A.53}$$

As we have arbitrary variation (dx, dy, dz), their respective coefficients must vanish.

$$\frac{\partial f}{\partial x} = \frac{\partial f}{\partial y} = \frac{\partial f}{\partial z} = 0.$$

Thus we have the familiar extremization condition of

$$\nabla f = 0. \tag{A.54}$$

However, if there is a constraint condition,

$$g(x, y, z) = c, \tag{A.55}$$

the variables (x, y, z) are not independent and one does not have arbitrary variations of all (dx, dy, dz). In principle, one can express one variable in terms of the others, say, $z = z(x, y)$, and with this smaller set of variables, work with only arbitrary (dx, dy). But this conventional approach is often difficult to implement; the method of Lagrangian multipliers is an alternative approach that is much simpler to work out.

A.4.1   The method

The constraint condition (A.55) reduces the number of independent variables. From this condition we have also

$$dg = \frac{\partial g}{\partial x} dx + \frac{\partial g}{\partial y} dy + \frac{\partial g}{\partial z} dz = 0. \tag{A.56}$$

Multiplying this equation by an arbitrary parameter λ, the **Lagrangian multiplier**, we can combine it with (A.53) to have the condition

$$d(f + \lambda g) = 0. \tag{A.57}$$

Writing this out, we have

$$\left(\frac{\partial f}{\partial x} + \lambda \frac{\partial g}{\partial x} \right) dx + \left(\frac{\partial f}{\partial y} + \lambda \frac{\partial g}{\partial y} \right) dy + \left(\frac{\partial f}{\partial z} + \lambda \frac{\partial g}{\partial z} \right) dz = 0.$$

Let λ be chosen so that

$$\frac{\partial f}{\partial z} + \lambda \frac{\partial g}{\partial z} = 0. \tag{A.58}$$

Since (dx, dy) are still arbitrary, we in fact have

$$\left(\frac{\partial f}{\partial x} + \lambda \frac{\partial g}{\partial x}\right) = \left(\frac{\partial f}{\partial y} + \lambda \frac{\partial g}{\partial y}\right) = \left(\frac{\partial f}{\partial z} + \lambda \frac{\partial g}{\partial z}\right) = 0$$

or, written compactly,

$$\mathbf{\nabla}(f + \lambda g) = 0. \tag{A.59}$$

In summary, in the presence of a constraint (A.55), the extremization condition, instead of (A.54), is given by (A.59). One can view this set of equations as three conditions; together with Eq. (A.55), they determine the four variables (x, y, z, λ).

A.4.2   Some examples

Let us look at two simple examples as illustrative applications of the method of Lagrangian multipliers.

Example 1   What rectangular parallelepiped with sides (x, y, z) has maximal surface area for a given volume V? Here we need to extremize the surface area $S = 2xy + 2yz + 2zx$, with the constraint of $xyz = V$. According to the method of Lagrangian multiplier (A.59), we have three extremization conditions $\partial_i(S + \lambda V) = 0$:

$$2(y + z) + \lambda yz = 0$$
$$2(z + x) + \lambda zx = 0$$
$$2(x + y) + \lambda xy = 0$$

which can be solved immediately to yield $x = y = z$. The parallelepiped of a given volume with the maximal surface area is a cube.

Example 2   For a cylinder with radius r and height h, what should be the ratio h/r so as to maximize the surface area for a given volume V? Here we need to extremize the surface area $S = 2\pi r^2 + 2\pi rh$, with the constraint $h\pi r^2 = V$. According to the method of Lagrangian multipliers we have two nontrivial extremization conditions that come from the differentiation of $(S + \lambda V)$ with respect to r and h:

$$4\pi r + 2\pi h + \lambda 2\pi rh = 0$$
$$2\pi r + \lambda \pi r^2 = 0,$$

which can be solved to yield $h = 2r$.

A.5 The Euler–Lagrange equation

In the above we studied the extremization of a function with respect to the variation of one or several variables. In this section we discuss the case of extremization of an integral (a functional) with respect to the variation of a whole function. The prototypical example is the variation of a curve between two fixed end-points so as to minimize its length. Such a curve is a **geodesic** (cf. Section 12.4.2). The geodesic curves of Euclidean space are straight lines, and of a spherical surface are great circles, etc. The differential equation for the function that satisfies such an extremization condition is the Euler–Lagrange equation.

A.5.1 Mechanics of a single particle

In physics one often encounters such a calculus of variations problem when one tries to understand the equation of motion as the Euler–Lagrange equation resulting from the minimization of the **action integral**:

$$S = \int_{\tau_i}^{\tau_f} L(x_k, \dot{x}_k) d\tau. \tag{A.60}$$

$x_j(\tau)$ is the trajectory of a particle with τ being the curve parameter, for example, the time variable $\tau = t$. We have also used the notation $\dot{x}_k \equiv dx_k/d\tau$. The integrand, called the **Lagrangian**, is the difference between the kinetic and the potential energy of the system:

$$L(x_k, \dot{x}_k) = T(\dot{x}_k) - V(x_k). \tag{A.61}$$

The **principle of minimal action** states that the action is a minimum with respect to the variation of the trajectory $x_k(\tau)$ with its end-points fixed at initial and final positions: $x_k(\tau_i)$ and $x_k(\tau_f)$. This minimization requirement can be translated into a partial differential equation as follows. The variation of the Lagrangian being

$$\delta L(x_k, \dot{x}_j) = \frac{\partial L}{\partial x_k} \delta x_k + \frac{\partial L}{\partial \dot{x}_j} \delta \dot{x}_j, \tag{A.62}$$

the minimization of the action integral becomes

$$0 = \delta S = \delta \int_{\tau_i}^{\tau_f} L(x_i, \dot{x}_j) d\tau = \int_{\tau_i}^{\tau_f} \left(\frac{\partial L}{\partial x_k} \delta x_k + \frac{\partial L}{\partial \dot{x}_j} \frac{d}{d\tau} \delta x_j \right) d\tau$$

$$= \int_{\tau_i}^{\tau_f} \left(\frac{\partial L}{\partial x_j} - \frac{d}{d\tau} \frac{\partial L}{\partial \dot{x}_j} \right) \delta x_j d\tau. \tag{A.63}$$

To reach the last expression we have performed an integration-by-parts on the second term

$$\int_{\tau_i}^{\tau_f} \frac{\partial L}{\partial \dot{x}_j} d(\delta x_j) = \left[\frac{\partial L}{\partial \dot{x}_j} \delta x_j(\tau) \right]_{\tau_i}^{\tau_f} - \int_{\tau_i}^{\tau_f} (\delta x_j) d\left(\frac{\partial L}{\partial \dot{x}_j} \right).$$

The first term on the RHS can be discarded because the end-point positions are fixed, $\delta x_k(\tau_i) = \delta x_k(\tau_f) = 0$. Since δI must vanish for arbitrary variations

$\delta x_k(\tau)$, the expression in parentheses on the RHS of (A.63) must also vanish. The result is the **Euler–Lagrange equation**:

$$\frac{d}{d\tau}\frac{\partial L}{\partial \dot{x}_k} - \frac{\partial L}{\partial x_k} = 0. \tag{A.64}$$

For the simplest case of $L = \frac{1}{2}m\dot{x}^2 - V(\mathbf{x})$, the Euler–Lagrange equation is just the familiar $\mathbf{F} = m\mathbf{a}$ equation as (A.64) yields $m\ddot{\mathbf{x}} + \nabla V = 0$.

A.5.2   Lagrangian density of a field system

In the above section, we discussed the understanding of the equation of motion for a particle as the Euler–Lagrange equation required by the minimization of the action $S = \int dt L(q, \dot{q})$, with q being the (generalized) coordinate. Thus a knowledge of the Lagrangian $L = L(q, \dot{q})$ is equivalent to knowing the equation of motion. Similarly a field can be viewed as a system having infinite degrees of freedom with its generalized coordinate q_i being the field itself, $q_i = \phi(x)$, where $\phi(x)$ is some generic field. (Namely, one can view the label x as a "continuous index" $q_i \to \phi_x$.) For such a continuum system, the field equation is the Euler–Lagrange equation (A.64) written in terms of the **Lagrangian density** $\mathcal{L}$ with $L = \int d^3x \mathcal{L}$ and an action $S = \int d^4x \mathcal{L}(x)$. The density is a function of the field and its spacetime derivatives $\mathcal{L} = \mathcal{L}(\phi, \partial_\mu \phi)$. We expect the generalization of (A.64) to be

$$\partial_\mu \frac{\partial \mathcal{L}}{\partial(\partial_\mu \phi)} - \frac{\partial \mathcal{L}}{\partial \phi} = 0. \tag{A.65}$$

A knowledge of the (Lorentz invariant) Lagrangian density $\mathcal{L}$ is equivalent to knowing the (Lorentz covariant) field equation.

It should be stressed that the physical content of the variational principle involving the Lagrangian density is no different from the case involving the Lagrangian. The former case is merely for systems with infinite degrees of freedom. We illustrate this point with a concrete example[4] of a 1D system of an infinite number of particles, coupled to their neighboring particles by a spring with identical spring constant k. We start with a collection of N particles and call their displacement from their respective equilibrium positions ϕ_i so that the Lagrangian (A.61) of the system is

$$L = \sum_i^N \left[\frac{1}{2}m\dot{\phi}^2 - \frac{1}{2}k\left(\phi_{i+1} - \phi_i\right)^2 \right]$$

$$= \sum_i^N a\frac{1}{2}\left[\frac{m}{a}\dot{\phi}^2 - ka\left(\frac{\phi_{i+1} - \phi_i}{a}\right)^2 \right] = \sum_i^N a\mathcal{L}_i, \tag{A.66}$$

where a is the distance separating the neighboring equilibrium positions. We can thus interpret $\mathcal{L}_i$ as the Lagrangian density (linear Lagrangian per unit length). We now go from this discrete system to a continuum system by letting $N \to \infty$ and $a \to dx$ with $m/a = \mu$ and $ka = Y$ fixed. Clearly in this way we also have $[(\phi_{i+1} - \phi_i)/a] \to d\phi/dx$ and

[4]See, for example, Sakurai (1967, Section 1–2).

$$L = \int \mathcal{L}(\phi, \partial_x\phi, \partial_t\phi) \quad \text{with} \quad \mathcal{L} = \frac{1}{2}\left[\mu\left(\frac{\partial\phi}{\partial t}\right)^2 - Y\left(\frac{\partial\phi}{\partial x}\right)^2\right]. \quad \text{(A.67)}$$

The variation of the action proceeds just as before, with the displacement field $\phi(x, t)$ being treated as a generalized coordinate:

$$0 = \delta S = \delta \int_{\tau_i}^{\tau_f} \mathcal{L} dx dt = \int_{\tau_i}^{\tau_f} \left(\frac{\partial\mathcal{L}}{\partial\phi}\delta\phi + \frac{\partial\mathcal{L}}{\partial(\partial_x\phi)}\frac{d}{dx}\delta\phi + \frac{\partial\mathcal{L}}{\partial(\partial_t\phi)}\frac{d}{dt}\delta\phi\right) d\tau$$

$$= \int_{\tau_i}^{\tau_f} \left(\frac{\partial\mathcal{L}}{\partial\phi} - \frac{\partial}{\partial x}\frac{\partial\mathcal{L}}{\partial\partial_x\phi} - \frac{\partial}{\partial t}\frac{\partial\mathcal{L}}{\partial\partial_t\phi}\right)\delta\phi d\tau. \quad \text{(A.68)}$$

Just as in (A.63) we have performed two integrations-by-parts to obtain the last expression. This condition then leads to the Euler–Lagrange equation in terms of the Lagrangian density

$$\frac{\partial\mathcal{L}}{\partial\phi} - \frac{\partial}{\partial x}\frac{\partial\mathcal{L}}{\partial\partial_x\phi} - \frac{\partial}{\partial t}\frac{\partial\mathcal{L}}{\partial\partial_t\phi} = 0. \quad \text{(A.69)}$$

When we plug in the expression (A.67) of the Lagrangian density into (A.69), we obtain the field equation

$$Y\frac{\partial^2\phi}{\partial x^2} - \mu\frac{\partial^2\phi}{\partial t^2} = 0. \quad \text{(A.70)}$$

We note that Eq. (A.69) may be written as (A.65) if we understand that the dummy index is summed over $0, 1$ with $\partial_0 = \frac{1}{c}\partial_t$.

Einstein's papers

B

B.1   Einstein's journal articles cited in the text

CPAE: One can find Einstein's writings in the *Collected Papers of Albert Einstein* (CPAE 1968–2012). So far 13 volumes have come out in print. His journal articles in German, up till 1922, are contained in Volumes 2, 3, 4, 6, and 7. Published separately are the English translations (by A. Beck for Volumes 2, 3 and 4, by A. Engel for Volume 6 and 7) that we have referred to below as CPAEe. Volumes 1, 5, 8, 9, 10, 12, and 13 are mostly Einstein's correspondence and other writings dealing with scientific matters. The reader may also wish to consult the original volumes of CPAE which contain many useful comments (in English) on the source text.

   The following is a list of journal article publications by Albert Einstein that are explicitly cited in the book. The translated English title is given in square brackets. We offer brief comments indicating the contents of each individual paper.

- (1902). *Annalen der Physik* **9**, 417. [Kinetic theory of thermal equilibrium and of the second law of thermodynamics. *CPAEe* **2**, 68.] Study of the equipartition theorem and the definitions of temperature and entropy. (Cited in Section 1.1.)
- (1903). *Annalen der Physik* **11**, 170. [A theory of the foundations of thermodynamics. *CPAEe* **2**, 48.] The problem of irreversibility in thermodynamics. Unaware of the work done by other researchers in the field, especially that by Willard Gibbs (1839–1903), Einstein in this and the previous paper rediscovered much fundamental content of statistical mechanics. (Cited in Section 1.1.)
- (1904). *Annalen der Physik* **14**, 354. [On the general molecular theory of heat. *CPAEe* **2**, 68.] Theory of fluctuations and a new determination of Boltzmann's constant. (Cited in Section 1.1 and Section 6.1.1.)
- (1905a). *Annalen der Physik* **17**, 132. [On a heuristic point of view concerning the production and transformation of light. *CPAEe* **2**, 86.] The "photoelectric paper" in which the idea of photons was proposed. (Chapters 3 and 4.)
- (1905b). *University of Zurich Dissertation*. [A new determination of molecular dimensions. *CPAEe* **2**, 104.] Einstein's doctoral thesis,

published in *Annalen der Physik* **19**, 289 (1906). Corrections, *ibid.,* **34**, 591 (1911). (Chapter 1.)

- (1905c). *Annalen der Physik* **17**, 549. [On the movement of small particles suspended in stationary liquids required by the molecular-kinetic theory of heat. *CPAEe* **2**, 123.] The Brownian motion paper. (Chapter 2.)

- (1905d). *Annalen der Physik* **17**, 891. [On the electrodynamics of moving bodies. *CPAEe* **2**, 140; see also Einstein *et al.* (1952).] The special relativity paper. (Section 9.4.1 and Chapter 10.)

- (1905e). *Annalen der Physik* **18**, 891. [Does the inertia of a body depend upon its energy content? *CPAEe* **2**, 172; see also Einstein *et al.* (1952).] The $E = mc^2$ paper. (Section 10.5.)

- (1906). *Annalen der Physik* **20**, 199. [On the theory of light production and light absorption. *CPAEe* **2**, 192.] Einstein's first comment on Planck's $E = h\nu$ relation. (Cited in the introduction section of Chapter 5.)

- (1907a). *Annalen der Physik* **22**, 180. [Planck's theory of radiation and the theory of specific heat. *CPAEe* **2**, 214.] Einstein's derivation of Planck's distribution and also his proposed quantum theory of specific heat. (Chapter 5.)

- (1907b). *Jahrbuch der Radioaktivität und Elektronik,* **4**, 411. [On the relativity principle and the conclusions drawn from it. *CPAEe* **2**, 252.] In the last section (titled "Principle of relativity and gravitation") of this review paper on special relativity, the EP was proposed. Its implications for electromagnetic phenomena were outlined. (Cited in Sections 12.2 and 12.3.)

- (1909a). *Physikalische Zeitschrift* **10**, 185. [On the present status of the radiation problem. *CPAEe* **2**, 357.] Fluctuations of radiation calculated. (Cited in Section 6.1.)

- (1909b). *Physikalische Zeitschrift* **10**, 817. [On the development of our views concerning the nature and constitution of radiation. *CPAEe* **2**, 379.] A photon carries momentum. (Cited in Sections 4.2 and 6.1.)

- (1911). *Annalen der Physik* **35**, 898. [On the influence of gravitation on the propagation of light. *CPAEe* **3**, 379; see also Einstein *et al.* (1952).] Einstein returned to the EP, first proposed in 1907, and carried out the detailed calculations of gravitational redshift, gravitational time dilation, and bending of light ray by the sun. (Section 12.3.)

- (1912a). *Annalen der Physik* **38**, 355. [The speed of light and the statics of the gravitational field. *CPAEe* **4**, 95.] First of the two papers in the continuing development of general relativity. These two papers are the last in which Einstein allows time to be warped while keeping space flat. He realizes that the Lorentz transformations of special relativity must be generalized and that the new theory of gravitation must be nonlinear. (Chapters 13 and 14.)

- (1912b). *Annalen der Physik* **38**, 443. [On the theory of the static gravitational field. *CPAEe* **4**, 107.] Second of the two GR papers in 1912. (Chapters 13 and 14.)

- (1913) with Marcel Grossmann. *Zeitschrift für Mathematik und Physik* **62**, 225 and 245. [Outline of a generalized theory of relativity and of a theory of gravitation. I. Physical Part by A. Einstein; II. Mathematical Part by M. Grossmann. *CPAEe* **4**, 151 and 172.] In this pioneering paper the single Newtonian scalar gravitational field is replaced by 10 fields, which are the components of the metric tensor. However, the field equation is still not correctly identified. (Section 12.4, Chapters 13 and 14.)

- (1914) *Preussische Akademie der Wissenschaften, Sitzungsberichte* p. 1030. [The formal foundations of the general theory of relativity. *CPAEe* **6**, 30.] In this paper Einstein derives the geodesic motion of point particles, and re-derives his 1911 results of the bending of light and gravitational redshift using the new metric tensor theory. (Chapter 13.)

- (1915a). *Preussische Akademie der Wissenschaften, Sitzungsberichte*, p. 778 and p. 799 (4 and 11 November 1915). [On the general theory of relativity. *CPAEe* **6**, 98 and 108.] Corrected a fundamental misconception in his GR theory. (Cited in Section 14.3.4.)

- (1915b). *Preussische Akademie der Wissenschaften, Sitzungsberichte*, p. 831 (18 November 1915). [Explanation of the perihelion motion of Mercury from the general theory of relativity. *CPAEe* **6**, 112.] Einstein presents his Mercury precession calculation and shows that the bending of starlight is twice as large as the EP result. (Section 14.5.1.)

- (1915c). *Preussische Akademie der Wissenschaften, Sitzungsberichte*, p. 844 (25 November 1915). [The field equations of gravitation, *CPAEe* **6**, 117.] Einstein presents his GR field equation. (Section 14.4.)

- (1916a). *Annalen der Physik* **49**, 769. [The foundation of the general theory of relativity. *CPAEe* **6**, 146; see also Einstein *et al.* (1952).] The formal publication of the general theory of relativity. (Section 12.1 and Chapters 13 and 14.)

- (1916b). *Verhandlungen der Deutschen Physikalischen Gesellschaft*, **18**, 318. [Emission and absorption of radiation in quantum theory. *CPAEe* **6**, 212.] (Section 6.3.)

- (1916c). *Mitteilungen der Physikalischen Gesellschaft, Zürich*, **16**, 47. [On the quantum theory of radiation. *CPAEe* **6**, 220.] (Section 6.3.)

- (1916d). *Preussische Akademie der Wissenschaften, Sitzungsberichte*, p. 1111 (26 October 1916). [Hamilton's principle and the general theory of relativity, *CPAEe* **6**, 240; see also Einstein *et al.* (1952).] (Section 17.1.4.)

- (1917). *Preussische Akademie der Wissenschaften, Sitzungsberichte*, p. 142. [Cosmological considerations in the general theory of relativity. *CPAEe* **6**, 421; see also (Einstein *et al.* 1952).] (Chapter 15.)

- (1918). *Preussische Akademie der Wissenschaften, Sitzungsberichte*, p. 154. [On gravitational waves. *CPAEe* **7**, 3.] (Section 14.4.2.)

- (1924). *Zeitschrift für Physik* **27**, 392. [The note appended to a paper by Bose entitled "Wärmegleichgewicht im Strahlungsfeld bei Anwesenheit von Materie" (Thermal equilibrium in the radiation field in the presence of matter)] Bose–Einstein statistics and condensation. (Section 7.2 and 7.4.)

- (1925) *Sitzungsberichte der Preussischen Akademie der Wissenschaften (Berlin)*, p. 3 and p. 18. [Quantum theory of the monatomic ideal gas] BE condensation, photons, and statistical mechanics. (Chapter 7.)
- (1935) with B. Podolsky and N. Rosen. Can a quantum-mechanical description of physical reality be considered complete? *Physical Review* **47**, 777. EPR paradox posited. Nonlocal feature of quantum mechanics was brought to the fore. (Chapter 8.)

Listing of Einstein's publications

Albert Einstein: Philosopher-Scientist. (ed. P.A. Schilpp, Harper and Brothers. Co. New York, 1949, 1951). The book, in two volumes, contains a complete listing of Einstein's publications.

List of Scientific Publications by Albert Einstein—Wikipedia. At this online site, one has an easy access to the list (last checked May 22, 2012).

B.2   Further reading

I consulted many books in the preparation of this book, but some are particularly helpful and are listed below.

- Pais, Abraham (1982). This is a physicist's biography written by a physicist. A proper reading requires a solid physics background. It is hoped that, after working through the present book, the reader will have a deeper understanding of this Einstein biography. The 2005 re-issue of the book also includes a new *Forward* by Roger Penrose that is, in my opinion, a particularly insightful appraisal of Einstein's contribution to physics.
- Schilpp, P.A., ed. (1949, 1951). This has a collection of essays by N. Bohr, M. Born, K. Godel, W. Pauli, and others on the occasion of Einstein's 70th birthday. They relate to Einstein's scientific and philosophical ideas. This book also includes his *Autobiographical Notes*. This brief essay is the closest Einstein ever came to writing an autobiography. Although a very personal account, it says little about his private life, and concerns entirely the development of his physics ideas. The account was also published in 1979 as a separate booklet by Open Court Pub. La Salle, IL.
- Stachel, John (1998). Here is another place where one can find easy access to Einstein's 1905 papers in English translation. Stachel also provides the historical context and insightful comments on these important papers.
- Stachel, John (2002). A collection of this Einstein scholar's many essays on all aspects of Einstein's life and physics.
- Landau, L.D. and E.M. Lifshitz (1959). The two chapters II and VI are particularly relevant for the discussion of viscous fluids and Brownian motion as given in our Chapters 1 and 2.
- Tomonaga, Sin-itiro (1962). The book is a clear and accessible presentation of the 'old quantum theory'.

- ter Haar, D (1967). This also contains a collection of the classics of the old quantum theory (by Planck, Einstein, Bohr, and others) in English translation.
- Huang, Kerson (2009). This concise introduction of the key concepts in statistical mechanics should be a helpful reference for the discussion contained in our Chapters 2, 3, and 7.
- Blundell, Stephen J. and Katherine M. Blundell (2009). A comprehensive introduction to thermal physics with many useful mathematical details.
- Cheng, Ta-Pei (2010). In this recent publication, one can find some of the calculational details of special and general relativity, discussed in Parts III and IV.
- Longair, Malcolm S. (2003). The physics is offered in its historical context. A valuable book for readers to consult on most of the topics covered in our presentation.
- Kennedy, Robert E. (2012). Differing from our presentation of Einstein's physics, Kennedy's book follows closely, equation by equation, the selected Einstein papers.
- Isaacson, Walter (2008). This is a popular and masterful biography of Einstein that covers many aspects of his life.

Answers to the 21 Einstein questions

Here are brief answers to the Einstein questions raised in the Preface. More details and pertinent context are given in the text with the relevant chapter and section numbers as shown.

1. In Einstein's doctoral thesis (Chapter 1), he derived two ways of relating N_A, Avogadro's number, to the viscosity and diffusion coefficient of a liquid with suspended particles. The second relation, the Einstein–Smoluchowski relation, also allowed N_A to be deduced from measurement in the Brownian motion (Chapter 2). Finally, from the blackbody radiation spectrum (Section 4.1.1) one could deduce, besides Planck's constant, also the Boltzmann constant $k_B = 1.380 \times 10^{-23}$ J K^{-1}, which led directly to $N_A = R/k_B = 6.02 \times 10^{23}$/mol, because the gas constant $R = 8.314$ J K^{-1}/mol was already known.

2. Even though Einstein conjectured that the motion he predicted in his 1905c paper, as discussed in Chapter 2, was the same as Brownian motion, he was prevented from being more definitive because he had no access then to any literature on Brownian motion. He was outside the mainstream academic environment and did not have the research tools typically associated with a university.

3. It is a common misreading of history that had Einstein's derivation of energy quantization in his 1905 study of blackbody radiation as a direct extension of Planck's work on the same problem in 1900. In fact Einstein's derivation of energy quantization was different from that of Planck's, and was by an approach that was, from the viewpoint of the then accepted physics, less problematic. But the important difference between Einstein and Planck was that Einstein, through his derivation by way of the equipartition theorem of the Rayleigh–Jeans law, was the first one to understand clearly the challenge that blackbody radiation posed for classical physics (cf. Section 4.1). Thus Einstein from the very beginning appreciated the fundamental nature of the break with classical physics this new proposal represented. Planck on the other hand had resisted the new photon idea for more than 10 years after its proposal in 1905 (cf. Chapter 3 and Section 5.1).

4. This can be understood most readily using Einstein's derivation of Planck's distribution law, as given in Section 5.1. Energy quantization implies that the step between energy levels becomes ever greater as the

radiation frequency increases. The Boltzmann factor of $\exp(-E/k_B T)$ would then suppress the ultraviolet contribution.

5. Before Einstein proposed his quantum theory, the success of the equipartition theorem in explaining the pattern of specific heat was very much confused. For instance, why the vibrational degrees of freedom must be ignored in the case of gases while they are the dominant components in solids. This led many to question the whole idea of the molecular composition of matter, as the counting of their degrees of freedom did not seem to match the observed result. See Section 5.2.

6. A field obeys a wave equation and its solution can be viewed as a collection of oscillators (Section 3.1). A quantum field is a collection of quantum oscillators. In the quantum mechanical treatment, the dynamical variables of oscillators are taken to be noncommuting operators, leading to the particle features of the system. The raising and lowering operators in the quantum formalism provide the natural language for the description of emission and absorption of radiation, and more generally, for the description of particle creation and annihilation. The surprising result of wave–particle duality discovered by Einstein in his study of fluctuations of radiation energy found its natural resolution in quantum field theory, when the fluctuation was calculated for these quantized waves with noncommuting field operators. More details are provided in Section 6.4.

7. Einstein advocated the local realist viewpoint that an object had definite attributes whether they had been measured or not. The orthodox interpretation of quantum mechanics (that measurement actually produces an object's properties) would imply that the measurement of one part of an entangled quantum state can instantaneously produce the value of another part, no matter how far these two parts have been separated. Einstein's criticism shone a light on this 'spooky action-at-a-distance' feature; its discussion and debate have illuminated the meaning of quantum mechanics. It led later to Bell's theorem showing that these seemingly philosophical questions could lead to observable results. The experimental vindication of the orthodox theory has sharpened our appreciation of the nonlocal features of quantum mechanics. Nevertheless, the counter-intuitive QM picture of objective reality still troubles many, leaving one to wonder whether quantum mechanics is ultimately a complete theory (Chapter 8).

8. The key idea of Einstein's special relativity is the new conception of time. Time, just like space, becomes a coordinate-dependent quantity. This, when augmented by the postulate of the constancy of light velocity, leads directly to the Lorentz transformation as the coordinate transformation among inertial frames of reference. This is in contrast to Lorentz's derivation based on a model of the aether–light interaction. While Einstein's derivation in this new kinematics implied its applicability to all of physics, Lorentz's specific dynamical theory, even if it were correct, was restricted to electrodynamics only (Section 10.3.1 and the final three sections of Chapter 9).

9. Stellar aberration, Fizeau's experiment, and Fresnel's formula can be viewed as lending important experimental support to what Einstein

needed in proposing a coordinate-dependent time—the key element of special relativity. See our discussion, in Sections 9.3 and 9.4, of their relations to Lorentz's 'local time'. Their straightforward derivation from special relativity is given in Section 10.6.

10. To obtain the length of an object one must find the positions of the front and the back of the object. The relativity of time comes into play in these two measurements. See Section 10.1.3.

11. Special relativity is 'special' because it restricts the invariance of physics laws to a special set of coordinate systems: the inertial frames of reference (Section 9.1), while general relativity allows all coordinate frames. Special relativity is not applicable to gravity because the concept of 'inertial frames' becomes meaningless in the presence of gravity (Section 12.2). The general theory of relativity is automatically a theory of gravitation because, according to the equivalence principle, any accelerated frame can be regarded as an inertial frame with gravity. General relativity in an 'interaction-free situation' is a theory of pure gravity; the GR version of any other interaction, say, electrodynamics, is the theory of that interaction in the presence of a gravitational field. See the introduction and final remarks in Section 13.4.

12. Although Minkowski's geometric formulation is a mathematical language that did not immediately lead to any new physical results in special relativity, it nevertheless supplies the framework in which the symmetry between space and time can be implemented in an elegant way. Einstein finally became appreciative of such a language when he realized that it provided him with just the avenue to extend special to general relativity. Einstein's greatest ability lay in his extraordinary physical instinct. It took him some time to truly value the connection between mathematics and new physics theory: some theoretical physics insights came about only when the necessary mathematical languages were available to facilitate such advances. The formulation of general relativity in the framework of Riemannian geometry is of course a glorious example. In this case Einstein was fortunate to have the assistance of his mathematician friend Marcel Grossmann. Still, Einstein had to struggle a great deal and, very much to his credit, he was finally able to find the correct GR field equation. We may speculate on the reason why Einstein was less successful in his unified field theory program. Besides his failure to take note of the new discoveries of the weak and strong forces as new fundamental interactions, he could possibly have made more progress had he been as great a mathematician as he was a great physicist. In this connection we have in mind the case of Newton who formulated his new theory of mechanics and gravitation that was greatly facilitated by his concurrent invention of calculus. See the discussion in Sections 11.1, 14.3.4, and 17.4.2.

13. The realization that gravity can be transformed away in a coordinate frame in free fall was called by Einstein 'my happiest thought'. It became the basis of the principle of equivalence between inertia and gravitation, which was used by Einstein as the handle to extend special to general relativity (Section 12.2.2). The moment of elation when Einstein found

out, in mid-November 1915, that he could correctly explain, from first principles in his new gravitational theory, the observed precession of the planet Mercury's orbit (Section 14.5.1) was, according to Pais, 'the strongest emotional experience in Einstein's scientific life'.

14. By a geometric theory, or a geometric description, of any physical phenomenon, we mean that the results of physical measurements can be attributed directly to the underlying geometry of space and time. Einstein started by studying the generalization of the equivalence between inertia and gravitation (first observed as the equality between inertial and gravitational masses) to electromagnetism. He showed that such a 'strong equivalence principle' implied a gravitational frequency shift, gravitational time dilation, and gravitational bending of a light ray (see Section 12.3). Such considerations led Einstein to the idea that the gravitational effect on a body can be attributed directly to some underlying spacetime feature. Thus, gravitational time dilation could be interpreted as the warping of spacetime in the time direction; a disk in a rotationally symmetric gravitational field has a non-Euclidean relation between its circumference and its radius, etc. (cf. Sections 13.1 and 13.2). In this way these partial GR results suggested to Einstein that 'a gravitational field is simply spacetime with curvature'. Such a description is clearly compatible with the EP result that any gravitational field can be transformed away locally, just as any curved space is locally flat. To what physical realm does Einstein's theory extend Newtonian gravity? It can be demonstrated that the GR equations, whether its equation of motion (the geodesic equation) or its field equation (the Einstein equation), reduce to their corresponding parts in the Newtonian theory when one takes the 'Newtonian limit': when particles move with nonrelativistic speed in a weak and static gravitational field. See Sections 13.2.2 and 14.4.1. This means that GR extends Newtonian gravity to the realm of a time-dependent gravitation field which is strong and allows for particles moving close to the speed of light.

15. The GR field equation, the Einstein equation, can be written as an equality between the spacetime curvature (the Einstein tensor) on the geometry side and the energy–momentum–stress tensor on the energy–matter side. The curvature being the nonlinear second derivatives of the metric, which is interpreted as the relativistic gravitational potential, is the relativistic version of the familiar tidal forces (Sections 14.3 and 14.4).

16. The observed changing rotation rate of the Hulse–Taylor binary pulsar system was found to be in agreement with the GR prediction over a time period of more than two decades (see Fig. 14.2 in Section 14.4.2).

17. The structure of the Schwarzschild spacetime is such that its metric elements

$$g_{00} = -\frac{1}{g_{rr}} = -\left(1 - \frac{r^*}{r}\right),$$

change sign when the radial distance r moves across the Schwarzschild radius r^*. In this way the various spacetime intervals ds^2 change from being space-like to time-like, and vice versa (cf. Section 11.3). This means a time-like or light-like worldline (as traced out by a material particle or a light ray), which always moves in the direction of ever increasing time when outside the black hole ($r > r^*$), once it crosses the event horizon (to the $r < r^*$ region), will be forced to move in the direction of $r = 0$. Pictorially we can represent this as 'lightcones tipping over across the $r = r^*$ horizon'. We say such features demonstrate the full power and glory of general relativity: Relativity requires space and time to be treated on an equal footing—as is best done by taking space-time as the physics arena. In special relativity the spacetime geometry is still flat, while general relativity involves a warped spacetime. In the case of a black hole when the radial size r is comparable to r^* the warpage of spacetime is so severe that the roles of space and time can be switched (Sections 11.3 and 14.5).

18. Each of these fundamental constants can be viewed as the 'conversion factor' that connects disparate realms of physics: Planck's constant h connects waves to particles; the light velocity c, between space and time; and Newton's constant G_N, between geometry and matter/energy. Einstein made pivotal contributions to all these connections through his discoveries in quantum theory, and special and general relativity (Sections 3.4.2, 6.1, 11.4, 14.4, and also 17.1.1).

19. As recounted in George Gamow's brief autography, *My Worldline*, Einstein apparently told Gamow that his introduction of the cosmological constant was 'the biggest blunder of my life'. But we now regard Einstein's discovery of this gravitational repulsion term Λ as a great contribution to modern cosmology: Λ is the crucial ingredient of inflationary cosmology, describing the explosive beginning of the universe, and in the present cosmic epoch, it is the 'dark energy' that constitutes 75% of the cosmic energy content and causes the universe's expansion to accelerate (Section 15.3).

20. The claim that Einstein's idea was of paramount importance in the successful creation of the Standard Model of particle physics is based on the fact that his teaching on the importance of symmetry principles in physics gave us the framework to understand particle interactions. Especially, the whole idea of gauge symmetry grew from the idea of spacetime-dependent transformations in the general theory of relativity. The Standard Model shows that all the principal fundamental interactions: electrodynamics, weak and strong interactions, are gauge interactions (Chapter 16, especially Sections 16.1 and 16.5.5).

21. The driving force behind Einstein's 20-year effort in the unified field theory program was his hope that such a unification would shed light on the quantum mystery. His motivation for new physics was often prompted by the promise of wider comprehension that a new synthesis would bring. While Einstein was not ultimately successful in this effort, his pursuit has inspired the research of others in this direction. In Chapter 17 we present the Kaluza–Klein (KK) theory as a shining example of Einstein's unification program. It not only unifies gravitation with electrodynamics in a GR theory with a 5D spacetime, but also suggests a possible interpretation of the internal charge space and gauge symmetry as reflecting the

existence of a compactified extra spatial dimension. On the other hand, the KK theory did not shed light on the origin of quantum mechanics; in fact it incorporates quantum field theory in order to have a self-consistent description. Nevertheless, the effort to incorporate quantum mechanics, in the form of the Standard Model, with gravity, in the form of general relativity, is a major forefront of modern theoretical physics research.

Glossary of symbols and acronyms

1 Latin symbols

a	Bohr radius, compact radius of extra dimension
$a(t)$	scale factor in Robertson–Walker metric
$\hat{a}_\pm$	raising and lowering operators creation and annihilation operators
A^μ, $\mathbf{A}$	EM 4-potential, 3-potential
$\mathbf{B}$	magnetic vector field
C	heat capacity (specific heat)
c	light speed
$c(x)$	light speed with respect to coordinate time
c_s	sound speed
D	diffusion coefficient, distance
D_μ	covariant derivative
$\mathrm{diag}(a_1, a_2)$	diagonal matrix $\begin{pmatrix} a_1 & 0 \\ 0 & a_2 \end{pmatrix}$
$\mathbf{E}$	electric vector field
E	energy
E_{Pl}	Planck energy
e	electric charge, eccentricity, electron field
$\mathbf{e}_\mu$	basis vectors in a manifold
$\mathcal{E}$	total nonrelativistic energy
F	force
f	energy flux (luminosity per unit area)
$f_\tau(\Delta)$	probability density at distance Δ at step τ
$F_{\mu\nu}$	electromagnetic field tensor Yang–Mills field tensor
$\tilde{F}_{\mu\nu}$	electromagnetic field dual tensor
g	determinant of metric matrix $g_{\mu\nu}$
g	number of states with same energy
$\mathfrak{g}, g$	gravitational acceleration, field
g_s	QCD strong coupling
$g_{\mu\nu}$	spacetime metric tensor
G_{N}	Newton's constant
G_μ	Yang–Mills field or gluon field
$G_{\mu\nu}$	Einstein curvature tensor
h	Planck's constant, $\hbar = h/2\pi$
H	Hubble's constant $\dot{a}(t)/a(t)$
H	Hamiltonian
I	electric current
j^μ, $\mathbf{j}$	EM current 4-vector, 3-vector

K	kinetic energy
K	Gaussian curvature
$\mathbf{k}, k^\mu$	wave 3-vector, 4-vector
k	wave number $k = |\mathbf{k}| = 2\pi/\lambda$
k	spring constant
k	curvature signature $(0, \pm 1)$
k_B	Boltzmann's constant
K^μ	covariant force 4-vector
$[L]^\lambda_\mu$	Lorentz transformation
l_P	Planck length
L	length (side of a volume)
L	Lagrangian
L	radiation energy
$\mathcal{L}$	Lagrangian density
m	mass
m_G	gravitational mass
m_I	inertial mass
M	molecular weight, mass (source)
m_P	Planck mass
n	number density, number of moles
n	integer number, index of refraction
n_i	component of a unit radial vector
N	number (e.g. in a frequency interval)
$\hat{n}$	number operator $= \hat{a}^\dagger_- \hat{a}^\dagger_+ /(\hbar\omega)$
N_A	Avogadro's number
$O(x)$	of the order of x
$[\hat{O}g]$	(derivative) operator acting on the metric
P	molecular radius, number of quanta
$\mathcal{P}$	pressure
p	pressure, probability, momentum magnitude
$\mathbf{p}$	momentum 3-vector
p^μ	components of momentum 4-vector
Q	heat function
q	charge, generalized position, quark field
r	radial coordinate
r^*	Schwarzschild radius
R	gas constant ($R = N_A k_B$)
R	Rydberg constant
R	radius of curvature, Ricci curvature scalar
R_0	radius in Robertson–Walker geometry
$[\mathbf{R}]_{ij}$	element of a rotational matrix
$R^\mu_{\lambda\alpha\beta}$	Riemann curvature tensor
$R_{\mu\nu}$	Ricci curvature tensor
S	surface area, entropy
S	action (time integral of Lagrange)
$SU(3)$	special unitary group in 3D
$S_{x,y,z}$	spin components
s	invariant spacetime interval
t_0	cosmic time of the present epoch age of the universe
t_H	Hubble time ($= 1/H_0$)
T	temperature (on absolute scale)
T_E	Einstein temperature

T_D	Debye temperature
$T_{\mu\nu}$	energy–momentum–stress 4-tensor (also $t_{\mu\nu}$)
T_{ij}	momentum stress 3-tensor
U	energy
U	internal space transformation matrix
u	energy density
U^μ	velocity 4-vector
$U(1)$	unitary group in 1D
V	volume, velocity magnitude
$\tilde{v}$	immersed small volume
v_i	velocity component
W	rate of energy transfer (power)
W	work function (energy threshold)
W	complexion, number of microstates
W_μ	weak vector boson field operator
x	spatial or spacetime position
Z	atomic number (number of protons in the nucleus)
Z_μ	neutral weak vector boson field
z	wavelength shift, redshift

2 Greek symbols

α	fine structure constant
α	parameter in Wien & Planck distributions
α_{ij}	velocity coefficient symmetric tensor
β	parameter in Wien & Planck distributions
β	v/c, velocity in unit of c
γ	the Lorentz factor $\left(1 - \beta^2\right)^{-1/2}$
γ_μ	Dirac gamma matrices
$\Gamma^\nu_{\lambda\rho}$	Christoffel symbols
δ_{ij}	Kronecker delta
∂_i, ∂_μ	(3D) del operator $\partial_i = \frac{\partial}{\partial x_i}$ (4D) with $\partial_0 = \frac{1}{c}\frac{\partial}{\partial t}$, etc.
∇	del operator with components $\left(\frac{\partial}{\partial x_1}, \frac{\partial}{\partial x_2}, \frac{\partial}{\partial x_3}\right)$
ε_{ijk}	Levi-Civita symbols (3D)
$\varepsilon_{\mu\nu\lambda\rho}$	Levi-Civita symbols (4D)
ϵ	energy, angular excess
η, η^*	viscosity, effective viscosity
$\eta_{\mu\nu}$	metric of (flat) Minkowski spacetime
θ	polar angle coordinate
$\theta(x)$	gauge function
κ	gravity strength $8\pi G_\mathrm{N}/c^4$
λ	wavelength
Λ	cosmological constant
$[\mathbf{\Lambda}]^\mu_\nu$	coordinate transformation matrix
μ	mobility, chemical potential
ν	frequency
ν, ν_e	neutrino field, electron neutrino field
ρ	radiation energy density per unit frequency
ρ	mass density, charge density
ρ	particle number density

ρ_c	cosmic critical density
σ	radial distance, area
σ_i	Pauli matrices
σ_{ij}	stress tensor due to viscosity
τ	proper time
τ_{ij}	stress tensor of an ideal fluid
φ	volume fraction
φ	entropy density per unit frequency
Φ	EM or gravitational potential
ϕ	phase of a wave
ϕ	azimuthal angle coordinate
$\phi(x)$	(generic) scalar field
ϕ^μ	force density
$\chi(x)$	gauge function
$\Psi(x)$	wavefunction or $\psi(x)$
ψ	rapidity parameter of relative frames
ω	angular frequency
$\omega(E)$	density of states having energy E
Ω	$d\Omega = \sin\theta\, d\phi\, d\theta$ element of solid angle
Ω	ratio of density to critical density (ρ/ρ_c)
Ω_B	baryonic matter density ratio
Ω_M	total mass density ratio
Ω_{DM}	dark matter mass density ratio
Ω_Λ	dark energy density ratio

3   Acronyms

1D, ... 4D	one dimensional, ... four dimensional
BEC	Bose–Einstein condensation
BBR	blackbody radiation
CMBR	cosmic microwave background radiation
CPAEe	Collected Papers of Albert Einstein (English translation)
DM	dark matter
DOF	degrees of freedom
ETH	Eidgenössische Technische Hochschule [(Swiss) Federal Polytechnic Institute]
EM	electromagnetic
EP	equivalence principle
EPR	Einstein–Podolsky–Rosen
EPT	equipartition theorem
FLRW	Friedmann–Lemaître–Robertson–Walker
GR	general relativity
GUT	grand unified theory
KK	Kaluza–Klein
ΛCDM	Lambda Cold Dark Matter
LHS	left-hand side
NR	nonrelativistic
QCD	quantum chromodynamics
QED	quantum electrodynamics
QM	quantum mechanics
RHS	right-hand side

SHO	simple harmonic oscillator
SM	Standard Model
SNe	supernovae
SR	special relativity
SSB	spontaneous symmetry breaking
SuppMat	supplementary material
WIMP	weakly interacting massive particle

4   Miscellaneous units and symbols

AU	astronomical unit (average distance between earth and sun)
GeV	giga electron volt
Gyr	giga (billion) years
K	degree kelvin
kpc	kiloparsec
MeV	million electron volt
Mpc	megaparsec
μ	micron $\left(10^{-6} \text{ m}\right)$
nm	nanometer $\left(10^{-9} \text{ m}\right)$
Å	angstrom $\left(10^{-10} \text{ m}\right)$
pc	parsec
$\lvert\psi\rangle$	ket vector, with its dual, the bra vector $\langle\phi\rvert$ and inner product $\langle\phi\,\lvert\psi\rangle$
$\odot$	symbol for the sun (e.g. $M_{\oplus}$ = solar mass)
$\oplus$	symbol for the earth (e.g. $R_{\oplus}$ = earth's radius)

Bibliography

Aspect, A. *et al.* (1981): Experimental tests of realistic local theories via Bell's theorem, *Physical Review Letters* **47**, 460.

Bell, J.S. (1964): On the Einstein Podolsky Rosen paradox, *Physics* **1**, 195.

Blundell, S. J. and Blundell, K.M. (2009): *Concepts in Thermal Physics*, 2nd ed. Oxford University Press.

Bohr, N. (1913): On the constitution of atoms and molecules, *Philosophical Magazine* **26**, 1.

Bohr, N. (1935): Can quantum-mechanical description of physical reality be considered complete?, *Physical Review* **48**, 696.

Boltzmann, L. (1884): Ableitung des Stefan'schen Gesetzes, betreffend die Abhängigkeit der Wärmestrahlung von der Temperatur aus der electromagnetischen Lichttheorie (Derivation of Stefan's law on the temperature dependence of the thermal radiation from the electromagnetic theory of light) *Annalen der Physik und Chemie* **22**, 291.

Born, M., Heisenberg, E., and Jordan, P. (1926): Zur Quantenmechanik (On quantum mechanics) *Zeitschrift für Physik* **35**, 557. Translated in (Van der Waerden 1968).

Bose, S. (1924): Plancks gesetz und lichtquantenhypothese (Planck's law and the light quantum hypothesis) *Zeitschrift für Physik* **26**, 178.

Bose, S. (1976): The beginning of quantum statistics: A translation of "Planck's law and the light quantum hypothesis", *American Journal of Physics* **44**, 1056.

Brown, L.M., Pais, A., and Pippard A.B. (eds) (1995): *Twentieth Century Physics Vol I, II, III*. IOP Publishing Bristol; AIP Press, New York.

Carroll, S. (2004): *Spacetime and Geometry: An Introduction to General Relativity*, Addison Wesley, San Francisco.

Cheng, T.P. (2010): *Relativity, Gravitation and Cosmology: A Basic Introduction*. 2nd ed. Oxford University Press.

Cheng, T.P. and Li, L.F. (1984): *Gauge Theory of Elementary Particle Physics*, Oxford University Press.

Cheng, T.P. and Li, L.F. (1988): Resource letter, Gauge invariance, *American Journal of Physics*, **56**, 586. This article also appears as the introduction in *Gauge Invariance, An Anthology with Introduction and Annotated Bibliography*, T. P. Cheng and L. F. Li (eds.), American Association of Physics Teachers, College Park MD, 1990.

Clauser, J.F. and Shimony, A. (1978): Bell's theorem: experimental tests and implications, *Reports on Progress in Physics* **41**, 1881.

Compton, A.H. (1923): A quantum theory of the scattering of X-rays by lights elements, *Physical Review* **21**, 483.

CPAE (1968–2012): *Collected Papers of Albert Einstein*, Volumes 1–13, J. Stachel, *et al.* (ed.), Princeton University Press.

de Broglie, L. (1924): *Recherches sur la théorie des quanta* (Research on the quantum theory). Doctoral thesis published in 1925.

Debye, P. (1912): Zur Theorie der spezifischen Waerme (On the theory of specific heat), *Annalen der Physik* **39**, 789.

Debye, P. (1923). Zerstreuung von Rontgenstrahlen und Quantentheorie (Scattering of X-rays and quantum theory), *Physikalische Zeitschrift* **24**, 161. Translated in *The Collected Papers of Peter J.W. Debye*. Interscience, New York, 1954.

d' Espagnat, B. (1979): The quantum theory and reality, *Scientific American* **241**(5), 158.

Dietrich, J.P. *et al.* (2012): A filament of dark matter between two clusters of galaxies, *Nature* **487**, 202.

Dirac, P.A.M. (1927): The quantum theory of the emission and absorption of radiation, *Royal Society Proceedings A* **114**, 243.

Duncan, A. and Janssen, M. (2008). Pascual Jordan's resolution of the conundrum of the wave–particle duality of light, *Studies in the History and Philosophy of Modern Physics* **39**, 634.

Einstein, A. (1902–35): Einstein's physics papers are listed separately in Appendix B (Einstein's journal articles).

Einstein, A. (1922): Kyoto Address: How I created the theory of relativity. Translated by S. Abeka in *Historical Study of Physical and Biological Sciences*, **30**, 1 (2000). http://www.jstor.org/stable/27757844

Einstein, A. (1949): *Autobiographical Notes*, in (Schilpp 1949, 1951); also, Open Court Pub. La Salle, IL. 1979.

Einstein, A. *et al.* (1952): *The Principle of Relativity: A collection of original memoirs on the special and general theory of relativity by H.A. Lorentz, A. Einstein, H. Minkowski and H. Weyl* (Translation by W. Perrett and G.B. Jeffery), Dover Publications, New York.

Fixsen, D.J. *et al.* (1996): The cosmic microwave background spectrum from the full COBE FIRAS data set, *Astrophysical Journal* **473**, 576.

Fock, V. (1926): Über die invariante Form der Wellen- und der Bewegungsgleichung für einen geladenen Massenpunkt (On the invariant form of the wave and the equation of motion for a charged mass point), *Zeitschrift für Physik* **39**, 226.

Friedmann, A. (1922): Über die Krümmung des Raumes (On space curvature), *Zeitschrift für Physik* **10**, 377.

Gamow, G. (1970): *My World Line—An informal autobiography*, Viking. New York, p. 44.

Gross, D. (1995): Oskar Klein and gauge theory, *Proceedings of the Oskar Klein Centenary Symposium*, U. Lindström, (ed.), World Scientific, Singapore.

Hafele, J.C. and Keating, R.E. (1972): Around-the-world atomic clocks: observed relativistic time gains, *Science*, **177**, 168.

Heilbron, J. (1977): *Lectures on the history of atomic physics 1900–1922*. in Weiner (1977), p. 40.

Hilbert, D. (1915): Die grundlagen der physik (The basics of physics) *Nachrichten von der Gesellschaft der Wissenschaften zu Göttingen, Mathematisch-Physikalische Klasse*, 395.

Huang, K. (2009): *Introduction to Statistical Physics*, 2nd. ed. Taylor and Francis, London.

Hubble, E. (1929): A relation between distance and radial velocity among extra-galactic nebulae, *Proceedings of the National Academy of Sciences USA* **15**, 168.

Isaacson, W. (2008). *Einstein: His Life and Universe*, Simon & Schuster, New York.

Jackson, J.D. and Okun, L.B. (2001): Historical roots of gauge invariance, *Review of Modern Physics* **73**, 663.

Kaluza, Th. (1921): Zum Unitätsproblem in der physik (On the unification problem in physics), *Preussische Akademie der Wissenschaften, Sitzungsberichte*, p. 966.

Kennedy, R. E. (2012). *A Student's Guide to Einstein's Major Papers*, Oxford University Press.

Kirchhoff, G. (1860): Über das verhältniss zwischen dem emissionsvermögen und dem absorptionsvermögen der körper für wärme and licht, *Annalen der Physik und Chemie* (Leipzig) **109**, 275–301. Translated by Guthrie, F. On the relation between the radiating and absorbing powers of different bodies for light and heat. *Philosophical Magazine* Series 4, **20**, 1.

Klein, M (1977): The Beginnings of quantum theory, in Weiner (1977), p. 1.

Klein, O. (1926): Quanten-theorie und 5-dimensionale relativitätstheorie (Quantum theory and the 5-dimensional relativistic theory), *Zeitschrift für Physik* **39**, 895; The atomicity of electricity as a quantum law, *Nature* **118**, 516.

Kuhn, T.S. (1978): *Blackbody Theory and the Quantum Discontinuity 1894–1912*, Oxford University Press.

Landau, L.D. and Lifshitz, E.M. (1959): *Fluid Mechanics*, Pergamon Press, London.

Lemaître, G. (1927, 1931): Un univers homogène de masse constante et de rayon croissant, rendant compte de la vitesse radiale des nébuleuses extra-galactiques, *Annales de la Societe Scientifique de Bruxelles* **47A**, 15. A related, but not identical, version with the same title was published later in 1931: A homogeneous universe of constant mass and increasing radius accounting for the radial velocity of extra-galactic nebulae, *Monthly Notices of the Royal Astronomical Society* **91**, 483.

London, F. (1927): Quantenmechanische deutung der theorie von Weyl (Quantum mechanical interpretation of the theory of Weyl), *Zeitschrift für Physik* **42**, 375.

London, F. (1938): The λ-phenomenon of liquid helium and the Bose–Einstein degeneracy, *Nature* **141**, 643.

Longair, M. S. (2003): *Theoretical Concepts in Physics. An Alternative View of Theoretical Reasoning in Physics*, 2nd ed. Cambridge University Press.

Lorentz, H.A. (1895): *Versuch einer theorie der elektrischen und optischen erscheinungen in bewegten körpern* (An attempt at a theory of electrical and optical phenomena in moving bodies), Leiden: E.J. Brill (see Pais 1982, pp. 124–126.) English translation of §§ 89–92 (on the topic of 'length contraction') can be found in Einstein *et al.* (1952).

Lorentz, H.A. (1904): Electromagnetic phenomena in a system moving with any velocity less than that of light, *Proceedings of the Academy of Amsterdam*, **6**, reprinted in Einstein *et al.* (1952).

Lummer, O. and Pringsheim, E. (1900): Über die strahlung der schwarzen körper für langer wellen (On the long wavelength of blackbody radiation), *Verhandlungen der Deutschen Physikalischen Gesellschaft* **2**, 163.

Mather, J.C. *et al.* (1994): Measurement by the FIRAS instrument on the COBE satellite showing that cosmic background radiation spectrum fitted the theoretical prediction of blackbody radiation perfectly, *Astrophysical Journal* **420**, 439.

Millikan, R.A. (1916): Einstein's photoelectric equation and contact electromotive force. *Physical Review* **7**, 18.

Millikan, R.A. (1949): Albert Einstein on his seventieth birthday. *Review of Modern Physics*, **21**, 343.

Minkowski, H. (1908): Space and time: An address delivered at the 60th Assembly of German Natural Scientists and Physicians, at Cologne, 21 September, 1908. English translation in Einstein *et al.* (1952).

Moore, W. (1989): *Schrödinger: life and thought*, Cambridge University Press.

Norton, J. D. (2004): Einstein's investigations of Galilean covariant electrodynamics prior to 1905. *Archive for History of Exact Sciences* **59**, 45.

Pais, A. (1982): *Subtle is the Lord. The Science and Life of Albert Einstein*, Oxford University Press.

Pauli, W. (1926): Über das Wasserstoffspektrum vom Standpunkt der neuen Quantenmechanik (On the hydrogen spectrum from the standpoint of the new quantum mechanics), *Zeitschrift für Physik* **36**, 336. Translated in Van der Waerden (1968).

Pauli, W. (1940): The connection between spin and statistics. *Physical Review* **58**, 716.

Peebles, P.J.E. (1984): Impact of Lemaître's ideas on modern cosmology. *The Big Bang and Georges Lemaître: Proceedings of a Symposium in Honour of G. Lemaître Fifty Years after his Initiation of Big-Bang Cosmology,* pp. 23–30. A. Berger (ed.), D. Reidel, Dordrecht, Netherlands.

Penzias, A.A., and Wilson, R.W. (1965): A measurement of excess antenna temperature at 4080 Mc/s, *Astrophysical Journal* **412**, 419.

Perlmutter, S. *et al.* Supernova Cosmology Project (1999): Measurements of omega and lambda from 42 high redshift supernovae, *Astrophysical Journal* **517**, 565.

Perrin, M.J. (1909): Brownian movement and molecular reality. *Ann. Chem. et Phys.* Translated by F. Soddy, reprinted in *The Question of the Atom: From the Karlsruche Congress to the Solvay Conference, 1860–1911,* M.J. Nye, (ed.) pp. 507–601, Tomash Publishers, Los Angeles, 1984.

Planck, M. (1900a): Über eine verbesserung der Wienschen spektralgleichung (On an improvement of Wien's equation for the spectrum), *Verhandlungen der Deutschen Physikalischen Gesselschaft* **2**, 202. Translated in ter Haar (1967).

Planck, M. (1900b): Über das gesetz der energieverteilung im normalspektrum (On the theory of energy distribution law of the normal spectrum), *Deutsche Physikalische Gesellschaft. Verhandlungen* **2**, 237. Translated in ter Haar (1967).

Planck, M. (1901): Über das Gesetz der Energieverteilung im Normalspektrum (On the law of energy distribution of the normal spectrum), *Annalen der Physik* **4**, 553. Translated in ter Haar (1967).

Planck, M. (1931): Letter from M. Planck to R.W. Wood. See Hermann, A., *The Genesis of Quantum Theory 1899–1923,* MIT Press, Cambridge, MA, 1971, p. 20.

Rayleigh, Lord (1900): Remarks upon the law of complete radiation, *Philosophical Magazine* **49**, 539. The paper is also reprinted and discussed in Longair (2003) pp. 320–328.

Riess, A.G. *et al.* High-z Supernova Search Team (1998): Observational evidence from supernovae for an accelerating universe and a cosmological constant, *Astronomical Journal* **116**, 1009.

Riess, A.G. (2000): The case for an accelerating universe from supernovae, *Publications of the Astronomical Society of the Pacific* **112**, 1284.

Rubens, H. and Kurlbaum, F. (1900): Über die Emission angwelliger wärmestrahlen durch den schwarzen Körper bei verschiedenen temperaturen (On the emission of long wave heat rays by a blackbody in various temperatures), *Sitzungsberichte der Königlich Preussischen Akademie der Wissenschaften zu Berlin*: 25 October, p. 929.

Sakurai, J.J. (1967): *Advanced Quantum Mechanics,* Addison-Wesley, Reading MA.

Schilpp, P.A. (ed.) (1949, 1951): *Albert Einstein: Philosopher-Scientist,* Vol I and II, Harper and Brothers Co., New York.

Schroeder, D.V. (2000): *An Introduction to Thermal Physics,* Addison-Wesley, San Francisco.

Schrödinger, E. (1926): Quantisierung als Eigenwertproblem (Quantization as an eigenvalue problem), *Annalen der Physik* **79**, 361 and 485. Translated in *Collected Papers on Wave Mechanics,* Chelsea Publishing Company, New York, 1982.

Shankland, R. S. (1963/73): Conversations with Einstein, *American Journal of Physics,* **31**(1963), pp. 47–57; **41** (1973), pp. 895–901.

Smoluchowski, M. (1906): Zur kinetischen Theorie der Brownschen Molekular-bewegung und der Suspensionen (The kinetic theory of Brownian motion and the suspensions), *Annalen der Physik* **21**, 756.

Sommerfeld, A. (1924): *Atombau und Spektralilinien* (Atomic structure and spectral lines) 4th ed., p. VIII. Vieweg, Braunschweig.

Stachel, J. (1995): *History of Relativity,* in Brown *et al.* (1995) p. 249.

Stachel, J. (1998): *Einstein's Miraculous Year: Five Papers that Changed the Face of Physics.* Princeton University Press.

Stachel, J. (2002): *Einstein from 'B' to 'Z'.* Birkhäuser, Boston.

Stefan, J. (1879): Über die Beziehung zwischen der Wärmestrahlung und der Temperatur (On the relation between the thermal radiation and temperature), *Sitzungsberichte der mathematisch-naturwissenschaftlichen Classe der kaiserlichen Akademie der Wissenschaften (Wien)* **79**, 391.

Stenger, J. *et al.* (1998): Optically confined Bose–Einstein condensate, *Journal of Low Temperature Physics* **113**, 167.

ter Haar, D. (ed.) (1967): *The Old Quantum Theory.* Pergamon Press, Oxford.

Tomonaga, S. (1962): *Quantum Mechanics.* North Holland, Amsterdam.

Uhlenbeck, G. and Goudsmit, S. (1925): *Naturwissenschaften* **47**, 953; Spinning electrons and the structure of spectra. *Nature* **117**, (1926) 264.

Van der Waerden, B.L. (ed.) (1968): *Sources of Quantum Mechanics.* Dover, New York.

Vikhlinin, A. *et al.* (2009): Chandra cluster cosmological project III: Cosmological parameter constraints, *Astrophysical Journal* **692**, 1060.

Weinberg, S. (1977): The search for unity: Notes for a history of quantum field theory, *Daedalus* **106**, 17.

Weiner, C. (ed.) (1977): *History of Twentieth Century Physics, Proceedings of the 1972 International School of Physics 'Enrico Fermi', Course 57.* Academic Press, New York.

Weisberg, J.M. and Taylor, J.H. (2003): The relativistic binary pulsar B1913+16, *Proceedings of Radio Pulsars*, Chania, Crete, 2002 (eds) M. Baileś, *et al.* (ASP Conference Series).

Weyl, H. (1918): Gravitation and electricity, in *Preussische Akademie der Wissenschaften, Sitzungsberichte.* Translated in Einstein *et al.* (1952).

Weyl, H. (1919): Eine neue erweiterung der relativitaetstheorie (A new extension of the relativity theory), *Annalen der Physik* **59**, 101

Weyl, H. (1928, 1931): *Gruppentheorie und quantenmechanik*, S. Hirzel, Leipzig 1928; *Theory of Groups and Quantum Mechanics*, 2nd ed. (transl. H.P. Robertson), Dover, New York, 1931.

Wien, W. (1893): Die obere grenze der wellenlängen, welche in der wärmestrahlung fester köper vorkommen können (The upper limit of the wavelength that can occur in the thermal blackbody radiation), *Annalen der Physik* **49**, 633.

Yau, S.T. and Nadis, S. (2010): *The Shape of Inner Space*, Basic Books, New York.

Zee, A. (2010). *Quantum Field Theory in a Nutshell*, 2nd ed. Princeton University Press.

Index